LASER 2006

LASER 2006

Proceedings of the 7th International Workshop on
Application of Lasers in Atomic Nuclei Research
"Nuclear Ground and Isometric State Properties" (LASER 2006)
held in Poznań, Poland, 29 May–1 June 2006

Edited by

Z. BŁASZCZAK

Adam Mickiewicz University, Poznań, Poland

B. MARKOV

Joint Institute for Nuclear Research, Dubna, Russia

and

K. MARINOVA

Joint Institute for Nuclear Research, Dubna, Russia

Reprinted from *Hyperfine Interactions*
Volume 171, Nos. 1–3 (2006)

Springer

A C.I.P. Catalogue record for this book is available from the Library of Congress.

978-3-540-71112-4

Published by Springer
P.O. Box 990, 3300 AZ Dordrecht, The Netherlands

Sold and distributed in North, Central and South America
by Springer
101 Philip Drive, Norwell, MA 02061, U.S.A.

In all other countries, sold and distributed
by Springer
P.O. Box 990, 3300 AZ Dordrecht, The Netherlands

Printed on acid-free paper

Table of Contents

Hyperfine Interact (2006) 171:1–2
DOI 10.1007/s10751-006-9499-7

Preface

**Zdzisław Błaszczak · Boris Markov ·
Krassimira Marinova**

Published online: 9 February 2007

The VII International Poznań Workshop "Nuclear Ground and Isomeric State Properties", organised by the Faculty of Physics Adam Mickiewicz University (*AMU*) and the Flerov Laboratory of Nuclear Reactions, Joint Institute for Nuclear Research (*JINR*), was held in Poznan, from May 29th to June 1st, 2006. The conferences belong to the cycle of workshops on the "Application of Lasers in Atomic Nuclei Research" and take place every second year in Poznan.

The host of the workshop is the Adam Mickiewicz University, whose traditions date back to the sixteenth century, founded in 1919 and at present belonging to the largest universities in Poland. The workshop sessions were held in a beautifully localised new campus of the Faculty of Physics, in nice and comfortable conditions.

The scientific cooperation between the Faculty of Physics AMU and the JINR has been developing for many years and is particularly intense with the Flerov Laboratory of Nuclear Reactions. The first workshop on the "Laser Spectroscopy of Atomic Nuclei", took place at the JINR in Dubna, from December 18th to 20th, 1990. After a break of a few years caused by changes in the political and economical situation in Central and Eastern Europe, another meeting was organised at the Faculty of Physics, Poznan, from May 29th to 31st, 1995. Each workshop was devoted to somewhat different topic (see Proceedings of the 6th Workshop, Hyperfine Interactions, vol. 162, 2005). The main aim of the meetings was to gather the specialist in the field at one place, present the recent achievements, review the

Z. Błaszczak
Institute of Nuclear Physics, ul. Radzikowskiego 152, 31-342 Krakow, Poland

B. Markov
Flerov Laboratory of Nuclear Reactions, Joint Institute for Nuclear Research,
141980 Dubna, Moscow region, Russia

K. Marinova (✉)
Laboratory of Nuclear Reactions, Joint Institute for Nuclear Research,
141980 Dubna, Moscow region, Russia
e-mail: laser2006@jinr.ru

 Springer

methods of research, discuss and correlate the directions of future projects. The scientific programme of the VII Workshop covered the following topics:

- Nuclear properties studied by the use of laser and storing techniques: spin, moments, hyperfine anomaly, radii, masses and nuclear half lives;
- Test of fundamental interactions and symmetries by laser technique;
- Development and applications of laser ion sources;
- Development and applications of gas catchers, ion guides and gas-jets;
- Cooling and trapping techniques for radioactive beams;
- Powerful pulsed lasers in nuclear research;
- Trace analysis by lasers.

The VII Workshop, gathered about 50 participants from many countries representing important scientific centres: Belgium (KU Leuven), Canada, (TRIUMF), CERN, Finland (IGISOL), France (Orsay), Germany (GSI and University of Mainz), Holland (KVI Groningen), JINR, Poland (AMU and TU), United Kingdom (Universities of Manchester and Glasgow) and USA (University of NY). The workshop was open to the academic staff and students from Poznan universities.

The sessions begin with an overview on the past developments and finished with a look in the future directions. The total number of the talks was 38, including 18 invited lectures. The high quality of the presented results, the participation of different generations of scientists, including many young physicists, demonstrated the importance and the validity of the discussed topics. It is not only our impression that the aim of the Workshop was fulfilled. The Workshop showed that the field under discussion is very attractive for the near future, that interesting questions are still open and that a common background can be found for future investigations

The Organizers acknowledged the decision of the steering committee of LASER-EURONS JRA (Joint Research Activity) to combine the 2006-meeting with the Workshop in Poznan. Many participants of the JR Activity took part in the laser-workshop and reported on their scientific progress. In addition, some members of LASER JRA took part in the discussion on the last day of the workshop. Thus, a fruitful interrelation with the Workshop was realized and the issues were discussed in depth. The second part of the discussion has been restricted to the outcomes of the Workshop, its future organisation and the scientific problems that should be included in the programme. It was pointed out that the efforts have to be focused on exotic species study, often but not necessary on-line related as well as on new fields such as laser oriented (atom or ion) traps for precision mass spectrometry, laser based ultratrace isotope determination etc. An idea was suggested to organize together with the future workshops a half day course for students to introduce them into the field.

Particular words of appreciation need to be directed to the EXAKT Group (friends of the Experimental Atomic and Nuclear Physics) from the University of Mainz and GSI, Darmstadt whose financial support permitted participation of 12 PhD and post doctoral students. Special thanks are due to Prof P. Van Duppen (*Leuven*) and Dr K. Wendt (*Mainz*) for the excellent organized discussions.

In the unanimous opinion of the participants the conference was an important European scientific meeting. It was a significant forum for exchange of information and experience. Its popularity among young scientists makes grounds for expectations that the conference will continue for many years.

Hyperfine Interact (2006) 171:3–40
DOI 10.1007/s10751-007-9513-8

Optical spectroscopy of radioactive atoms

H. Henry Stroke

Published online: 27 March 2007

Abstract An epitome of optical methods used in atomic spectroscopy of radioactive atoms is presented. The overview addresses a number of results in atomic structure and hyperfine structure, and the implications in the study of electric and magnetic properties of nuclei. An *aperçu* is given of the concomitant development of the experimental methods, from simple optical techniques to laser spectroscopy, and from use of "off-line" experiments to ones using ISOLDE-type facilities.

Keywords Radioactive atom spectroscopy · Isotope shifts · Hfs ·
Laser spectroscopy · Nuclear multipole moments

1 Introduction

A brief historical accounting, paralleling the presentation at the VII International Workshop Laser 2006 in Poznan, Poland, is made of the spectroscopy of radioactive atoms, from its beginning to the present day. It is forcibly sketchy, relying largely on personal experience and acquaintance, certainly not exhaustive, and, in the process, it omits a great number of important contributions of colleagues, for which I apologize. Nonetheless, it hopefully presents a picture – highlights – which should give an appreciation of the richness of the physics results obtained and the effort of many, over nearly a century, in developing more and more sensitive and precise

H. H. Stroke (✉)
Department of Physics, New York University, 4 Washington Place,
New York, NY 10003, USA
e-mail: henry.stroke@nyu.edu

H. H. Stroke
CERN, 1211 Geneva 23, Switzerland

Fig. 1 Jacques Pinard
(courtesy of J.P.)

experimental techniques to reach the results. I give a sampling of the literature, from old to current, which should provide some indication of these efforts [1–12].[1] This presentation was aided substantially by a recent review that was made in collaboration with Jacques Pinard, Fig. 1, of the Laboratoire Aimé Cotton in Orsay [13].

As an introduction, the extension of "classical" spectroscopy to radioactive atoms is discussed. By classical we refer to the light sources or absorption cells, not to the methods: these were in fact extended radically from early dispersive grating and Fabry–Pérot techniques to Fourier spectroscopy [14–16],[2] which allows an entire spectrum to be recorded with high resolution. The first radioactive atom spectroscopy is in fact that of ^{209}Bi, in 1926, which has been found to be an α emitter only recently, with a half-life of 1.9×10^{19} years [17]. In the early 1930s, less exotic radioisotopes, of radium and radon, were measured, and can be found listed in Charlotte Moore's *Atomic Energy Levels III* table.

I digress for an instant. In 2005 the world celebrated the *annus mirabilis* of Einstein's monumental papers. The decade, starting in the vicinity of 1925, constituted for physics in general, and for a number of subjects of concern here, veritable *anni mirabiles*, both for theory and experiment. In 1924, Wolfgang Pauli [18] suggested the existence of a nuclear spin and hyperfine structure in the atomic

---

[1] Refs. [7, 8] include the recent application at the Jyväskylä IGISOL facility of the "cooler-buncher" technique, which will represent a crucial advance for radioactive atom spectroscopy.

[2] I recall that Fourier spectroscopy is based on the Michelson interferometer where the fringe visibility is measured as one of the mirrors is displaced. This actually gives the Fourier transform of the spectrum. Michelson constructed an analogue device to obtain the inverse transform; this appears to have discouraged further spectroscopic applications, until digital techniques were introduced (see references in [15, 16]).

Fig. 2 Sam Goudsmit (courtesy American Institute of Physics, Emilio Segrè Visual Archives)

spectrum, then measured and interpreted for bismuth by Ernest Back and Sam Goudsmit (Fig. 2) [19, 20]. This would have been the accepted chronology, but for the correction brought to light recently by Takashi Inamura [21]. He brings to light the early studies of isotope shifts and hyperfine structure by Hantaro Nagaoka [22]. This, Inamura recounts, was recognized by Pauli but not in most of the atomic physics books with which we were educated. I note, however, that the "bible," Condon and Shortley [23], does give a reference to some of Nagaoka and co-workers isotope shift studies, as do Back and Goudsmit [19] to both this and earlier work. If I can be forgiven for a purely provincial outlook at New York University, I note the first theoretical paper on volume-dependent isotope shifts and the effect of a distributed nuclear charge distribution on atomic hyperfine structure by Jenny Rosenthal (Fig. 3) and Gregory Breit [24] (the so-called *Breit–Rosenthal–Crawford– Schawlow correction*), the calculation in 1931 by Herman Yagoda of the wavelengths of the resonance lines of the then not yet discovered unstable element, francium [25, 26], a subject to which I return, and, parenthetically, to finish with NYU, the first experiment on β-decay asymmetry, reported by Richard Cox, et al., though certainly unexplained at that time [27]! Before touching on some of the milestones, I call attention to the largely ignored paper by Victor Weisskopf [28] that later laid the foundation for level-crossing spectroscopy [29, 30] with fruitful applications in the study of hfs spectra of radioisotopes.

2 Measurable quantities

We divide the optical spectroscopies of radioisotopes into two broad areas of interest:

A.   Atomic physics and interactions
B.   Electron–nuclear interactions

Fig. 3 Jenny Rosenthal Bramley at her 1929 NYU commencement: the first woman to earn a PhD in physics (courtesy NYU Alumni Today)

2.1 Atomic physics

2.1.1 Actinides

This series of radioisotopes, that corresponds to the lighter lanthanides, has been studied over many years (Jean-François Wyart, Jean Blaise) at the Laboratoire Aimé Cotton, long under the direction of Pierre Jacquinot (Fig. 4).

A number of other laboratories, Livermore, Argonne, Lund, Amsterdam, …, have been actively engaged in the measurement and analysis of these spectra. The results are available on the site http://www.lac.u-psud.fr/LAC/data/database.htm. The classification of the atomic spectra relies on the Rydberg–Ritz principle, aided in the configuration assignments by observation of hfs and isotope shifts.

2.1.2 Electron correlations

Isotope shifts, discussed below, are the primary interest for nuclear structure studies. The extraction of this "volume effect" of the nuclear charge distribution is limited by mass-dependent effects. These depend in turn on the correlation of the motion of all the electrons [23] – and is thus a nuisance many-body problem, unfortunately important in light and middle-A nuclei (A, mass number).

 Springer

Fig. 4 Pierre Jacquinot
(courtesy Laboratoire Aimé
Cotton, Centre National de la
Recherche Scientifique)

2.1.3 Spectrum of francium

Theoretical work by Yagoda [25, 26] on the spectrum of francium (called eka-cesium until its discovery in 1939 by Marguerite Perey [31]) (Section 1) relied on analyses of isoelectronic series. His predictions for the resonance lines proved to be closer to the experimental values [32, 33] than a number of more recent, sophisticated, calculated values. Although a finely tuned laser is not the best adapted tool for searching for an atomic transition over large wavelength ranges, experiments were successful for several resonance lines (Fig. 5).

2.2 Electron–nuclear hyperfine interaction

In his 1924 paper [18], Pauli encompasses a good part of the physics of hfs, including the effect of the extended charge distribution on the electron-nuclear interaction, *i.e.* the volume-dependent isotope effect. As mentioned in Section 1, in order to account for observed structures, he also introduced the existence of a nuclear spin, and notes that the electron-nuclear interaction is much smaller than the electron-electron interaction. Goudsmit and Robert Bacher [34] applied the resulting interval rule to fit measured spectra with this magnetic dipole interaction, but encountered difficulties: the electric quadrupole moment of the nucleus was not yet known. The inclusion of its interaction with the gradient of the atomic electric field at the position of the nucleus was calculated in the prize essay of Hendrik B.G. Casimir [35]. A

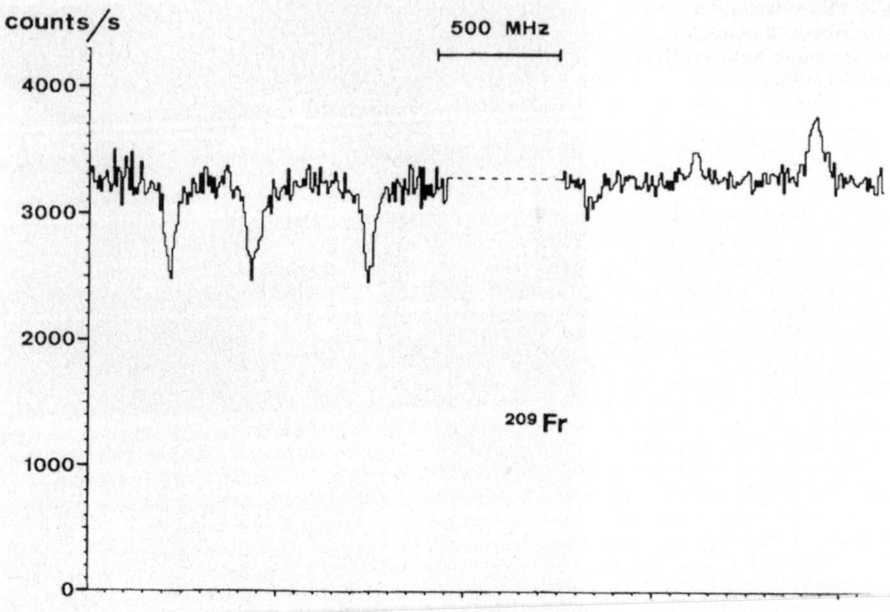

Fig. 5 Francium D2 line, from Ref. [33] (courtesy ©American Physical Society, http://prola.aps.org/abstract/PRA/v22/i6/p2732_1)

few years later, he also calculated [36] the next term in the electron-nuclear hfs interaction, caused by a possible nuclear octupole moment, which we detected a dozen years later [37].

2.2.1 Isotope shifts

In the preceding paragraph we discussed the *multipole* electron-nuclear interaction, but omitted the *monopole* term, the point Coulomb interaction. From the so-called "center-of-gravity" or "center-of-mass" theorem [38] we know that measurements of the hyperfine interactions (multipolarity k>0) give no information about the center of gravity of the atomic level, determined by k=0. The latter, on the other hand, is dependent on the spherically averaged value of the nuclear charge distribution, which, in turn, gives rise to the volume-dependent part of isotope shifts.

2.2.2 Bohr–Weisskopf effect

The magnetic counterpart of the isotope shift is the Bohr–Weisskopf (Fig. 6) effect, also known as the "hfs anomaly" [39]. It reflects the influence on the magnetic dipole hfs interaction of the spatially-distributed nuclear magnetization – both spin and orbit. For ordinary atoms the effect is observed via the comparison of isotopic ratios of hfs interaction constants with those of independently-measured nuclear g-factors. For atoms, this is a relatively small effect, for muonic atoms a large effect. But the enormously greater precision of atomic measurements makes the atomic spectroscopy much more sensitive to these nuclear effects.

 Springer

Fig. 6 Viki Weisskopf at the "Bohr-Weisskopf" atomic beam apparatus at ISOLDE (photo by author)

As a historical aside, we reproduce (Fig. 7) an introductory page of the first edition of the book of Hans Kopfermann [40].

What he neglects is in fact the Bohr–Weisskopf effect, which in many cases is only a fraction of one percent! One can forgive Kopfermann: his book was the first important compendium of this field, and he educated a whole generation of post-war physicists in Germany, many of whom became leaders in atomic physics in their own right. Kopfermann was also probably the first (or was among the first) German physicist to be invited after World War II to visit MIT, where he gave a very stimulating set of lectures on hfs. An English edition of his book was published later [41].

3 Experimental methods

3.1 Atomic beam techniques

The experiments of Otto Stern and Walther Gerlach led first to non-resonance atomic beam experiments, e.g. the "zero-moment" method for determining nuclear spins and hfs [42]: it consists, for alkali atoms, in determining in the Zeeman effect

B. Magnetische Wechselwirkungen

In ganz entsprechender Weise läßt sich die magnetische Wechselwirkung zwischen Kern und Elektronenhülle durch ein Kernvektorpotential $\mathfrak{A}$ und die Stromdichte $\vec{\sigma}$ der Hüllenelektronen darstellen:

$$V_{magn} = -\frac{1}{c}\int(\mathfrak{A},\vec{\sigma})\,d\tau. \qquad (1,4)$$

Man kann das Kernvektorpotential außerhalb des Kerns nach Potenzen von $\frac{1}{r}$ entwickeln, wobei das Vektorpotential in erster Näherung das eines magnetischen Kerndipols $\mathfrak{A} = \frac{[\vec{\mu_I},\vec{r}]}{r^3}$ wird. Höhere magnetische Pole, welche sicher nur als kleine Störungen anzusehen sind, und für deren Existenz bisher keine experimentellen Anzeichen vorliegen, sollen unberücksichtigt bleiben. Dann wird:

$$V_{magn} = -\frac{1}{c}\int\frac{([\vec{\mu_I},\vec{r}]\vec{\sigma})}{r^3}\,d\tau = -\frac{1}{c}\int\left(\vec{\mu_I}, \frac{[\vec{r},\vec{\sigma}]}{r^3}\right)d\tau. \qquad (1,5)$$

Die Größe $\frac{1}{c}\int\frac{[\vec{r},\vec{\sigma}]}{r^3}d\tau$ stellt das Magnetfeld $\mathfrak{H}(0)$ dar, welches die Elektronen am Kernort erzeugen. Man erhält

$$V_{magn} = -(\vec{\mu_I}, \mathfrak{H}(0)) = -\mu_I H(0)\cos(\vec{\mu_I}, \mathfrak{H}(0)). \qquad (1,6)$$

Der Betrag $\delta H(0)$, der von dem innerhalb des Kernvolumens fließenden Teil der Elektronenstromdichte stammt, und der in (1, 5) nicht erfaßt ist, dürfte im allgemeinen unmerklich klein sein. Selbst bei den s-Elektronen der schweren Atome, welche dem Kern sehr nahekommen, kann die Vernachlässigung der magnetischen Wechselwirkung innerhalb des Kernvolumens nur einen Fehler von wenigen Prozenten in (1, 6) ausmachen.

Fig. 7 From the first edition of Kopfermann's book, Kernmomente [40]: he neglects the influence of the penetration of the electron inside the nuclear volume, which would lead to an error of only a few percent (courtesy of Akademische Verlagsgesellschaft, Berlin)

of the hfs the magnetic fields, B, for which the effective magnetic moment $\mu_{eff} = -\partial E/\partial B=0$: such atoms are detected after traversing undeflected an inhomogeneous magnetic field. E is the energy of a particular hyperfine level. An example for the 2.1-y ^{134}Cs [43] is shown in Fig. 8.

This was followed by the atomic (or molecular) beam magnetic resonance (ABMR) method of Rabi and co-workers [44], Fig. 9. Here, two inhomogeneous magnets, A and B, with their field gradients in opposite directions, were separated by a homogeneous magnetic field, C, that also contained an rf loop. In the absence of rf transitions, the atoms are refocused on the detector by the B magnet. At the resonant frequency, atoms make transitions from positive to negative μ_{eff} or *vice versa*, producing a small decrease in detected atom intensity.

It was soon recognized by Jerrold Zacharias that having the magnetic-field gradients of the A and B magnets in the same direction would increase the signal-to-noise

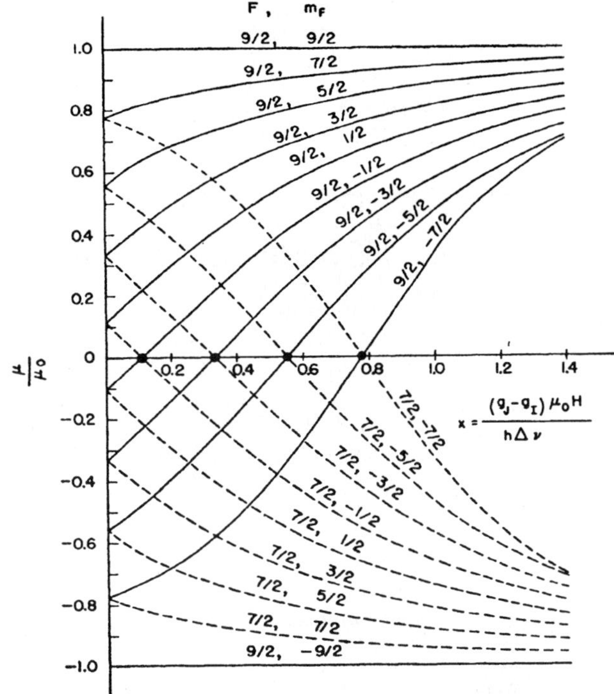

Fig. 8 The effective magnetic moment of the energy levels for the atom with electron angular momentum $J = \frac{1}{2}$ and nuclear angular momentum $I = 4$. The solid circles represent the magnetic fields at which μ_{eff} vanishes. From [43] (courtesy ©American Physical Society, http://prola.aps.org/abstract/PR/v87/i4/p676_1)

Fig. 9 Some of the key figures in ABMR and resonance physics. *Back row*: Norman Ramsey, Charles Townes, Vernon Hugues, Ed Purcell, William Nierenberg. *Front row*: Jerrold Zacharias, I.I. Rabi, Julian Schwinger, Gregory Breit (courtesy American Institute of Physics, Emilio Segrè Visual Archives)

ratio considerably: one looks for a small signal on essentially zero background in the absence of resonance. This was implemented by Zacharias in the measurement [45] of 1.25×10^9-y ^{40}K, found in nature with an abundance of 0.0117%. This was the beginning of radioactive ABMR work on alkali atoms at MIT [46], Fig. 10. (It ended in 1954 with a major radioactivity spill by the author!) I note that some 5×10^9 Bq of ^{134}Cs were typically loaded into the ABMR source (10 times less activity for a mixture of ^{137,135}Cs).

A number of other laboratories around the world started work on radioactive ABMR. This was reviewed by William Nierenberg [1]. I only point out work on the first nuclear isomer, 2.9-h ^{134m}Cs, by Victor Cohen and Donald Gilbert at Brookhaven National Laboratory [47], and the development of a focussing ABMR,

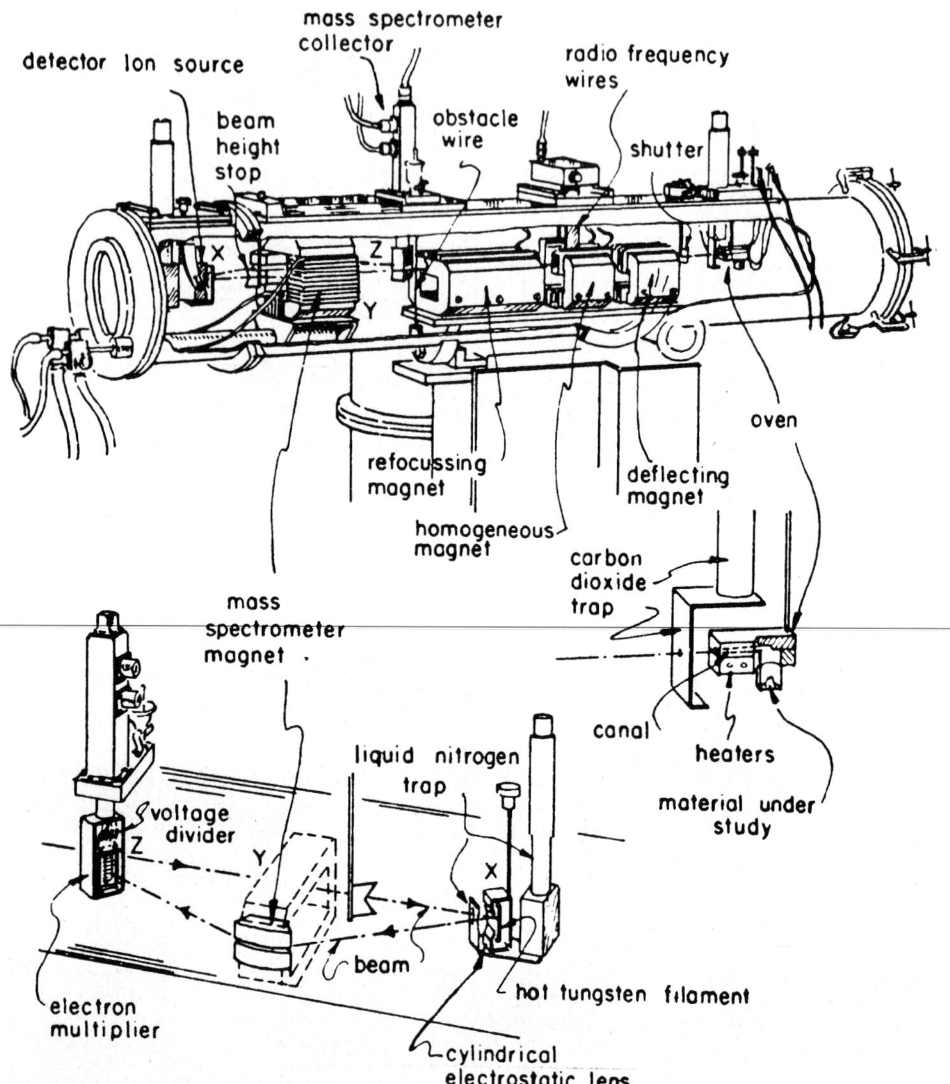

Fig. 10 Atomic beam apparatus used at MIT for radioactive-atom experiments. Note the use of a mass spectrometer in the detecting system [46] (courtesy ©American Physical Society http://prola.aps.org/abstract/PR/v76/i8/p1068_1)

Fig. 11, at Princeton University, under Donald Hamilton [48], with which isotopes in the range of minutes could be measured, "off-line," the cyclotron used for the production being just some 50 m away, Fig. 12.

Obtaining a resonance curve was not trivial because of instabilities in the electron-bombardment atomic beam oven. The circular detector buttons were split into 8 sectors. Each sector was exposed for 20 s at a particular frequency, rotated to the next sector for a background exposure, the frequency changed to the following value,

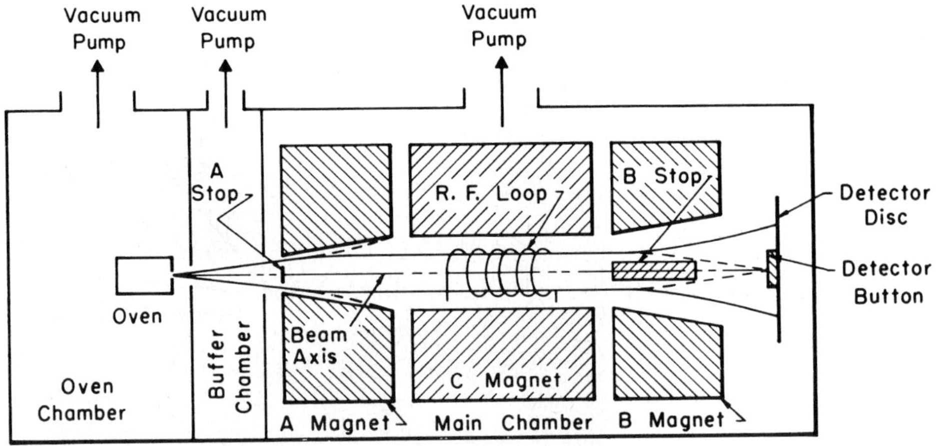

Fig. 11 Princeton focussing ABMR schematic. The A and B magnets have a six-pole geometry. The atoms are collected on detector buttons, which are removed from the apparatus for measuring the collected radioactivity

Fig. 12 Photograph of Princeton ABMR. The port through which the detector buttons are inserted into the apparatus is seen *above the middle* of the end plate

 Springer

Fig. 13 Francis Bitter
in his MIT laboratory
with his technical assistant
(with permission, © courtesy
MIT Museum, [51])

Fig. 14 Jean Brossel (courtesy
Laboratoire Kastler-Brossel,
Ecole Normale Supérieure)

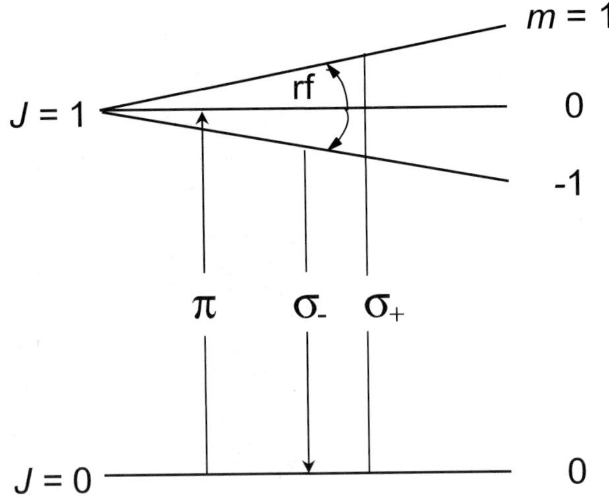

Fig. 15 Double resonance with polarization detection

etc. for a total accumulation time of 5 min on each sector. This was signal averaging at its most primitive!

3.2 Optical methods

3.2.1 Optical double resonance

After a false start [49] by Francis Bitter, Fig. 13, he and Jean Brossel [50], Fig. 14, developed the "Double-Resonance" method for studying the structure of excited atomic states with great sensitivity [51].

The double resonance consists of an optical resonance followed by an rf resonance, the detection of the latter relying on a change of polarization of the re-emitted light. This is illustrated in Fig. 15: in this example, atoms in a magnetic field in $J=0$ are excited to a $J=1$, $m=0$ state with use of π-polarized light. Following rf transitions, $\sigma_{\pm}$ light is emitted.

Much of the work in Bitter's laboratory at MIT for over one and a half decades was devoted to the hfs spectroscopy of stable and radioactive mercury isotopes, ranging from mass number, $A = 192$ to 204. I will touch on the several experimental techniques. An elaboration of the schematic of the polarization double resonance method is shown in Fig. 16 from the work on 23.8-h ^{197m}Hg of Henry Hirsch [52, 53].

These experiments were done long before the advent, in 1966, of tunable dye lasers [54, 55]. If you look at Fig. 16 carefully, you will note that the tunable light source to produce the first (optical) resonance, either for double resonance or level-crossing experiments, is a ^{198}Hg isotopic lamp placed in a magnetic field. Such "Zeeman" tuning was already used in 1929 by Marcel Schein [56], and was rediscovered by Bitter to become known as the"Bitter magnetic scanning" technique. The energy levels of ^{197m}Hg in a magnetic field and the transitions used in the double resonance and level-crossing experiments are indicated in Fig. 17.

 Springer

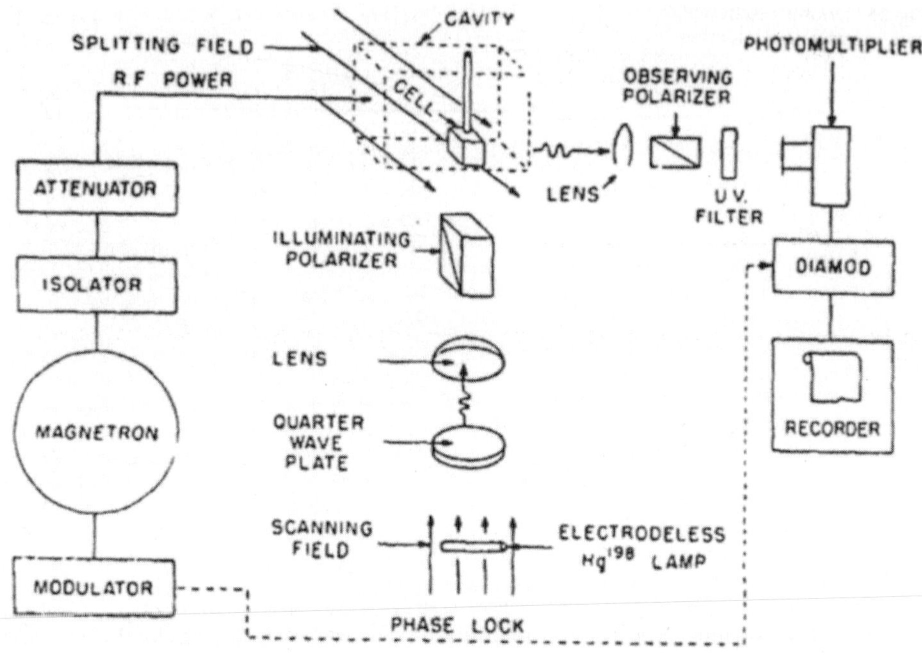

Fig. 16 Double-resonance apparatus schematic (from [52, 53], courtesy of the Optical Society of America)

Fig. 17 Zeeman hfs levels of ^{197m}Hg (from [52, 53], courtesy of the Optical Society of America)

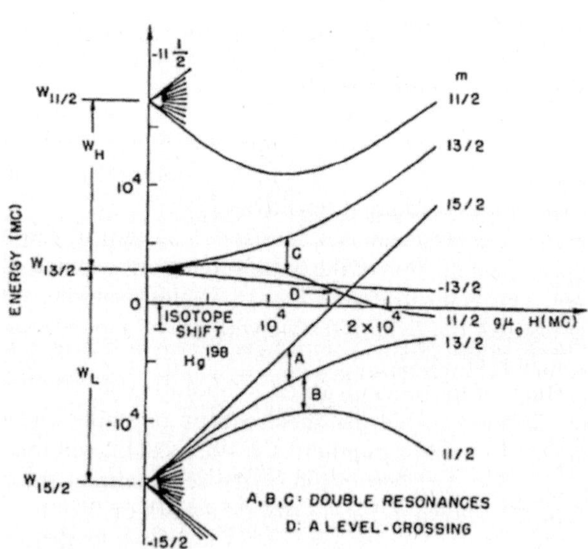

3.2.2 Double resonance by frequency change

For special spectra, e.g. in mercury, Robert Kohler [57, 58] introduced a new double-resonance method that depends on a change in frequency in a transition rather than

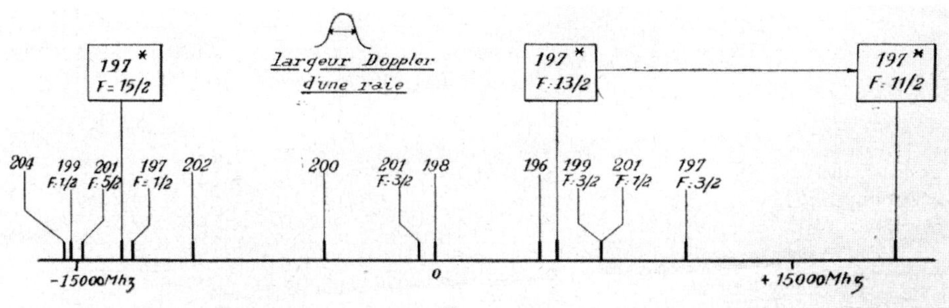

Fig. 18 Hyperfine components in mercury spectrum (from [59], courtesy of the Société Française de Physique)

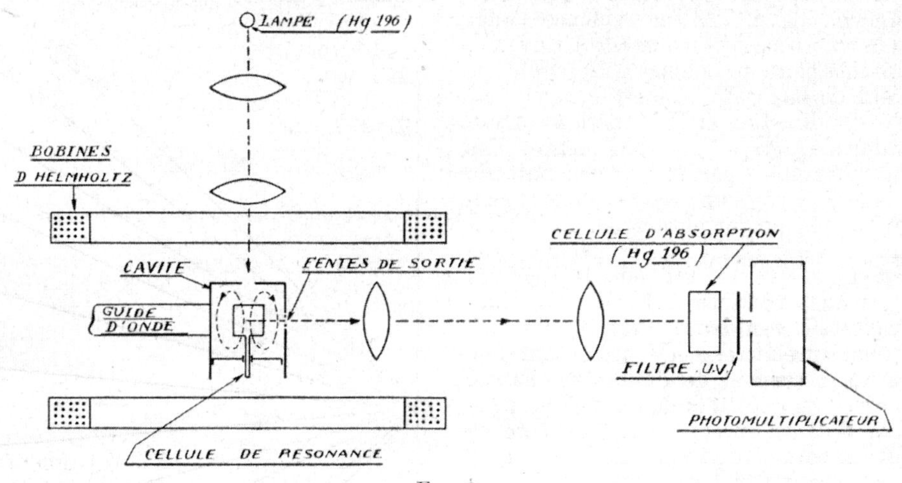

Fig. 19 Experimental setup in the Kohler-type double resonance experiment for ^{197m}Hg hfs measurement (from [59], courtesy of the Société Française de Physique)

a change in the polarization of the light. The hfs separations have to be much larger than the Doppler widths of the transitions. The scheme was applied by Claude Brot [59], then at MIT, to a precision measuremennt of the hfs separations of ^{197m}Hg, shown in Fig. 18.

The scheme was possible here because of the near coincidence of the F=13/2 component of ^{197m}Hg and the 0.15 percent stable isotope ^{196}Hg. The latter had to be enriched both for the light source and the absorption cell, Fig. 19.

3.2.3 Level-crossing spectroscopy

The scheme for level-crossing spectroscopy is not far removed from Fig. 17, except that changes in the observed intensity of re-emitted light, rather than of the polarization of the light, are observed. We have already cited the theoretical basis, [28, 29]. Simply, if transition amplitudes from two excited levels to a common lower

Fig. 20 Willis Lamb, Alfred Kastler, and Hans Kopfermann, with George Series in the background *on the right* (Photograph by Speck, Heidelberg, courtesy American Institute of Physics, Emilio Segré Visual Archives and International Conference on Optical Pumping, 24–26 April, 1962)

Fig. 21 Optical pumping transitions for mercury isotopes with nuclear spin $I=\frac{1}{2}$, e.g. ^{197}Hg, ^{199}Hg

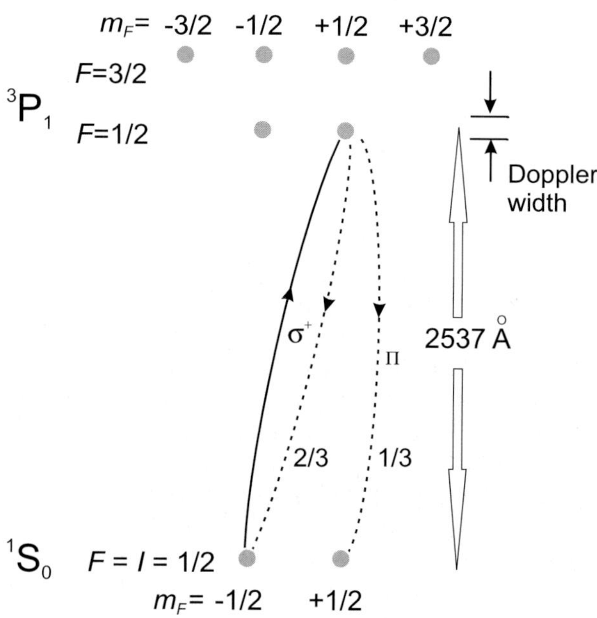

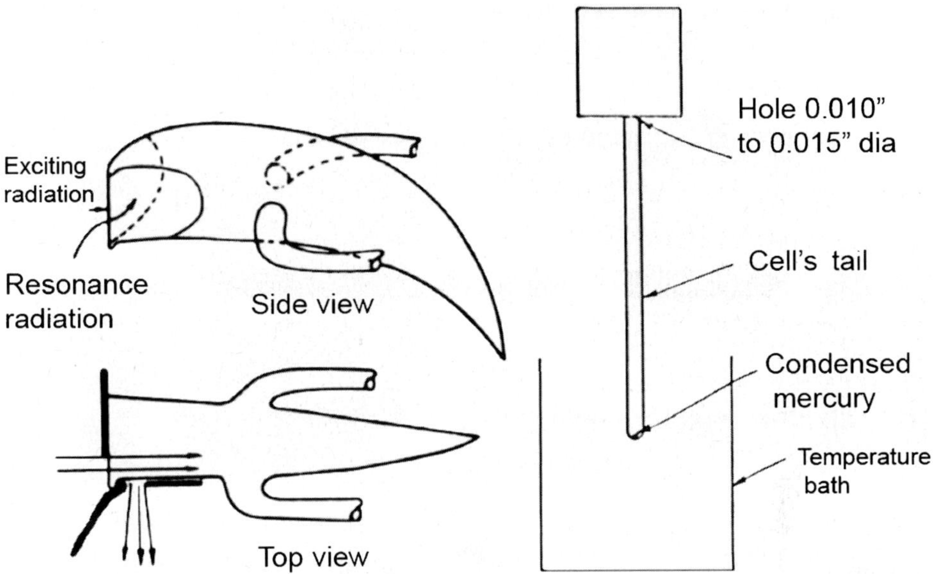

Fig. 22 Left: *Wood's horn; right*, resonance cell in which optical pumping was successfully observed (see, HHS in [51], p. 289, courtesy MIT Press)

levels are a and b, away from the level crossing the intensity is $a^2 + b^2$, while at the level-crossing magnetic field it is $(a + b)^2$, leading to an observable interference term. The value of the crossing fields, from which hfs interactions can be obtained, are determined without being limited by Doppler broadening. The method has been applied to several radioactive isotopes, e.g. ^{197m}Hg and ^{203}Hg.

3.2.4 Optical pumping

The *optical pumping* method also dates to this period. It originated with Alfred Kastler, (Fig. 20), in 1950 [60]. We describe below its importance to the measurement of nuclear magnetic moments, in particular radioisotopes.

The basic scheme is shown in Fig. 21.

The mercury is excited from the ground ^{1}S$_0$, $F = \frac{1}{2}$ state with circularly polarized light to the excited ^{3}P$_1$, $F = \frac{1}{2}$ state. The selection rule $\Delta m = +1$ allows only the $m = -\frac{1}{2}$ state to be excited. The spontaneous re-radiation allows transitions with $\Delta m = \pm 1$, thereby populating the $m = +\frac{1}{2}$ sublevel, where the atoms are blocked. Since the ground-state angular momentum is due entirely to the nuclear spin, a corresponding nuclear orientation is produced. Attempts at observing it met with initial failure [61]. The cause was eventually found in the resonance cells that were used. As when work on gas lasers first began, the above optical work also relied heavily on the book of Allan Mitchell and Mark Zemansky (the former at New York University) [62], in particular the resonance cells. The type of cell that was used at first was the "Wood's horn" (named after R.W. Wood), shown in Fig. 22.

In the *horn*, a large surface of mercury liquid is seen by the oriented atoms and condensation rapidly equalizes the two magnetic substates, m. In the cell shown on

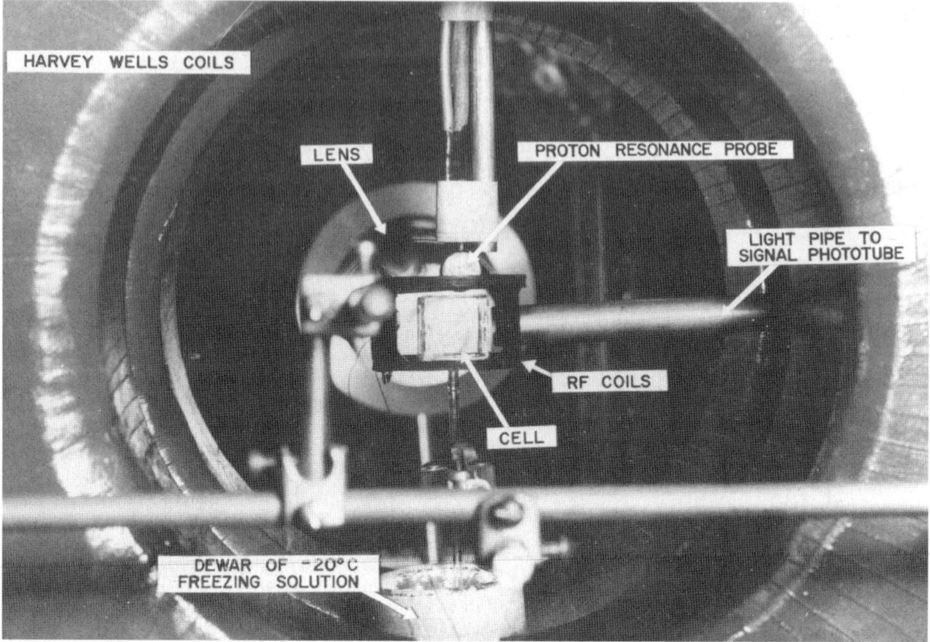

Fig. 23 Optical pumping setup for measurement of ^{195,197}Hg (courtesy W.T. Walter [66])

the right, the atoms do not see the mercury reservoir and relaxation is negligible. An optical pumping experiment with stable mercury isotopes by Bernard Cagnac thus became successful [63–65] and the first measurement of radioisotopes, ^{195}Hg and ^{197}Hg were made [66, 67]. Figure 23 shows William Walter's experimental setup.

At the onset of this section, the importance of the optical pumping experiments to the measurement of nuclear moments with great sensitivity was stated. In Section 2.2.2 the requirement of nuclear magnetic moments, or *g-factors*, was brought out. For stable isotopes, these can be obtained by nmr in bulk samples, obviously not possible for minute quantities of radioisotopes. Optical pumping, as described in particular for mercury, permits the measurement: while a resonance cell is being pumped it absorbs light. When orientation is complete the vapor becomes transparent. The repopulation of the *m*-levels can be re-established by the nmr transition, and detected by the change in transparency. I should mention that another possibility to obtain the magnetic moment, especially in alkali atoms, is to measure the term in the hfs Hamiltonian that depends directly on the interaction of the magnetic moment with the applied magnetic field. This was used in obtaining the moments in a series of cesium radioisotopes [68].

Optical pumping orientation was successful early on in alkali atoms. Asymmetry in beta decay of oriented ^{21}Na was measured by Otten, Fig. 24, and co-workers [69], a forerunner of a an entire program of measurements on radioactive nuclei at ISOLDE.

Springer

Fig. 24 Ernst Otten: the pioneer in radioactive-atom spectroscopy at ISOLDE (photo by author)

3.2.5 Radioactive source preparations

I concentrate on the experiences at MIT and some of our later work: these should serve to give a picture of "off-line" source preparations, and increase our appreciation of doing spectroscopy at ISOL facilities! In most experiments transmutation reactions were used in the production, so as to avoid large stable-isotope backgrounds. For a number of experiments the production was at a cyclotron with use of a (p, xn) nuclear reaction. For the mercury isotopes, this was straightforward with the use of a gold target. After irradiation, a cell or electrodeless discharge lamp was produced with a system as shown in Fig. 25.

In a few cases a nuclear reactor was used to produce the radioisotope, but this demanded subsequent mass separation. For other elements, radiochemistry was required, and for thallium the development of a liquid mercury target acceptable to cyclotron establishments! Typical lamps and cells are shown in Fig. 26. In use, the lamps frequently necessitated to be heated with the use of a propane torch to drive the few atoms back from the quartz surface into the electric discharge.

3.2.6 Grating optical spectroscopy

Most of the optical spectroscopy of the radioactive isotopes was done with the use of a 10-m focal length (slightly shorter at NYU) Czerny–Turner two-mirror monochromator, Fig. 27.

A 25-cm wide ruled diffraction grating, 300 lines/mm, blazed at 60^0, was the heart of the instrument. A photograph is shown in Fig. 28.

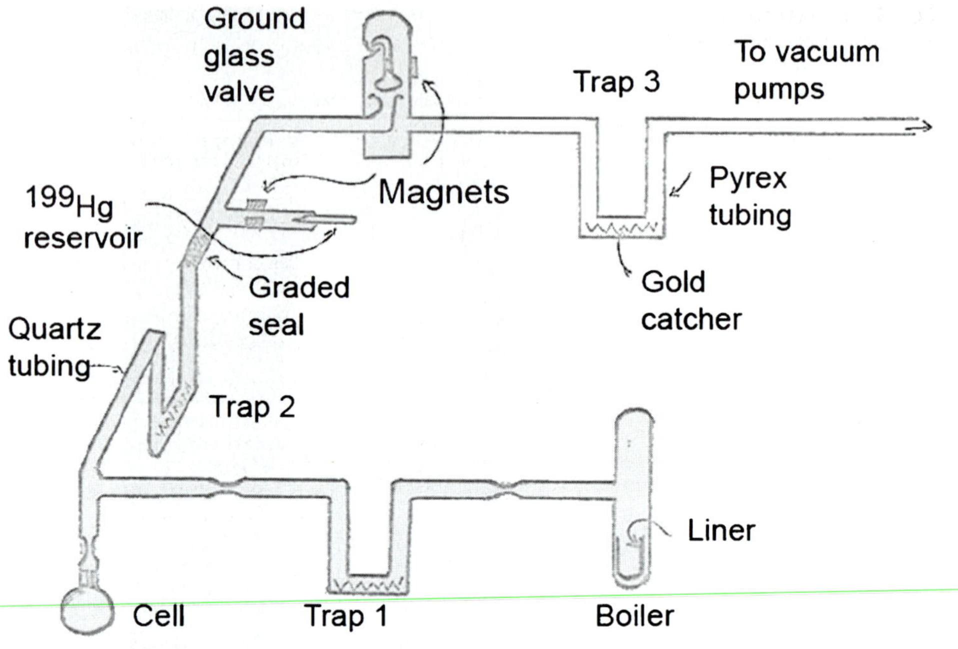

Fig. 25 The mercury in the irradiated gold foil is melted with an induction heater, driven to gold catchers, and finally to the resonance cell or lamp

Fig. 26 Optical-pumping cell and spherical electrodeless lamps. Scale is in cm

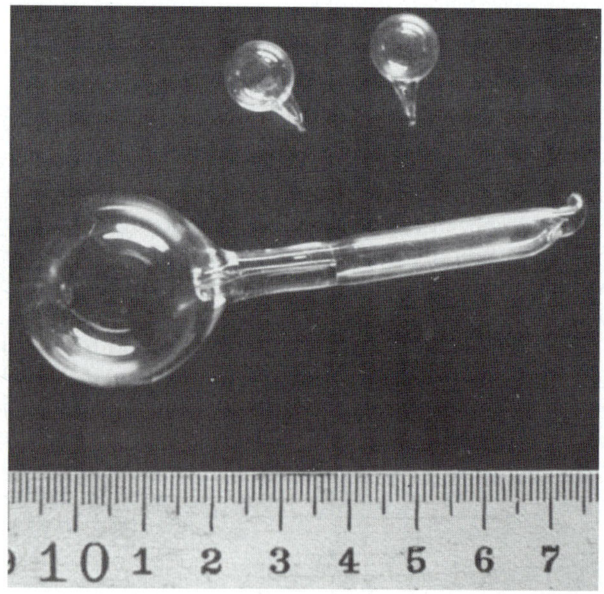

Photographic or photoelectric detection was used, as well as an image intensifier. Since the preparation time for the fabrication of a spectral lamp took about two weeks, and the lamps had a limited lifetime (not just radioactively, but frequently

Fig. 27 Grating spectrometer, schematic. The grating spacing is denoted by d

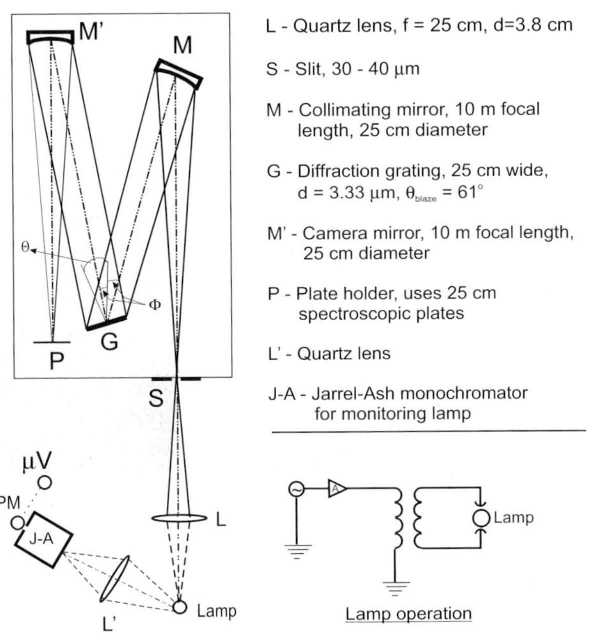

Fig. 28 1-slit, 2-diffraction grating, 3-photographic plate holder

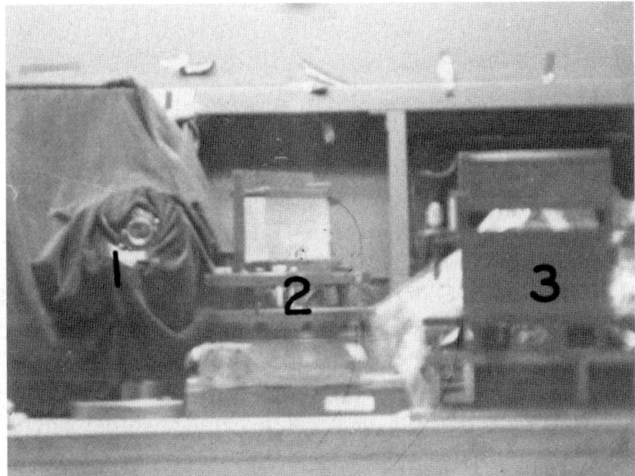

they may have lasted only from a few minutes to hours because of the atom "clean-up" in the discharge), additional camera mirrors and detectors were frequently set up to record simultaneously the spectra of several wavelengths of interest . Absorption spectroscopy was advantageous in atom conservation. A spectrum is shown in Fig. 29.

As a transition to laser spectroscopy, one can bring out the instrumental broadening of grating spectroscopy by comparing its spectrum of 32.9-y ^{207}Bi to one obtained with use of laser scanning, Fig. 30.

 Springer

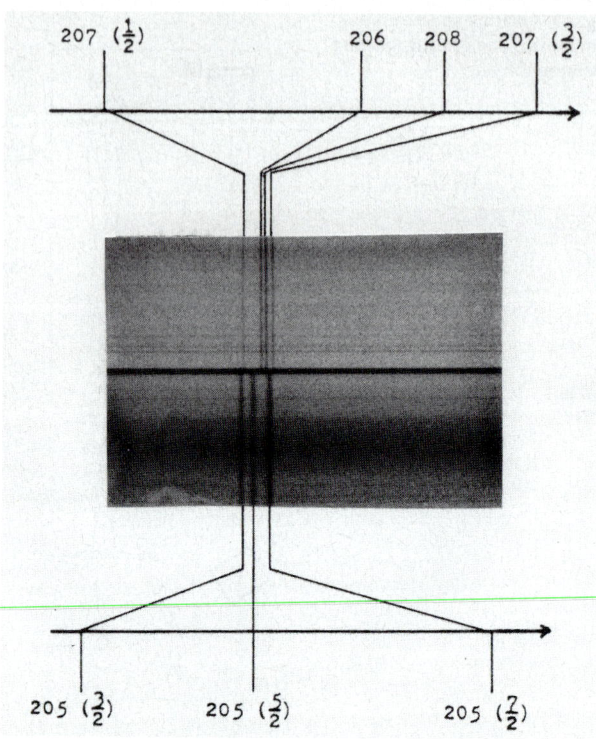

Fig. 29 Photographic absorption spectrum of ^{205}Pb and of natural lead

3.3 On-line experiments

I will restrict this overview to experiments at ISOLDE (CERN) as they are representative of developments elsewhere in the world. Work in laser-type spectroscopy is active in many laboratories: TRIUMF (Canada), GANIL and ORSAY (France), GSI, Mainz (Germany), Manchester (UK), ORNL, ANL, Stony Brook (USA), JINR (Dubna), Gatchina, Troitzk (Russia), Jyväskylä (Finland), Louvain-la-Neuve, Leuven (Belgium). New facilities are on the way at Lanzhou (China), INS (Japan). Others, that made early contributions to radioactive-atom spectroscopy, are no longer active in this field (Berkeley, Brookhaven, Harwell, Karlsruhe, Sussex). A major facility, EURISOL, is presently under study. There are comprehensive reviews, e.g. Otten, Kluge, Billowes and others [2, 4–8]. "On-line" refers to experiments done on isotopes directly as they are produced at the accelerator – though the isotopes may sometimes be first collected, e.g. in resonance cells.

3.3.1 RADOP

Nuclear orientation by optical pumping and detection of asymmetry in beta decay were already discussed in Section 3.2.4. This work was intensified at ISOLDE by Jochen Bonn, Gerhard Huber, Jürgen Kluge, and Ernst Otten in a series of experiments named RADOP (radioactive detection of optical pumping) [70], which allowed the detection of sudden deformations in neutron-deficient isotopes of

 Springer

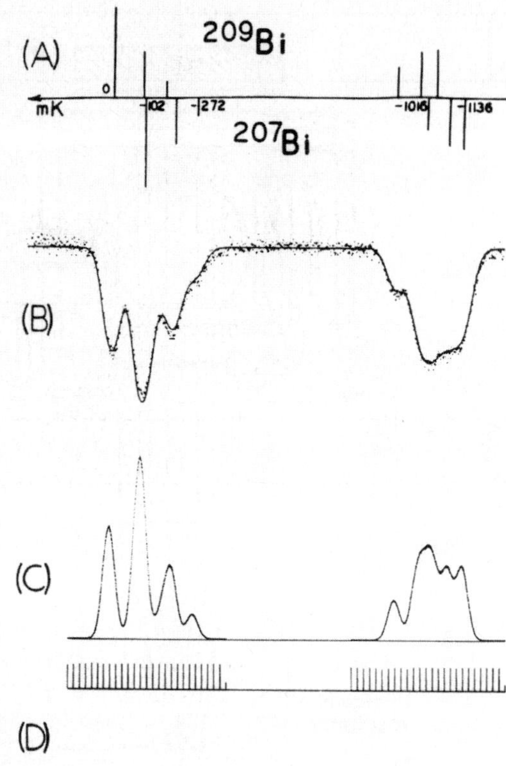

Fig. 30 A position of hfs components of ^{207}Bi and ^{209}Bi. **B** Dispersive spectrum. **C** Laser excitation-fluorescence spectrum. **D** Frequency markers (Zuyun Fang, PhD Thesis, NYU, 1988)

mercury [71]. In these early experiments, a Zeeman-tuned spectral lamp was still used for the optical pumping, Fig. 31.

3.3.2 Atomic beams

An extensive series of ABMR, non-laser experiments on cesium isotopes was done by Curt Ekström to obtain hfs systematics [72]. Use of the tuned laser as the spectrometric instrument appeared in atomic beam experiments by the Orsay group [73] on a long series of sodium isotopes. They could take pride in the measurement by this method of the first optical spectrum of the element francium [74]. Their experimental setup [75] is shown in Fig. 32.

The last of the atomic beam experiments, Fig. 33, is again of the magnetic resonance type, but in which the laser is used for state selection. It was designed for the measurement of the Bohr–Weisskopf effect in a long series of cesium isotopes [76], in an analogous way to those in [68]. Fig. 34 is a photo of H.T. Duong, a prime designer of this experiment.

As with other on-line atomic beam experiments, the ions from ISOLDE have to be neutralized and thermalized. While this was done successfully in prior experiments by stopping, neutralizing the ISOLDE ion beam on an yttrium surface, and re-evaporating to produce the thermal atomic beam, in this case the surface effect

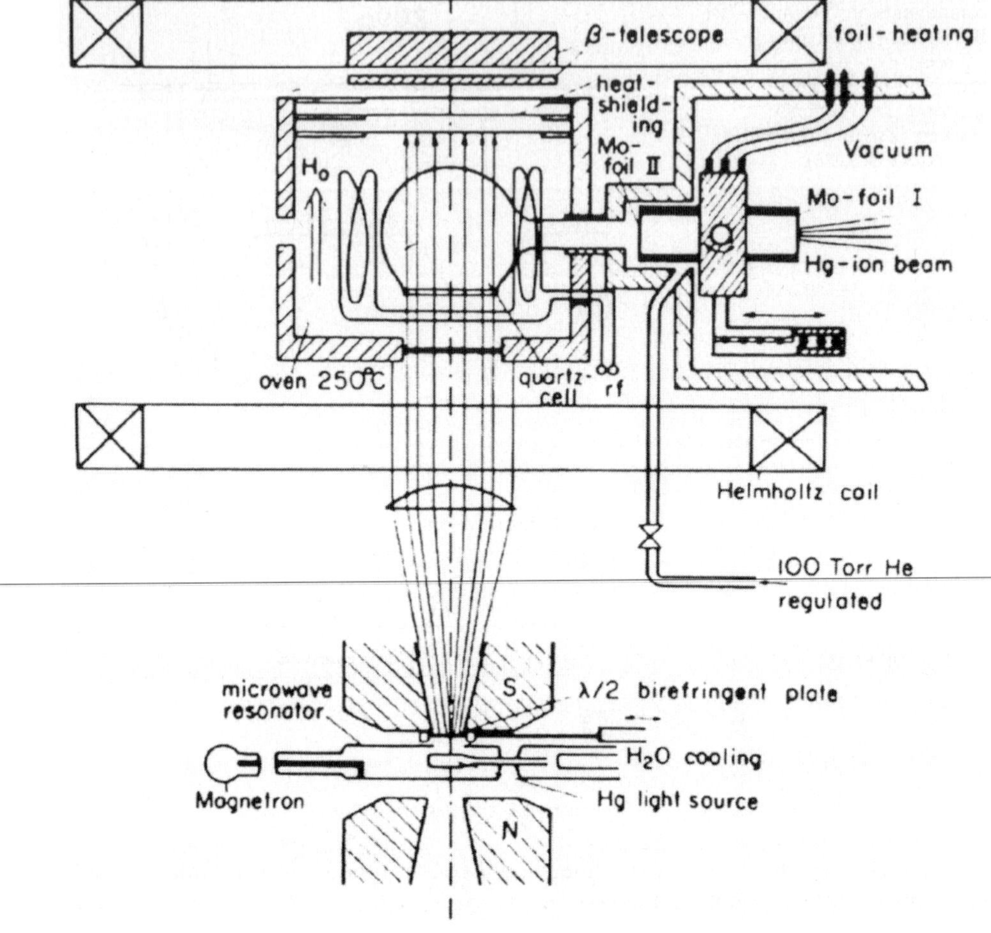

Fig. 31 RADOP setup for measurement of 2.4-m ^{187}Hg (from *Optical Pumping of Neutron Deficient* ^{187}Hg [70], ©1971, with permission of Elsevier)

was not reliable. After considerable efforts the neutralizer, shown in Fig. 35, finally allowed the precision measurement of the hfs of 1.64-m ^{126}Cs. This Bohr-Weisskopf study awaits the future.

3.3.3 Collinear, RIS

Collinear laser spectroscopy and resonance ionization (mass) spectroscopy (RI(M)S [77]) laid the groundwork for many experiments, a number of which are reported at the VII International Workshop Laser 2006. Collinear laser spectroscopy originated in Mainz, where the initial experiments were performed [78]. A schematic of the experiment as set up at ISOLDE [79] is shown in Fig. 36.

In the acceleration process, the velocity spread of the atom source is reduced to give Doppler-free spectroscopy. With a fixed-frequency laser, the hyperfine spectrum

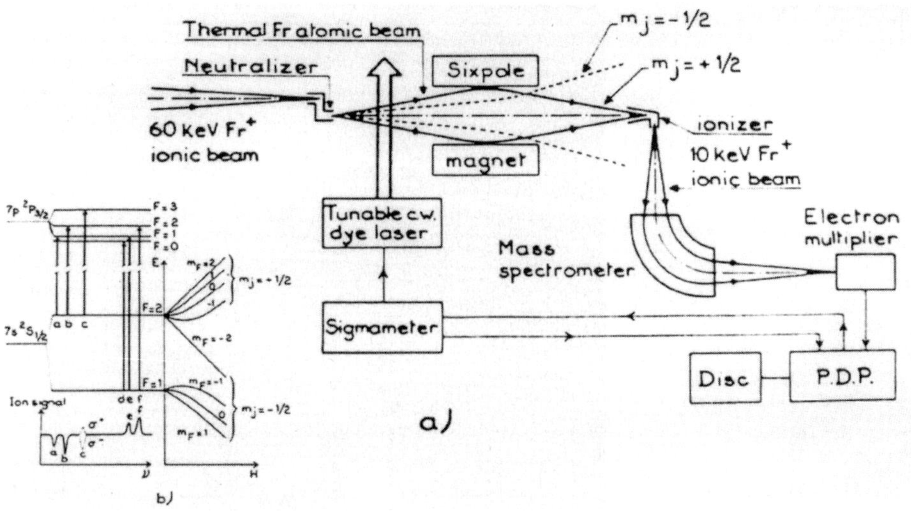

Fig. 32 **a**) Schematic of apparatus for the D2 francium-line measurement. **b**) Signals, exhibiting optical pumping effects in transitions (adapted from [75]; courtesy ©American Physical Society, http://prola.aps.org/abstract/PRC/v23/i6/p2720_1)

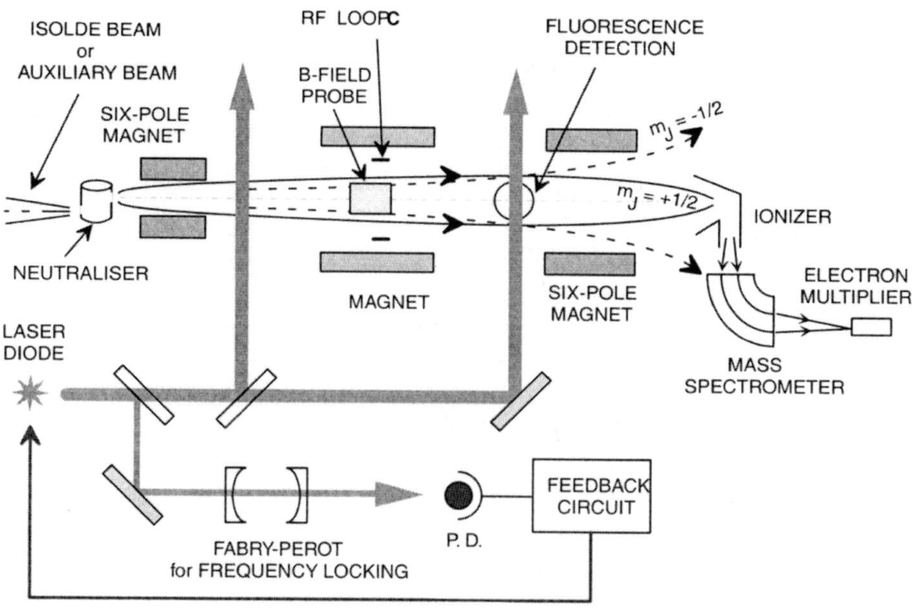

Fig. 33 ABMR apparatus with laser state selection (from *Precision hfs of* $^{126}Cs(T_{1/2} = 1.63m)$ *by ABMR* [76]; ©2005, with permission of Elsevier)

can be tuned across that frequency, and, in the figure shown, the fluorescence detected. An offshoot is the COLLAPS collaboration experiment in which the collinear setup ends with a crystal in which the selected atomic beam is embedded. Beta-decay asymmetry is then used to detect the resonance frequencies. Magda

Fig. 34 H.T. Duong (courtesy Jacque Pinard)

Fig. 35 Orthotropic source: ions which are not neutralized can be returned to the yttrium surface for more chances of neutralization (following design of T. Dinneen, A. Ghiorso, H. Gould, Rev. Sci. Instrum. 67(1996)752; from Ref.[76], © with permission of Elsevier, as in Fig. 33)

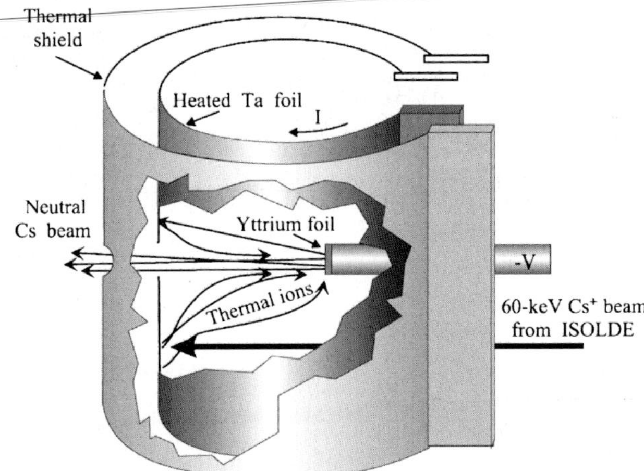

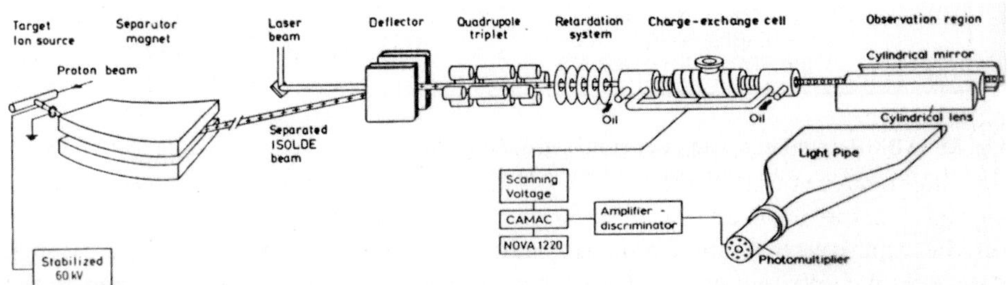

Fig. 36 Collinear laser spectroscopy schematic (adapted from Ref. [79], courtesy of Elsevier)

Fig. 37 Rainer Neugart, a pillar of ISOLDE laser spectroscopy! Standing on the *left*, Ken Ledingham, and in the background, Paul Campbell (courtesy H.-J. Kluge)

Kowalska and Kieran Flanagan are presenting results of experiments with these techniques in this Workshop. I add one more collinear laser spectroscopy experiment: the measurement of the francium D1 line [80]. This was another example of a difficult laser experiment: searching for the wavelength of an optical transition over a wide wavelength region with a fine-tooth comb. In this respect, the old-fashioned diffraction grating spectroscopy has the advantage of displaying lines in large spectral regions! At this point it is good to give particular mention of Rainer Neugart, Fig. 37, who has been intimately associated with the success of a good many of these experiments.

RI(M)S, which was initiated independently at ISOLDE, Orsay, and Gatchina, developed later at CERN into the COMPLIS (COllaboration for spectroscopy Measurement using a Pulsed Laser Ion Source) experiments for the study of refractory elements not produced directly as an ISOL beam. The atoms studied are radioactive daughters of beam ions implanted in graphite, desorbed by a high-power laser pulse, followed by resonance ionization spectroscopy, and time-of-flight detection, Fig. 38. These were substantially efforts by Mainz, McGill, Orsay, Gatchina, Troitzk, and

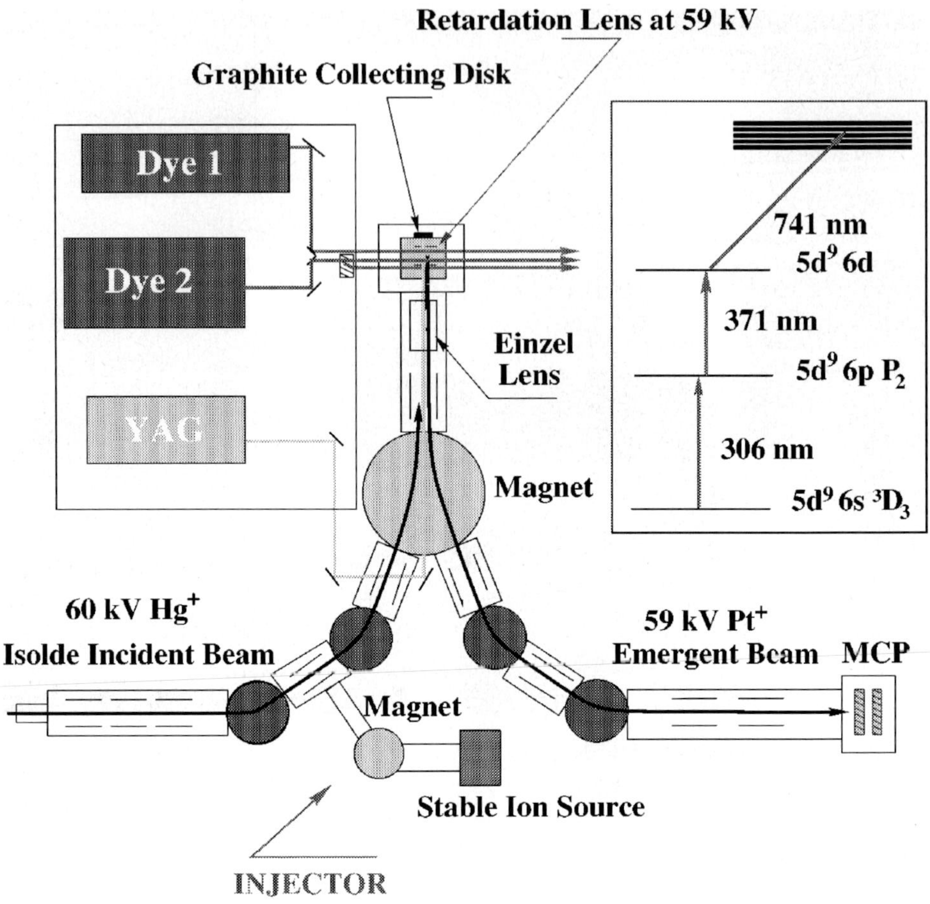

Fig. 38 COMPLIS experiment (from Ref. [86]). The transitions shown are for platinum

ISOLDE [81–86]. Rosa Sifi, Fig. 39, presents current work on tellurium isotopes at this Workshop.

4 Selected results

Atomic spectroscopy contributed many early data on nuclear spins, nuclear size, magnetic dipole and electric quadrupole moments which were important in the development of nuclear models and interactions. Although we are treating here mainly radioactive atoms we want at least to mention spectroscopic experiments that had *fundamental* implications: the quadrupole moment of the deuteron in revealing the tensor force in nuclei, the Lamb shift and the value of the gyromagnetic ratio of the electron in showing effects of quantum electrodynamics. But many of the spectroscopic experiments that we discuss led more to systematic properties from which one hopes to extract detailed nuclear data.

Fig. 39 From the most senior in the field of radioactive spectroscopy, Gerhard Huber, to the youngest, PhD candidate Rosa Sifi (courtesy H.-J. Kluge)

4.1 Isotope shifts and nuclear charge distributions

It is work with radioactive isotopes that has permitted systematic studies over large ranges of isotopes. A spectacular result was the shape staggering in the nuclear radii of mercury isotopes [87], shown in Fig. 40.

Another systematic study is of isotope shifts crossing magic numbers. A typical example, for cesium [88],[3] is shown in Fig. 41.

As we point out in Section 4.2, an effect of the magic-neutron isotope, ^{137}Cs, also appears in the Bohr–Weisskopf effect.

4.1.1 Odd–even staggering

It has been observed since the early days of isotope shift studies that the addition of a single neutron to an even-number neutron isotope generally produces less than one

---

[3]See, also, Otten [4].

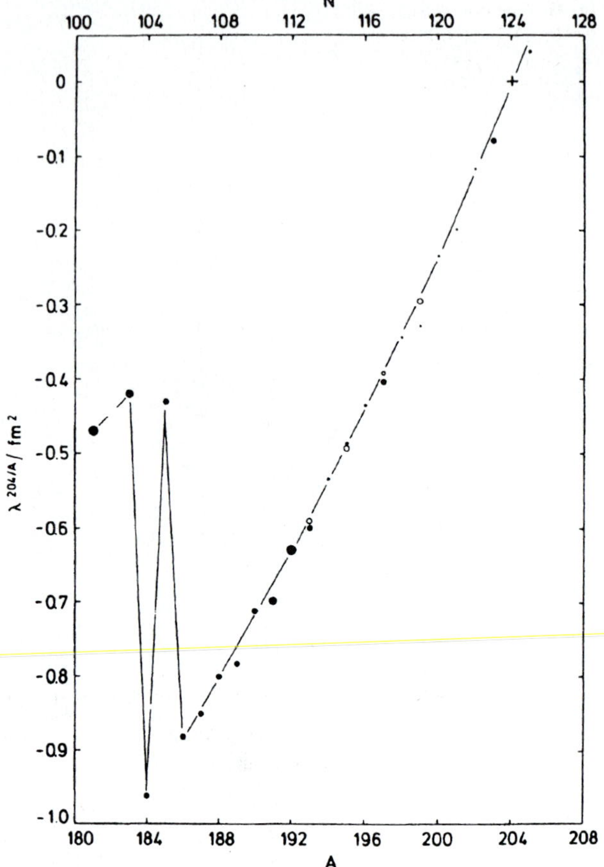

Fig. 40 Isotope shifts in mercury isotopes reflecting changes in the mean square radii (from Ref. [87]). The sudden changes for the lightest isotopes have been found to correspond to changes in deformation (courtesy ©American Physical Society, http//prola.aps.org/abstract/PRL/v39/i4/p180_1)

half of the shift caused by the addition of a neutron pair. We have described this by a "staggering parameter",

$$\gamma \equiv \frac{<r^2>_{N+1} - <r^2>_N}{\frac{1}{2}[<r^2>_{N+2} - <r^2>_N]}.$$

N refers to an even-neutron number nucleus. The parameter γ is more sensitive to details in the variations of the isotope shifts than what we can see in a plot, such as Fig. 41. In fact, with Jürgen Kluge and Dieter Proetel [89], we found possible to account for the trend of the odd-even staggering parameters, γ, for the mercury nuclear isomers. Fig. 42 is a photo of Kluge. He has been a prime mover in many important developments of modern radioactive-atom spectroscopy, and now in experiments on precision mass measurements with traps.

We have approached isotope shifts from the point of view of *relative shifts*, which allows us to compare *isotonic shifts*, as shown in Fig. 43, from [90]

This picture has been extended recently to radium and radon, with continuing isotonic similarities. The possibility of obtaining insight into the nuclear-neutron interaction from these measurements is being studied with Kieran Flanagan.

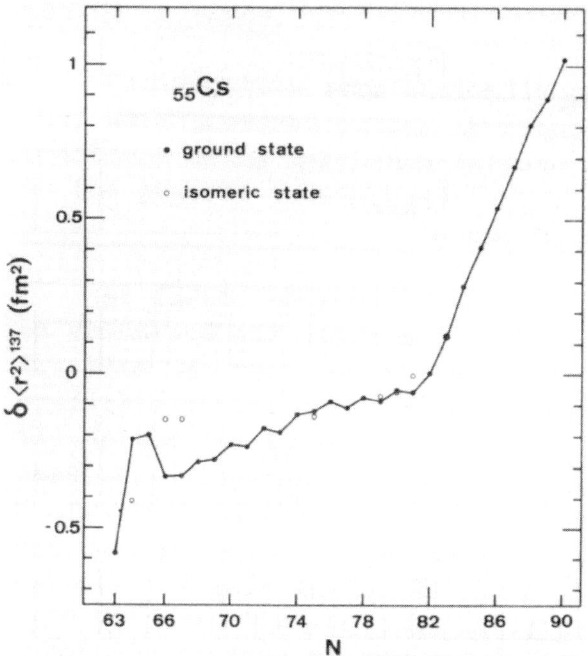

Fig. 41 Cesium isotope shifts: note break in $\Delta < r^2 >$ at the magic number $N=82$ (from [88])

Fig. 42 Jürgen Kluge (photo by author)

There are other systems which I have to pass over, such as isomer shifts in fission isomers and shifts in lithium halo nuclei. The physics that one can expect to obtain from isotope shifts is not exhausted!

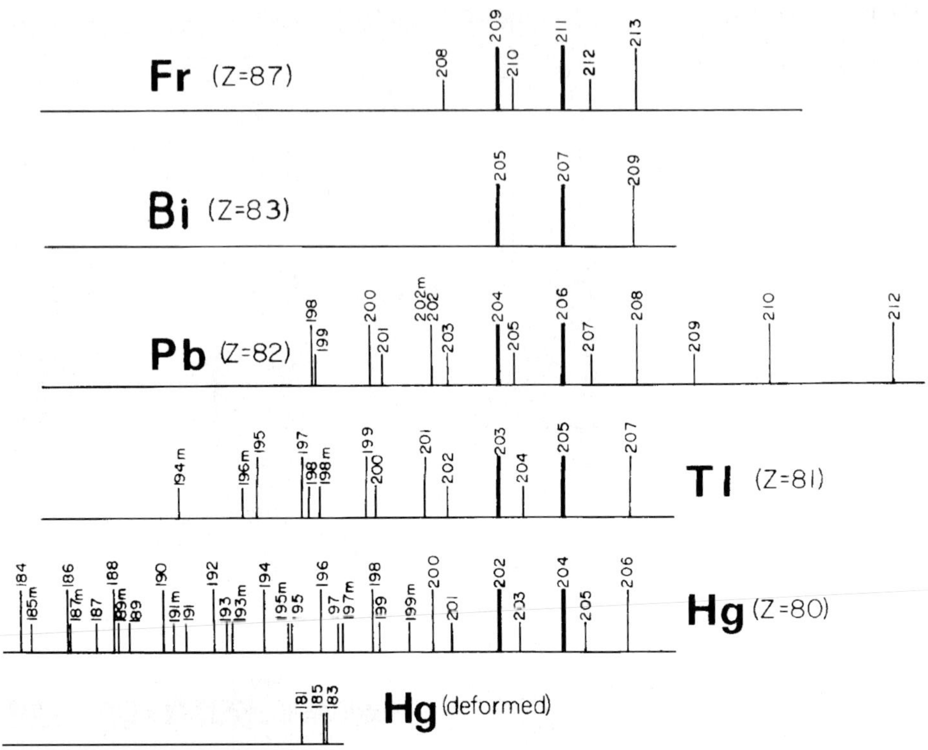

Fig. 43 Isotone shifts in the region of the doubly-magic nucleus, ^{208}Pb (from [90])

Fig. 44 "Hfs anomalies", or Bohr–Weisskopf effect in cesium radioisopes (from [68], courtesy ©American Physical Society, http://prola.aps.org/abstact/PR/v105/i2/p590_1)

| Isotopes | | Isotope 1 | | $-\Delta_{12}$ |
1	2	I	$\mu(\mu_N)$	percent
^{135}Cs	^{133}Cs	7/2	2.7290(3)	+0.037±0.009
^{137}Cs	^{133}Cs	7/2	2.8382(3)	+0.009±0.010
^{137}Cs	^{135}Cs			-0.020±0.009
^{134}Cs	^{133}Cs	4	2.9900(9)	+0.169±0.030
^{133}Cs		7/2	2.57893(3)	

4.2 Extended nuclear magnetization

In Section 2.2.2 this effect, additional to the simple magnetic dipole electron-nuclear interaction was introduced. Experimentally, it is revealed by the difference, Δ, between the ratio of the hfs interaction constants, a, for two isotopes and the ratio

Springer

Fig. 45 Hartmut Backe. Valentin Fedoseev can be recognized in the background (courtesy H.-J. Kluge)

of the nuclear gyromagnetic ratios, g. We have $a = a_{point-nucleus}(1 + \varepsilon)$, $\Delta_{12} = \varepsilon_1 - \varepsilon_2$ and take $\left(\frac{a_1}{a_2}\right)_{point} = \frac{g_1}{g_2}$. Then

$$\frac{a_1}{a_2}\frac{g_2}{g_1} - 1 \equiv \Delta.$$

For a set of cesium radioisotopes [68] it was found that while the magnetic moments change monotonically for three nuclei with identical spins, Δ changes signs between successive pairs, Fig. 44. This would not be the case were ε simply proportional to the nuclear magnetic moment.

While, qualitatively, a magic neutron number effect could be invoked because of N=82 in ^{137}Cs, the results required a picture that takes into account details of the nucleon configurations. This motivated a more quantitative approach, which involved nuclear configuration mixing theory [91].

5 Conclusion

Working back, magnetic moments, together with the Bohr-Weisskopf effect, can give information on neutron wave functions in nuclei. As suggested, isotope shifts can also provide such insights from systematic studies of relative shifts. Additionally, these

Fig. 46 The two Klaus's: Wendt, *left*, Jungmann, *right* (courtesy H.-J. Kluge)

can provide a handle on the neutron-nuclear interaction. Magnetic moments and the Bohr–Weisskopf effect also find importance in extracting quantum–electrodynamic effects from the hfs of atoms highly stripped of electrons. These are studied extensively at GSI, Darmstadt, and are reported at this Workshop by Hartmut Backe (Fig. 45).

Knowledge of neutron wave functions is also required in the interpretation of isotopically-differential parity non-conservation experiments. Such fundamental problems were discussed at the Workshop by Klaus Jungmann, Fig. 46.

Magnetic moments can, in particular nuclei, provide information on pion-exchange contributions. And then we can venture far afield in applications, such as the relation of the neutron-rich skin of heavy nuclei and neutron star structure, isotope-shift data for space-time variation of the fine structure constant, and surely a good number of exciting, still undiscovered new endeavors!

Fig. 47 Krassimira Marinova, moving spirit of the Workshop, *first row*. *Second row*, left to right, Boris Markov, Yury Gangsky; *third row*, Valentin Fedoseev, Hartmut Backe, Werner Lauth; *fourth row*, Serguei Zemlyanoi, Nikolai Tarantin, Zheng-Tian Lu; *fifth row*, Dana Borremans, Klaus Blaum, and two seats over to the right, Wilfried Nörteshäuser; Takashi Inamura is seen at upper right hand corner (courtesy R. Jarzebinska)

Acknowledgements I am grateful to the organizers of the Workshop, in particular Jürgen Kluge and Krassimira Marinova, Fig. 47, for the invitation to present at this Workshop this (somewhat biased toward personal experience) sketchy rendition of radioactive atom spectroscopy. I also thank the Physics Faculty of the Adam Mickiewicz University, Poznan, for their hospitality. I appreciate Tak Inamura's enlightening me on the early contributions of Nagaoka to hfs and isotope shift studies. I also want to acknowledge long-term collaborations, and friendly and stimulating interactions with many colleagues, among them Olav Redi, with them much of my work was carried out, Jacques Pinard and H.T. Duong, Laboratoire Aimé Cotton, Jürgen Kluge, Gerhard Huber, Rainer Neugart, Ernst Otten, Mainz and GSI, and many others whose names would fill a good number of pages. They built on the foundations of I.I. Rabi, Hans Kopfermann, Pierre Jacquinot, Jerrold Zacharias, Francis Bitter, Jean Brossel, Alfred Kastler, to name a few, and, in turn, trained a whole new generation of young physicists, many of whom are presenting their contributions at this Workshop.

References

1. Nierenberg, W.A.: Annu. Rev. Nucl. Sci. **7**, 349 (1957)
2. Jacquinot, P., Klapisch, R.: Rep. Prog. Phys. **42**, 773 (1979)
3. Bemis Jr., C.E., Carter, H.K.: Lasers in Nuclear Physics. Harwood Academic, Chur (1982)
4. Otten, E.-W.: In Treatise on Heavy-Ion Science, vol. 8, Bromley, D.A. (ed.), Plenum, New York, pp. 517 (1989)
5. Vergnes, M., Goutte, D., Heenen, P.H., Sauvage, J. (eds.): Nuclear shapes and nuclear structure at low excitation energies, Editions Frontières, Gif-sur-Yvette, (1994)
6. Kluge, H.-J., Nörteshäuser, W.: Spectrochim. Acta, Part B: Atom. Spectrosc. **58**, 1031 (2003)
7. Billowes, J.: Nucl. Phys., A **752**, 309c (2005)
8. Billowes, J.: Hyperfine Interact. **162**, 63 (2005)

9. Rao, G.N. (ed.): Modern Optics, Lasers and Laser Spectroscopy, Hyperfine Interactions, vol. 27 and 38 (1987)
10. Inamura, T.T., Wakasugi, W. (eds.): Lasers in Nuclear Physics, Hyperfine Interactions, vol. 74 (1992)
11. Schweikhard, L., Kluge, H.-J. (eds.): Atomic Physics at Accelerators: Laser Spectroscopy and Applications (APAC99), Hyperfine Interactions, vol. 127 (2000)
12. Blaszczak, Z., Markov, B., Marinova, K. (eds.): Laser 2004: Application of Lasers in Atomic Nuclear Research, ed. Hyperfine Interactions, vol. 162 (2005)
13. Pinard, J., Stroke, H.H.: Adv. Atom. Molec. Opt. Phys. **51**, 273 (2005)
14. Michelson, A.A.: Studies in Optics. Phoenix-University of Chicago, Chicago, p.34 (1927)
15. Françon, M.: Optical Interferometry. Academic, New York, 1966
16. Colloques Internationaux du CNRS, No. 161, 1966, Editions du CNRS, Paris, 1967 (references to work by Connes, P., Connes, J., Jacquinot, P., Pinard, J., et al.)
17. de Marcillac, P., Coron, N., Dambier, G., Leblanc, J., Moalic, J.-P.: Nature, **422**, 876 (2003)
18. Pauli, W.: Naturwiss. **12**, 741 (1924)
19. Back, E., Goudsmit, S.: Z. Phys. **43**, 321 (1927)
20. Back, E., Goudsmit, S.: Z. Phys. **47**, 174 (1928)
21. Inamura, T.T.: Hyperfine Interact. **127**, 31 (2000)
22. Nagaoka, H., Mishima, T.: Proc. Imp. Acad. **2**, 249 (1926)
23. Condon, E.U., Shortley, G.H.: Theory of Atomic Spectra. University Press, Cambridge, pp. 420 (1935)
24. Rosenthal, J.E., Breit, G.: Phys. Rev. **41**, 459 (1932)
25. Yagoda, H.: Phys. Rev. **38**, 2298 (1931)
26. Yagoda, H.: Phys. Rev. **40**, 1017 (1932)
27. Cox, R.T., McIlwraith, C.G., Kurrelmeyer, B.: Proc. Natl. Acad. Sci. U.S.A. **14**, 544 (1928)
28. Weisskopf, V.F.: Ann. Phys. (Leipzig) **9**, 23 (1931) (Eq. 54)
29. Colgrove, F.D., Franken, P.A., Lewis, R.R., Sands, R.H.: Phys. Rev. Lett. **3**, 420 (1959)
30. Bitter, F.: Appl. Opt. **1**, 1 (1962)
31. Perey, M.: C.R. Acad. Sci. **208**, 97 (1939)
32. Liberman, S., Pinard, J., Duong, H.T., Juncar, P., Vialle, J.-L., Jacquinot, P., Huber, G., Touchard, F., Büttgenbach, S., Pesnelle, A., Thibault, C., Klapisch, R.: C.R. Acad. Ser. B, **286**, 253 (1978)
33. Liberman, S., Pinard, J., Duong, H.T., Juncar, P., Pillet, P., Vialle, J.-L., Jacquinot, P., Touchard, F., Büttgenbach, S., Thibault, C., de Saint-Simon, M., Klapisch, R., Pesnelle, A., Huber, G.: Phys. Rev., A **22**, 2732 (1980)
34. Goudsmit, S., Bacher, R.F.: Phys. Rev. **34**, 1501 (1929)
35. Casimir, H.B.G.: In: Archives du Musée Teyler, Ser. III, VIII, Martinus Nijhoff, The Hague, pp. 201 (1936)
36. Casimir, H.B.G., Karreman, G.: Physica **9**, 494 (1942)
37. Jaccarino, V., King, J.G., Satten, R.A., Stroke, H.H.: Phys. Rev. **94**, 1798 (1954)
38. de-Shalit, A., Talmi, I.. Nuclear Shell Theory. Academic, New York, pp. 215 (1963)
39. Bohr, A., Weisskopf, V.F.: Phys. Rev. **77**, 94 (1950)
40. Kopfermann, H.: Kernmomente, Akademische Verlagsgesellschaft, Leipzig (1940)
41. Kopfermann, H.: Nuclear Moments. Academic, New York, (1958)
42. Cohen, V.W.: Phys. Rev. **46**, 713 (1934)
43. Jaccarino, V., Bederson, B., Stroke, H.H.: Phys. Rev. **87**, 676 (1952)
44. Rabi, I.I., Zacharias, J.R., Millman, S., Kusch, P.: Phys. Rev. **53**, 318(L) (1938)
45. Zacharias, J.R.: Phys. Rev. **61**, 270 (1942)
46. Davis Jr., L., Nagle, D.E., Zacharias, J.R.: Phys. Rev. **76**, 1068 (1949)
47. Gilbert, D.A., Cohen, V.W.: Phys. Rev. **95**, 569 (1955)
48. Lemonick, A., Pipkin, F.M., Hamilton, D.R.: Rev. Sci. Instrum. **26**, 1112 (1955)
49. Bitter, F.: Phys. Rev. **76**, 833 (1949)
50. Brossel, J., Bitter, F.: Phys. Rev. **86**, 308 (1952)
51. Bitter, F.: Selected papers and commentaries, Erber T., Fowler, C.M. (eds.), MIT, Cambridge, Massachusetts (1969)
52. Hirsch, H.R.: PhD Thesis, MIT (1960)
53. Hirsch, H.R.: Opt. J. Soc. Am. **51**, 1192 (1961)

54. Sorokin, P.P., Lankard, J.P.: IBM J. Res. Dev. **10**, 162 (1966)
55. Schäfer, F.P., Schmidt, W., Volze, J.: Appl. Phys. Lett. **9**, 306 (1966)
56. Schein, M.: Suppl. 1, Helv. Phys. Acta **2**, 73 (1929)
57. Kohler, R.H.: PhD Thesis, MIT, (1961)
58. Kohler, R.H.: Phys. Rev. **121**, 1104 (1961)
59. Brot, P.C.: J. Phys. Radium **22**, 412 (1961)
60. Kastler, A.: J. Phys. Radium **11**, 255 (1950)
61. Bitter, F., Brossel, J.: Phys. Rev. **85**, 1051 (1952)
62. Mitchell, A.C.G., Zemansky, M.W.: Resonance Radiation and Excited Atoms. University Press, Cambridge (1934)
63. Cagnac, B., Barrat, J.P.: C.R. Acad. Sci. **249**, 534 (1959)
64. Cagnac, B.: PhD Thesis, Paris (1960)
65. Cagnac, B.: Ann. Phys. (Paris) **6**, 467 (1961)
66. Walter, W.T.: PhD Thesis, MIT (1962)
67. Walter, W.T.: Bull. Am. Phys. Soc. **7**, 295 (1962)
68. Stroke, H.H., Jaccarino, V., Edmonds Jr., D.S., Weiss, R.: Phys. Rev. **105**, 590 (1957)
69. Besch, H.J., Köpf, U., Otten, E.W.: Phys. Lett., B **25**, 120 (1967)
70. Bonn, J., Huber, G., Kluge, H.-J., Köpf, U., Kugler, L., Otten, E.W.: Phys. Lett., B **36**, 41 (1971)
71. Bonn, J., Huber, G., Kluge, H.-J., Otten, E.W.: Phys. Lett., B **38**, 308 (1972)
72. Ekström, C.: Adv. Quant. Chem. **30**, 361 (1998)
73. Huber, G., Touchard, F., Büttgenbach, S., Thibault, C., Klapisch, R., Duong, H.T., Liberman, S., Pinard, J., Vialle, J.L., Juncar, P., Jacquinot, P.: Phys. Rev., C **18**, 2342 (1978)
74. Liberman, S., Pinard, J., Duong, H.T., Juncar, P., Vialle, J.-L., Jacquinot, P., Huber, G., Touchard, F., Büttgenbach, S., Pesnelle, A., Thibault, C., Klapisch, R.: CERN Collaboration, C. R. Acad. Ser. B **286**, 253 (1978)
75. Thibault, C., Touchard, F., Büttgenbach, S., Klapisch, R., de Saint-Simon, M., Duong, H.T., Jacquinot, P., Juncar, P., Liberman, S., Pillet, P., Pinard, J., Vialle, J.L., Pesnelle, A., Huber, G.: Phys. Rev., C **23**, 2720 (1981)
76. Pinard, J., Duong, H.T., Marescaux, D., Stroke, H.H., Redi, O., Gustaffson, M., Nilsson, T., Matsuki, S., Kishimoto, Y., Kominato, K., Ogawa, J., Shibata, M., Tada, M., Persson, J.R., Nojiri, Y., Momota, S., Inamura, T.T., Wakasugi, M., Juncar, P., Murayama, T., Nomura, T., Koizumi, M.: ISOLDE collaboration. Nucl. Phys., A **753**, 3 (2005)
77. Kluge, H.-J.: Hyperfine Interactions **37**, 347 (1987)
78. Anton, K.-R., Kaufman, S.L., Klempt, W., Moruzzi, G., Neugart, R., Otten, E.W., Schinzler, B.: Phys. Rev. Lett. **40**, 642 (1978)
79. Mueller, A.C., Buchinger, F., Klempt, W., Otten, E.W., Neugart, R., Ekström, C., Heinemeier, J.: Nucl. Phys., A **403**, 234 (1983)
80. Duong, H.T., Juncar, P., Liberman, S., Mueller, A.C., Neugart, R., Otten, E.W., Peuse, B., Pinard, J., Stroke, H.H., Thibault, C., Touchard, F., Vialle, J.L., Wendt, K., and the ISOLDE Collaboration: Europhys. Lett. **3**, 175 (1987)
81. Wallmeroth, K., Bollen, G., Dohn, A., Egelhof, P., Grüner, J., Lindenlauf, F., Krönert, U., Campos, J., Rodriguez-Yunta, B., Borge, M.J.G., Venugopalan, A., Wood, J.L., Moore, R.B., Kluge, H.-J.: Phys. Rev. Lett. **58**, 1516 (1987)
82. Lee, J.K.P., Savard, G., Crawford, J.E., Thekkadath, G., Duong, H.T., Pinard, J., Liberman, S., Le Blanc, F., Kilcher, P., Obert, J., Oms, J., Putaux, J.C., Roussière, B., Sauvage, J.: ISOCELE Collaboration. Nucl. Instrum. Methods, B **34**, 252 (1988)
83. Alkhazov, G.D., Barzakh, A.E., Berlovich, É.I., Denisov, V.P., Dernyatin, A.G., Ivanov, V.S., Zherikhin, A.N., Kompanets, O.N., Letokhov, V.S., Mishin, V.I., Fedoseev, V.N.: JETP Lett. **37**, 274 (1983)
84. Fedoseyev, V.N., Letokhov, V.S., Mishin, V.I., Alkhazov, G.D., Barzakh, A.E., Denisov, V.P., Dernyatin, A.G., Ivanov, V.S.: Opt. Commun. **52**, 24 (1984)
85. See, also, Hurst, G.S., Payne, M.G., Kramer, S.D., Young, J.P.: Rev. Mod. Phys. **51**, 767 (1979)
86. Sauvage, J., Boos, N., Cabaret, L., Crawford, J.E., Duong, H.T., Genevey, J., Girod, M., Huber, G., Ibrahim, F., Krieg, M., Le Blanc, F., Lee, J.K.P., Libert, J., Lunney, D., Obert, J., Oms, J., Péru, S., Pinard, J., Putaux, J.-C., Roussière, B., Sebastien, V., Verney, D., Zemlyanoi, S., Arianer, J., Barré, N., Ducourtieux, M., Forkel-Wirth, D., Le Scornet, G., Lettry, J., Richard-Serre, C., Vernon, C.: Hyperfine Interact. **129**, 303 (2000)

87. Kühl, T., Dabkiewicz, P., Duke, C., Fischer, H., Kluge, H.-J., Kremmling, H., Otten, E.-W.: Phys. Rev. Lett. **39**, 180 (1977)
88. Liberman, S., Pinard, J., Duong, H.T., Juncar, P., Vialle, J.L., Pillet, P., Jacquinot, P., Huber, G., Touchard, F., Büttgenbach, S., Thibault, C., Klapisch, R., Pesnelle, A., In: Walther, H., Rothe, K. (eds.): Laser Spectroscopy IV. Springer, Berlin Heidelberg New York, pp. 527 (1979)
89. Stroke, H.H., Proetel, D., Kluge, H.-J.: Phys. Lett., B **82**, 204 (1979)
90. Barboza-Flores, M., Redi, O., Stroke, H.H.: Z. Phys., A **321**, 85 (1985)
91. Stroke, H.H., Blin-Stoyle, R.J., Jaccarino, V.: Phys. Rev. **123**, 1326 (1961)

Hyperfine Interact (2006) 171:41–55
DOI 10.1007/s10751-007-9510-y

Tests of fundamental symmetries and interactions – using nuclei and lasers

Klaus Peter Jungmann

Published online: 28 February 2007

Abstract State of the art laser technology and modern spectroscopic methods allow to address issues of fundamental symmetries and fundamental interactions in atoms with high precision experiments. In particular the discrete symmetries Parity (P), Charge Conjugation (C), Time Reversal (T) as well as their combinations CP and CPT are in the center of interest at present. Actual projects are concerned with Parity Violation in atoms, Time Reversal Violation in β-decays and searches for permanent Electric Dipole Moments (EDMs), and tests of CPT conservation in particle-antiparticle properties, in particular antiprotonic atoms.

Keywords Fundamental interactions · Fundamental symmetries · Precision measurements · Magnetic anomalies · β-decays · Electric dipole moments · Antiprotons · Radioactive beam facilities

PACS 11.30.-j · 11.30.Er · 06.20.Jr · 24.40.Bw

1 Introduction

The Standard Model (SM) in particle physics describes accurately all observations in this field. It appears that even recent spectacular observations in neutrino experiments can be included with moderate modifications. This far ranging theoretical framework lacks, however, a deeper and more satisfactory explanation for many of the facts which it describes so precisely. Among the open questions are the

---

This work is supported in part by the Dutch Stichting voor Fundamenteel Onderzoek der Materie (FOM) under programme number 48 (TRIμP).

K. P. Jungmann (✉)
Kernfysisch Versneller Instituut, University of Groningen, Zernikelaan 25, 9747 AA Groningen, The Netherlands
e-mail: jungmann@kvi.nl

large number of some 30 free parameters in the SM, the hierarchy of fundamental fermion masses, the number of three particle generations and the origin of Parity (P) Violation and combined Charge Symmetry (C) and Parity violation, i.e. CP-violation. If the SM is combined with Standard Cosmology, the dominance of mater over antimatter in the universe presents a serious unsolved puzzle. In order to provide answers to such intriguing questions speculative extensions have been constructed, such as supersymmetry, left–right symmetry, technicolor and many others. However, they have despite their elegancy no status in physics, yet, unless they can be experimentally verified by an observation of a unique prediction of one of these models.

We know two conceptually different approaches for confirming the SM and also to find New Physics beyond it:

– The direct observation of new particles or processes and
– Precise measurements of quantities, which can be calculated to sufficient accuracy within the SM, and where New Physics appears in a significant difference between theory and experiment.

Whereas the first approach typically is carried out in high energy physics, the second route uses experiments at low energies. Precision measurements at low energies offer indeed various possibilities to confirm the SM at a high level, to find new physics and to determine accurate values of important fundamental constants [1–4]. Stringent tests of the SM arise in particular through exploiting one of the recent achievements in atomic physics: cooling and storing of ions, atoms and molecules. This includes, e.g., precise measurements of magnetic anomalies [5–8], precision studies of nuclear β-decays [2–4, 9] and searches for permanent electric dipole moments of particles, nuclei, atoms and molecules [10] as well as spectroscopy of antiprotonic atoms [11].

In the recent years, several experiments have reported a few standard deviations differences between theoretical predictions and the measurements. Among those are experiments on the muon magnetic anomaly [5], the unitarity of the Cabbibo–Kobayashi–Maskawa matrix [12], nuclear β-decay [13], atomic parity violation [14] and many others. In some cases the differences disappeared completely after refinement of theory. However, not all of them [15]. Further work is needed to clarify the situation.

2 Fundamental symmetries and interactions

Symmetries play an important and central role in physics. Whereas global symmetries relate to conservation laws, local symmetries yield forces. Today four fundamental interactions are known in physics: (a) Electromagnetic, (b) Weak Interactions, (c) Strong Interactions, and (d) Gravitation. These forces are considered fundamental, because all observed dynamic processes in nature can be traced back to one or a combination of them. Together with fundamental symmetries they form the framework on which all physical descriptions ultimately rest. The Standard Model (SM) is a most remarkable theory. Electromagnetic, Weak and many aspects of Strong Interactions can be described to astounding precision in one single coherent picture. It is a major goal in modern physics to find a unified quantum field theory which includes all the four known fundamental forces. To achieve this, a satisfactory

 Springer

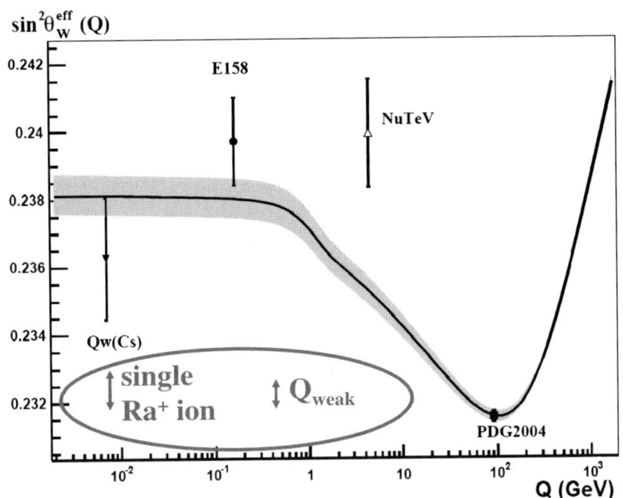

Fig. 1 Running of the weak mixing angle (see [16]). The present test of SM theory is not fully satisfactory. The estimated accuracy of future experiments is indicated in the inserted *ellipse*

quantum description of gravity remains yet to be found. This is a lively field of actual activity where string- and M-theories may offer a route to a better understanding of this aspect.

3 Discrete symmetries

3.1 Parity

The observation of Atomic Parity Violation was crucial for the acceptance of the SM as a unified electro-weak theory with a validity over several orders of magnitude in momentum transfer. The most recent completed experiment used Cs atoms [14] and it allows to extract a precise value of the weak mixing angle ($\sin^2 \Theta_W$). The running with energy [16] of this quantity is such that there is a minimum at the Z-pole and it is higher at lower and at higher energies due to the Abelian respectively non-Abelian character of QED and QCD. The experimental verification of that running is rather moderate (see Fig. 1).

Therfore, new precision experiments are indicated. Whereas the deep inelastic scattering experiment Q_{weak} at the Jefferson Laboratory (USA) will cover the intermediate energy range, new possibilities have emerged at atomic energies. In Fr atoms the weak effects are some 18 times bigger than in Cs. This advantage is planned to be exploited in experiments at Legnaro, Italy, and at Stony Brook, USA. These projects are being prepared by a number of auxiliary experiments, which measured important quantities such as transition frequencies and lifetimes [17–21]. Also trapping of Fr was achieved. The main hurdle to overcome will be the low number of Fr atoms available at the Legnaro and Stony Brook accelerators. Moving the experiments to the ISAC facility at TRIUMF, Canada, or the ISOLDE facility of CERN, Switzerland, may offer the necessary numbers of atoms.

It has been suggested to employ single heavy alkali-like ions and to observe light shifts in n ^{2}S$_{1/2}$-(n-1) ^{2}D$_{3/2}$ transitions. The detection scheme works via a sequence

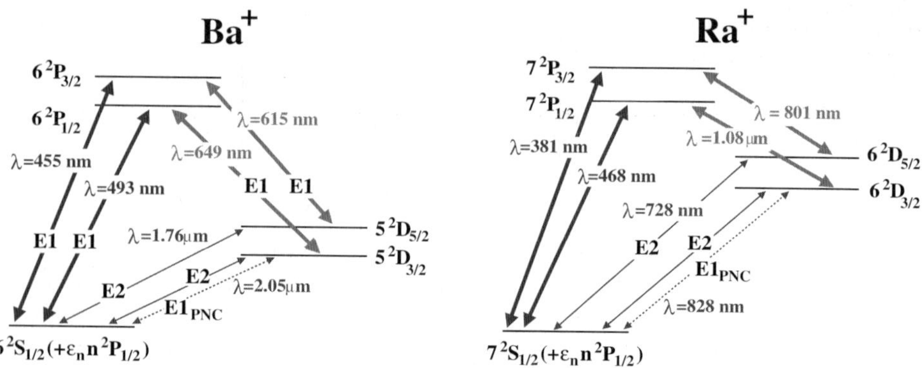

Fig. 2 Energy levels of Ba$^+$ and Ra$^+$. The atomic parity effect arises from admixtures of n $^2P_{1/2}$ states to the $^2S_{1/2}$ ground state of the ion owing to Z^0-boson exchange in the atomic system

of atomic excitations and relies on the technique of electron shelving. At the University of Washington promising preparatory experiments using Ba$^+$ have been performed [22]. Ra$^+$ ions have not only some 20 times larger parity effects, their relevant transitions also are easier accessible with all solid state laser systems. Such an experiment is considered at the Kernfysisch Versneller Instituut in Groningen, Netherlands. The availability of several isotopes with different nuclear spins would allow to study nuclear spin dependence and nuclear anapole moments. Because parity effects will be extracted through an interference of electromagnetic and weak forces, better knowledge of atomic wave functions at the sub-percent level will be mandatory – a posed challenge for atomic theorists. This challenge holds in particular for all atomic parity experiments with heavy elements (Fig. 2).

An interesting new development in this field comes from a suggestion of a hydrogen atomic parity experiment using a spin echo technique and Casimir forces near solid state surfaces (Nachtmann and DeKiewiet, private communication, 2006) [23]. The extremely high sensitivity of spin echo techniques may be sufficient to overcome the disadvantage of the smallness of atomic parity effects in hydrogen due to the low nuclear charge. The advantage of hydrogen lies in the fact that the theoretical calculations of the weak effects are straightforward and can be performed to high precision.

3.2 CP and T-violation

3.2.1 β-decays

In standard theory the structure of weak interactions is V-A, which means there are vector (V) and axial-vector (A) currents with opposite relative sign causing a left handed structure of the interaction and parity violation [24]. Other possibilities like scalar, pseudo-scalar and tensor type interactions which might be possible would be clear signatures of new physics. So far they have been searched for without positive result. However, the bounds on parameters are not very tight and leave room for various speculative possibilities. The double differential decay probability

$d^2W/d\Omega_e d\Omega_\nu$ for a β-radioactive nucleus is related to the electron and neutrino momenta **p** and **q** through

$$\frac{d^2W}{d\Omega_e d\Omega_\nu} \sim 1 + a\,\frac{\mathbf{p}\cdot\mathbf{q}}{E} + b\,\sqrt{1-(Z\alpha)^2}\,\frac{m_e}{E} + <\mathbf{J}>\cdot\left[A\,\frac{\mathbf{p}}{E} + B\,\mathbf{q} + D\,\frac{\mathbf{p}\times\mathbf{q}}{E}\right]$$

$$+ <\sigma>\cdot\left[G\,\frac{\mathbf{p}}{E} + Q\,\mathbf{J} + R\ <\mathbf{J}>\times\frac{\mathbf{q}}{\mathbf{E}}\right]$$

where m_e is the β-particle mass, E its energy, σ its spin, and **J** is the spin of the decaying nucleus. The coefficients D and R are studied in a number of experiments at this time and they are T violating in nature. Here D is of particular interest for further restricting model parameters. It describes the correlation between the neutrino and β-particle momentum vectors for spin polarized nuclei. The coefficient R is highly sensitive within a smaller set of speculative models, since here exist some already well established constraints, e.g., from searches for permanent electric dipole moments [24].

From the experimental point of view, an efficient direct measurement of the neutrino momentum is not possible. The recoiling nucleus can be detected instead and the neutrino momentum can be reconstructed using the kinematics of the process. Of course, the problem can also be reformulated using the kinematic variables of the recoil nucleus and the β-particle only [25]. Since the recoil nuclei have typical energies in the few 10 eV range, precise measurements can only be performed, if the decaying isotopes are suspended using extreme shallow potential wells. Such exist, for example, in atom traps formed by laser light, where many atomic species can be stored at temperatures below 1 mK. An overview over actual activities can be found in [9]. Such research is being performed at a number of laboratories worldwide.

In a recent measurement at Berkeley, USA, the asymmetry parameter a in the β-decay of ^{21}Na has been measured for optically trapped atoms [13]. The value differs from the present SM value by about 3 standard deviations. In order to explore whether this could be an indication of new physics reflected in new interactions in β-decay, the $(\beta + \gamma)/\beta$ decay branching ratio was re-measured at Texas A&M and at KVI, because some 5 measurements existed which in part disagreed significantly. The new values of 4.74(4)% [26] and 4.85(12)% (Achouri, private communication, 2006) agree well and do not affect the SM prediction in a significant way. The still remaining difference may be explained by Na dimer formation in the trap (Scielzo, private communication, 2006). The most stringent limit on scalar interactions for β-neutrino correlation measurements comes from an experiment on the pure Fermi decay of ^{38m}K at TRIUMF, where a was extracted to 0.5% accuracy and in good agreement with standard theory [27].

3.2.2 Permanent electric dipole moments (EDMs)

An EDM of any fundamental particle violates both parity and time reversal (T) symmetries. With the assumption of CPT invariance a permanent dipole moment also violates CP. EDMs for all particles are caused by CP violation as it is known from the K systems through higher order loops. These are at least 4 orders of magnitude below the present experimentally established limits. Indeed, a large number of speculative models foresees permanent electric dipole moments which could be as large as the present experimental limits just allow. Historically the non-observation

Table 1 Actual limits on permanent electric dipole moments [30–33]

Particle	Limit/measurement [e-cm]	method
e	$< 1.6 \times 10^{-27}$	Tl atomic beam (Berkeley)
μ	$< 2.8 \times 10^{-19}$	Muon g-2 storage ring (Brookhaven)
n	$< 3.0 \times 10^{-26}$	Stored cold neutrons (Grenoble)
Hg-atom	$< 2.1 \times 10^{-28}$	Hg vapour cell (Seattle)

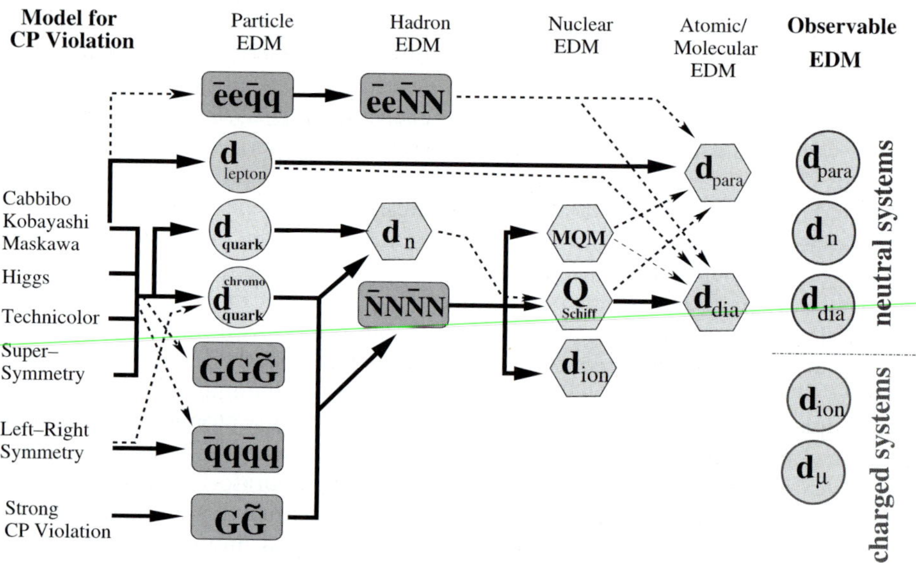

Fig. 3 A variety of theoretical speculative models exists in which an EDM could be induced through different mechanisms or a combination of them into fundamental particles and composed systems for which an EDM would be experimentally accessible. Up to now very sensitive experiments were only carried out for composed neutral systems. A novel technique may allow to sensitively access EDMs also for charged fundamental particles and ions

of permanent electric dipole moments has ruled out more speculative models than any other experimental approach in all of particle physics. EDMs have been searched for in various systems with different sensitivities (Table 1). In composed systems such as molecules or atoms fundamental particle dipole moments of constituents may be significantly enhanced [28]. Particularly in polarizable systems there can exist large internal fields.

There is no system which can be preferred over other to search for an EDM on any sound theoretical basis or insights. Only if any of the (not yet proven) scenarios of physics beyond the SM is chosen, one or the other experimental search may be favoured (see Fig. 3) [10, 29]. In fact, many systems need to be examined, because depending on the underlying process different systems have in general quite significantly different susceptibility to acquire an EDM through a particular mechanism. An EDM may be found an 'intrinsic property' of an elementary particle as we know them, because the underlying mechanism is not accessible at present.

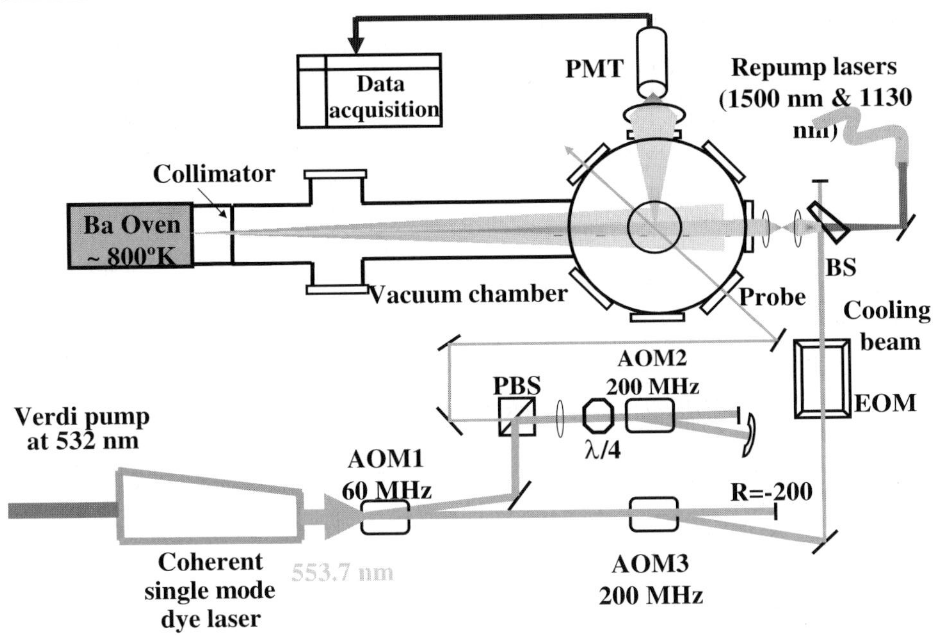

Fig. 4 Laser system to demonstrate cooling of a Ba atomic beam

However, it can also arise from CP-odd forces between the constituents under observation, e.g. between nucleons in nuclei or between nuclei and electrons. Such EDMs could be much larger than such expected for elementary particles originating within the popular, usually considered beyond standard theory models. No other constraints are known. It is therefore important to see all EDM experiments in a common context.

This highly active field of research has recently received a boost from a number of novel developments. One of them concerns the Ra atom, which has rather close lying $7s7p\,^3P_1$ and $7s6d\,^3D_2$ states. Because they are of opposite parity, a significant enhancement has been predicted for an electron EDM [34, 35], much higher than for any other atomic system. Further more, many Ra isotopes are in a region where (dynamic) octupole deformation occurs for the nuclei, which also may enhance the effect of a nucleon EDM substantially, i.e. by some two orders of magnitude. From a technical point of view the Ra atomic levels of interest for en experiment are well accessible spectroscopically and a variety of isotopes can be produced in nuclear reactions. The advantage of an accelerator based Ra experiment is apparent, because EDMs require isotopes with spin and all Ra isotopes with finite nuclear spin are relatively short-lived. For a successful experiment polarized Ra atoms need to be trapped in an atom trap where they can sufficiently long interact with an external electric field. Two Ra EDM experiments are at present prepared at the Argonne National Laboratory, USA, and at the Kernfysisch Versneller Instituut, Netherlands. Whereas the Argonne group uses the weaker $7s^2\,^1S_0$-$7s7p\,^3P_1$ intercombination line for cooling and trapping. Trapping of some 10 Ra atoms has recently been reported. The KVI approach employs the strong $7s^2\,^1S_0$-$7s7p\,^1P_1$ transition. The disadvantage of the second method lies in the strong branching into ^1D and ^3D states, where

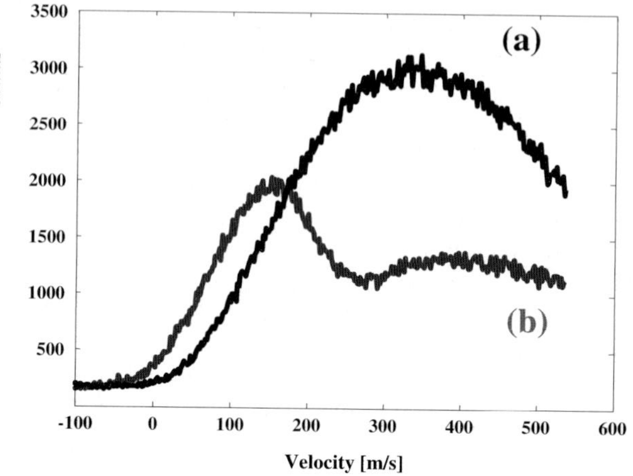

Fig. 5 **a** Maxwell–Boltzmann velocity distribution. **b** With properly adjusted cooling and re-pumping lasers the velocity distribution can be significantly shifted towards smaller velocities as a first step towards trapping of significant quantities of heavy alkali earth elements

the atomic electron can remain shelved for times too long for efficient cooling or trapping. Repumping involving three infrared lasers is needed. The discomfort is rewarded by much higher expected yields and long storage times. In order to develop the method stable Ba atoms are used (see Figs. 4 and 5). In those a significant and efficient modification of the Maxwell–Boltzmann velocity distribution, i.e. significant cooling has been achieved in an experimental arrangement using three repumping lasers in addition to a main cooling laser on the resonance line. In this way for the first time a heavy two electron system was cooled, in which a number of states exist and where the electron can be shelved with the consequence of an interruption of the cooling process.

A very novel idea was introduced recently for measuring an EDM of charged particles. The high motional electric field is exploited, which charged particles at relativistic speeds experience in a magnetic storage ring. In such an experiment the Schiff theorem can be circumvented (which had excluded charged particles from experiments due to the Lorentz force acceleration) because of the non-trivial geometry of the problem [28]. With an additional radial electric field in the storage region the spin precession due to the magnetic moment anomaly can be compensated, if the effective magnetic anomaly a_{eff} is small, i.e. $a_{eff} << 1$. The method was first considered for muons. For longitudinally polarized muons injected into the ring an EDM would express itself as a spin rotation out of the orbital plane. This can be observed as a time dependent (to first order linear in time) change of the above/below the plane of orbit counting rate ratio. For the possible muon beams at the future J-PARC facility in Japan a sensitivity of 10^{-24} e cm is expected [36]. In such an experiment the possible muon flux is a major limitation. For models with nonlinear mass scaling of EDMs such an experiment would already be more sensitive to some certain new physics models than the present limit on the electron EDM . An experiment carried out at a more intense muon source could provide a significantly more sensitive probe to CP violation in the second generation of particles without strangeness.

The deuteron is the simplest known nucleus. Here an EDM could arise not only from a proton or a neutron EDM, but also from CP-odd nuclear forces. It was shown very recently [37] that the deuteron can be in certain scenarios significantly

more sensitive than the neutron. In Eq. 3.2.2 this situation is evident for the case of quark chromo-EDMs: $d_{\mathcal{D}} = -4.67\,d_d^c + 5.22\,d_u^c$, $d_n = -0.01\,d_d^c + 0.49\,d_u^c$. It should be noted that because of its rather small magnetic anomaly the deuteron is a particularly interesting candidate for a ring EDM experiment and a proposal with a sensitivity of beyond 10^{-27} e cm exists. In this case scattering off a target will be used to observe a spin precession [38]. One can expect to extend the method to systems with arbitrary magnetic anomaly, if a relevant parameter on which the signal depends is modulated with the spin anomaly frequency and phase sensitive detection. Modulation of velocity is one option [39].

The method of searching EDMs in magnetic storage rings is particularly suited also for highly charged heavy ions. In this case polarization and the detection of spin rotation can be envisaged using laser light [38].

The highly active field of EDM searches includes at present a variety of experiments on the neutron and the electron EDM. Whereas in the neutron case basically the experiments follow the concepts of earlier measurements, novel approaches characterize the search for an electron EDM. Searches in Hg are continued and a new search in liquid Xe. Further, there are projects on molecules such as PbO, or molecular ions such as ThF^+ or condensed matter such as garnets, where in all cases one relies on the huge predicted enhancements due to local fields [28].

3.3 CPT tests with antihydrogen

$\overline{H}$ was produced first at CERN in 1995 [40] . The atoms were fast as the production mechanism required e^+e^- pair creation when $\overline{p}$s were passing near heavy nuclei. A small fraction of the e^+ form a bound state with the $\overline{p}$. The experiment was an important step forward showing that a few $\overline{H}$ could be produced. Unfortunately, the speed of the atoms does not allow any meaningful spectroscopy. Later a similar experiment was carried out at FERMILAB [41].

The successful production of slow $\overline{H}$ was first reported by the ATHENA collaboration in 2002 [42] and shortly later also by the ATRAP collaboration [43]. Both experiments use combined Penning traps in which first e^+ and $\overline{p}$ are stored separately and cooled. The atoms form when both species are brought into contact by proper electric potential switching in the combined traps. The detection in ATHENA relies on diffusion of the neutral atoms out of the interaction volume and the registration of πs which appear when the atoms annihilate on contact with matter walls of the container. In ATRAP the hydrogen atoms are re-ionized in an electric field and the $\overline{p}$s are observed using a capture Penning trap.

Most of the atoms are in excited states ($n > 15$) which can be seen from the fact that their physical size is above 0.1 μm [44–46]. For spectroscopy the atoms need to be in states with low n, preferentially the ground state. The production of such states is a major goal of the community for the immediate future. The kinetic energy of the produced $\overline{H}$ atoms has been measured to be of order 200 meV corresponding to a velocity of 6×10^4 m/s (Fig. 6). This is a factor of 400 above the value where neutral atom traps can hold them. Therefore cooling such atoms or identifying a production mechanism for colder $\overline{H}$ are a central topic.

For laser cooling of $\overline{H}$ a continuous laser at the II Lyman α frequency for $\overline{H}$ cooling has recently been developed [47]. One hopes to achieve the photo-recoil limit of 1.3 mK.

 Springer

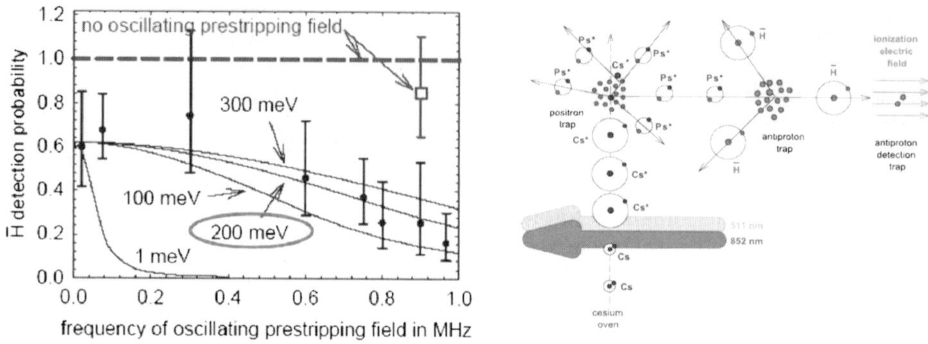

Fig. 6 *Right*: The $\overline{\text{H}}$ atoms produced up to date have typical kinetic energies of order 200 meV. Kinetic energies below 0.5 meV are needed to trap the atoms. *Left*: A new method to produce $\overline{\text{H}}$ uses charge exchange with PS atoms [44–46]

Recently a promising new method was demonstrated to obtain antihydrogen. It uses resonant charge exchange with excited positronium to obtain $\overline{\text{H}}$ atoms with essentially the same velocities as the $\overline{p}$s in the trap which can be made rather low by cooling [44–46, 48] (see Fig. 6).

A main motivation to perform precision spectroscopy on $\overline{\text{H}}$ is to test CPT invariance. There are two electromagnetic transitions which offer a high quality factor and therefore promise high experimental precision when H and $\overline{\text{H}}$ are compared: the 1S–2S two-photon transition at frequency $\Delta\nu_{1s-2s}$ and the ground state hyperfine splitting $\Delta\nu_{HFS}$, which both have within the SM in addition to the leading order contributions from QED, nuclear structure, weak and strong interactions,

$$\Delta\nu_{1s-2s} = \frac{3}{4} \times R_\infty + \varepsilon_{QED} + \varepsilon_{nucl} + \varepsilon_{weak} + \varepsilon_{strong} + \varepsilon_{CPT} \tag{1}$$

$$\Delta\nu_{HFS} = const \times \alpha^2 \times R_\infty + \varepsilon^*_{QED} + \varepsilon^*_{nucl} + \varepsilon^*_{weak} + \varepsilon^*_{strong} + \varepsilon^*_{CPT}. \tag{2}$$

In the Eqs. 1 and 2 it is assumed that only CPT violating contributions exist from interactions beyond the SM. If one assumes that ε_{CPT} and ε^*_{CPT} are of the same order of magnitude, the relative contribution is then larger by order α^{-2} for $\Delta\nu_{HFS}$. Further one can speculate that a new interaction may be of short range (contact interaction), which also favors measurements of $\Delta\nu_{HFS}$. Such an experiment has been recently proposed. It utilizes a cold $\overline{\text{H}}$ atom beam and has sextupole state selection magnets in a Rabi type atomic beam experiment [49]. For both experiments the atoms should be as cold as possible and one should use as many as possible atoms.

3.3.1 Gravitational force on $\overline{\text{H}}$

One of the completely open questions in physics concerns the sign of gravitational interaction for antimatter. It can only be answered by experiment. A proposal [50] exists (see Fig. 7) in which the deflection of a horizontal cold beam is measured in the earth's gravitational field. The experiment plans on a number of modern state of the art atomic physics techniques like sympathetic cooling of $\overline{\text{H}}^+$ ions by, e.g., Be$^+$ ions in an ion trap to achieve the necessary low temperatures of some 20 μK . After

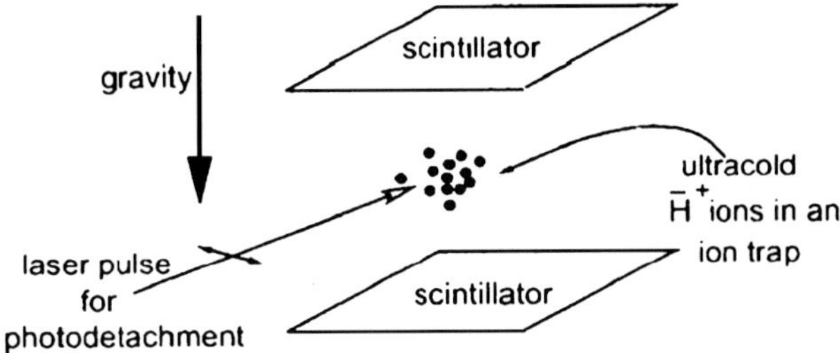

Fig. 7 The sign of the gravitational force on atomic $\overline{\mathrm{H}}$ needs to be determined by experiment, as there is no experimental evidence yet that antimatter and matter show identical behaviour concerning gravity [50]

pulsed laser photo-dissociation of the ion into $\overline{\mathrm{H}}$ and a e^+ the neutral atoms can then leave the trap. The atom's ballistic path can be measured.

3.3.2 Antiprotonic helium

Antiprotonic helium atoms and ions have received most attention in the past decade of all exotic atomic systems investigated with laser spectroscopy. The potential of antiprotonic helium for precision measurements in the field of fundamental interaction research was realized shortly after it had been discovered that $\overline{p}$s stopped in liquid or gaseous helium exhibit long lifetimes and do not rapidly annihilate with nucleons in the helium nucleus [51]. This can be explained, if one assumes that the $\overline{p}$s are captured in metastable states of high principal quantum number n and high angular momentum l, with $l \approx n$ (see Fig. 8 [52]. The capture happens at typically at $n \approx \sqrt{M^*/m_e} \approx 38$, where M^* is the reduced mass of the $(\overline{p}\mathrm{He})$ bound system.

With laser radiation the $\overline{p}$s in these atoms can be transferred into states where Auger de-excitation can take place. In the resulting H-like system Stark mixing with s-states results in nuclear $\overline{p}$ absorption and annihilation which is signaled by emitted pions. This way a number of transitions could be induced and measured with continuously increasing accuracy over the past decade. A precision of 6×10^{-8} has been reached for the transition frequencies [53], which has been stimulating for improving three-body QED calculations (see Fig. 9). It should be noted that with the high principal quantum numbers for the $\overline{p}$ the system shows also molecular type character [54, 55].

Among the spectroscopic successes the laser-microwave double resonance measurements of hyperfine splittings of $\overline{p}$ transitions could be measured [59]. There is agreement with QED theory [57, 58] at the 6×10^{-5} level and the measurement can be interpreted as a measurement of the antiprotonic bound state g-factor to this accuracy. In principle, hyperfine structure measurements in antiprotonic helium offer the possibility to measure the magnetic moment of the $\overline{p}$.

3.3.3 CPT tests with antiprotonic helium ions

The very good agreement of the QED calculations with the measurements of several transitions can be exploited to extract a limit on the equality of the ratio of the

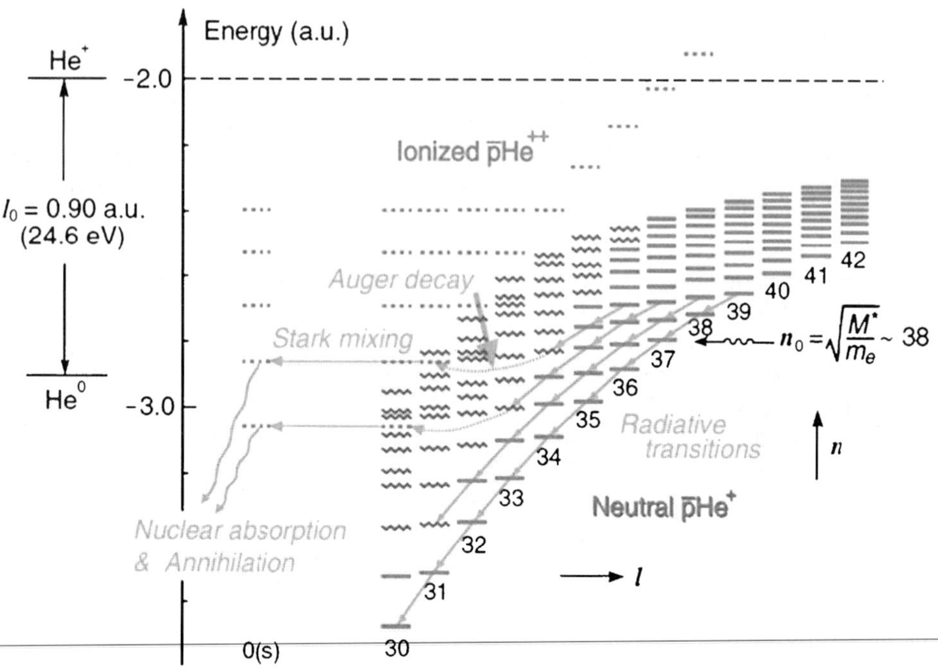

Fig. 8 In the formation process of antiprotonic helium the capture of the $\bar{p}$ into a state with quantum numbers $n \approx 38$ and maximal l is very likely. Such states are rather stable against $\bar{p}$ annihilation [51]

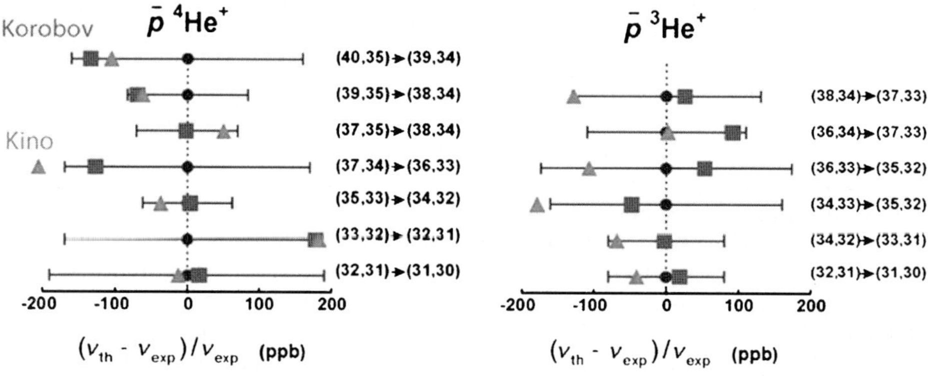

Fig. 9 Accurate QED calculations have been performed for antiprotonic $\overline{P}\,^4$He$^+$ and $\overline{P}\,^3$He$^+$. The differences between the results from independent calculations are below the accuracy achieved in experiments. These differences may be regarded as a an indication of the size of systematic uncertainties in the theoretical approaches chosen [60, 61]

squared charge and mass for proton and $\bar{p}$. Combined with the results of cyclotron frequency measurements [56] in Penning traps one can conclude that masses and charges of proton and $\bar{p}$ are equal within 6×10^{-8} in full agreement with expectations based on the CPT theorem [60, 61]. The collaboration estimates that a test down to the 10 ppb level should be possible.

 Springer

4 Facilities

At present experiments on fundamental interactions and symmetries using laser technology and exploiting nuclear properties are either performed in table top laboratory experiments with stable particles in several university laboratories worldwide or at a small number of accelerator laboratories where radioactive beams are made available. In the latter case the availability of sufficient beam time to debug precision experiments and to study systematic effects with the indicated care is a constant problem. Therefore new facilities are most welcome. The latest commissioned facility is the TRIμP facility at KVI, Netherlands, where fundamental interaction research is foreseen with a significant share of beamtime [62–64].[1]

Future possibilities for precision experiments in a large variety of radioactive atoms may include the ISAC facility at TRIUMF, Canada, provided a new target will be installed, the Spiral II facility of GANIL in Caen, France, the radioactive beams at the new RIKEN cyclotrons in the Tokyo area, Japan, and the FAIR facility of GSI in Darmstadt, Germany, where in particular one can look forward to the most intense source of slow antiprotons (FLAIR at FAIR), and . For the expected progress in precision measurements, which use lasers and radioactive nuclei to investigate fundamental symmetries and forces, it will be of crucial importance that the new facilities can be operated to give a significant share of beam time to this research.

5 Conclusions

Laser techniques and modern spectroscopy offer a variety of possibilities to investigate fundamental interactions and symmetries and to determine most accurate values of fundamental constants. The present high level of important results can be expected to be even improved in ongoing and planned experiments at existing and next generation facilities, providing information complementary to high energy physics.

Acknowledgements We would like to thank the organizers of Laser 2006 and in particular K. Marinova, a wonderful workshop with plenty of substantial discussions around an interdisciplinary field. We thank for their hospitality and their support.

References

1. Akesson, T., et al.: Towards the European strategy for particle physics: the Briefing Book. hep-ph/0609216 (2006)
2. Jungmann, K.: Fundamental symmetries and interactions. Nucl. Phys., A **751**, 87c (2005)
3. Jungmann, K.: Fundamental symmetries and interactions – some aspects. Eur. Phys. J., A **25**, 677 (2005)
4. Behr, J.A., et al.: Weak interaction symmetries with atom traps. Eur. Phys. J., A **25**, 685 (2005) and references therein

---

[1] An example of the achievable clean secondary beams was demonstrated in an experiment on the β decays of $A{=}12$ isotopes in excited states of ^{12}C, which may themselves decay into $3\,\alpha$ particles. Spectra obtained with beam ions implanted in a Si detector matrix – a kind of trap – are of relevance to the ^{12}C production process in stars [65].

 Springer

5. Bennett, G.W., et al.: Final report of the E821 muon anomalous magnetic moment measurement at BNL. Phys. Rev., D **73**, 072003 (2006) and references therein
6. Gabrielse, G., et al.: New determination of the fine structure constant from the electron g-value and QED. Phys. Rev. Lett. **97**, 030802 (2006)
7. Odom, B., et al.: New Measurement of the electron magnetic moment using a one-electron quantum cyclotron. Phys. Rev. Lett. **97**, 030801 (2006) and references therein
8. van Dyck, R.S., et al.: New high-precision comparison of electron and positron g-factors. Phys. Rev. Lett. **59**, 26 (1987)
9. Severijns, N., Beck, M., Naviliat-Cuncic, O.: Tests of the standard electroweak model in beta decay. nucl-ex/0605029 (2006)
10. Jungmann, K.: Searches for permanent electric dipole moments: some recent developments. In: Aulenbacher, K., Bradamante, F., Bressan, A., Martin, A. (eds.) Spin 2004, pp. 108–116. World Scientific Singapore (2005); see also: Symposium Lepton Moments III, Cape Cod. http://g2pc1.bu.edu/lept06/program.html (2006)
11. Jungmann, K.: Low energy antiproton experiments: a review. AIP Conf. Proc. **793**, 18 (2005)
12. Abele, H., et al.: Quark mixing, CKM unitarity. Eur. Phys. J., C **33**, 1 (2004)
13. Scielzo, N.D., et al.: Measurement of the β-ν correlation using magneto-optically trapped ^{21}Na. Phys. Rev. Lett. **93**, 102501 (2004)
14. Bennett, S.C., Wieman, C.E.: Measurement of the 6S $\rightarrow$ 7S transition polarizability in atomic cesium and an improved test of the standard model. Phys. Rev. Lett. **82**, 2484 (1999)
15. Eidelman, S.: Internat. Conf. High Energy Phys. Moscow (2006) hep-ex/0608025
16. Czarnecki, A., Marciano, W.J.: Electrons are not ambidextrous. Nature **435**, 437 (2005)
17. Aubin, S., et al.: Lifetime measurement of the 9s level of atomic francium. Opt. Lett. 2055 (2003)
18. Zhao, W.Z., et al.: Measurement of the 7p ^{2}P$_{3/2}$ level lifetime in atomic francium. Phys. Rev. Lett. 4169 (1997)
19. Gomez, E., et al.: Spectroscopy with trapped francium: advances and perspectives for weak interaction studies. Rep. Prog. Phys. **69**, 79 (2006)
20. Stancari, G., et al.: Francium sources at Laboratori Nazionali di Legnaro: design and performance. Rev. Sci. Instrum. **77**, 03A701 (2006)
21. Autov, S.N., et al.: Production and trapping of francium atoms. Nucl. Phys., A **746**, 421C (2004)
22. Sherman, J.A., et al.: Precision measurement of light shifts in a single trapped Ba$^+$ ion. Phys. Rev. Lett. **94**, 243001 (2004)
23. Bergmann, T., Nachtmann, O.: Atomic beam spin echo and parity violation. DPG, Frühjahrstagung, Frankfurt (2006)
24. Herczeg, P.: Beta decay beyond the standard model. Prog. Part. Nucl. Phys. **46**, 413 (2001)
25. Sohani, M.: TRIμP – a new facility to produce and trap radioactive isotopes. Acta Phys. Pol., B **37**, 231 (2006)
26. Iacob, V.E., et al.: Branching ratios for the β-decay of ^{21}Na. Phys. Rev., C **74**, 015501 (2006)
27. Gorelov, A., et al.: Scalar interaction limits from the β-ν correlation of trapped radioactive atoms. Phys. Rev. Lett. **94**, 142501 (2005)
28. Sandars, P.G.H.: Electric dipole moments of charged particles. Contemp. Phys. **42**, 97 (2001) and references therein
29. Liu, C.P., et al.: Schiff theorem and the electric dipole moments of hydrogen-like atoms. nucl-th/0601025 (2006)
30. Regan, B.C., et al.: New limit on the electron electric dipole moment. Phys. Rev. Lett. **88**, 071805 (2002)
31. McNabb, R., et al.: An improved limit on the electric dipole moment of the muon. hep-ex/0407008
32. Baker, C.A., et al.: An improved experimental limit on the electric dipole moment of the neutron. hep-ex/0602020 (2006)
33. Romalis, M.V., et al.: A new limit on the permanent electric dipole moment of Hg-199. Phys. Rev. Lett. **86**, 2505 (2001)
34. Ginges, J.S.M., Flambaum, V.V.: Violations of fundamental symmetries in atoms and tests of unification theories of elementary particles. Phys. Rep. **397**, 63 (2004) and references therein
35. Bieron, J., et al.: Lifetime and hyperfine structure of the D-3(2) state of radium. J. Phys., B **37**, L305 (2004)
36. Farley, F.J.M., et al.: New method of measuring electric dipole moments in storage rings. Phys. Rev. Lett. **93**, 052001 (2004)
37. Liu, C.P., Timmermans, R.G.E.: P- and T-odd two-nucleon interaction and the deuteron electric dipole moment. Phys. Rev., C **70**, 055501 (2004)

38. Jungmann, K.: In: Kluge, H.J., Jungmann, K., Khriplovich, I.B. (eds.) Proceedings of Workshop on Permanent Electric Dipole Moments. GSI (1999)
39. Onderwater, C.J.G.: Light ion EDM search in storage rings. Hyperfine Interact. (in press) and AIP Conf. Proc. **842**, 784 (2006)
40. Baur, G., et al.: Production of antihydrogen. Phys. Lett., B **368**, 251 (1996)
41. Blanford, G., et al.: Observation of atomic antihydrogen. Phys. Rev. Lett. **80**, 3037 (1998)
42. Amoretti, M., et al.: Production and detection of cold antihydrogen atoms. Nature **419**, 456 (2002)
43. Gabrielse, G., et al.: Background-free observation of cold antihydrogen with field-ionization analysis of its states. Phys. Rev. Lett. **89**, 213401 (2002)
44. Gabrielse, G.: Atom made entirely of antimatter: two methods produce slow antihydrogen. Adv. At. Mol. Opt. Phys. **50** (2004)
45. Amoretti, M., et al.: Positron plasma diagnostics and temperature control for antihydrogen production. Phys. Rev. Lett. **91**, 0055001-1 (2003)
46. Amoretti, M., et al.: Dynamics of antiproton cooling in a positron plasma during antihydrogen formation. Phys. Lett., B **590**, 133 (2004)
47. Eikema, K.S.E., et al.: Continuous coherent Lyman- ? Excitation of atomic hydrogen. Phys. Rev. Lett. **86**, 5679 (2001)
48. Storry, C.H., et al.: First laser-controlled antihydrogen production. Phys. Rev. Lett. **93**, 263401 (2004)
49. Wildman, E., et al.: Study of the hyperfine structure of antiprotonic helium. Nucl. Instrum. Methods, B **214**, 89 (2004)
50. Walz, J., Hänsch, T.: A proposal to measure antimatter gravity using ultracold antihydrogen atoms. Gen. Relativ. Gravit. **36**, 561 (2004)
51. Yamazaki, T., et al.: Antiprotonic helium. Phys. Rep. **366**, 183 (2002)
52. Nakamura, S.N., et al.: Delayed annihilation of antiprotons in helium gas. Phys. Rev., A **49**, 4457 (1994)
53. Hori, M., et al.: Sub-ppm laser spectroscopy of antiprotonic helium and a CPT-violation limit on the antiprotonic charge and mass. Phys. Rev. Lett. **87**, 093401 (2001)
54. Yamazaki, T.: In: Karshenboim, S.G., et al. (eds.) The Hydrogen Atom, p. 246. Springer, Berlin Heidelberg New York (2001)
55. Korobov, V.I.: Antiprotonic helium "atomcule": relativistic and QED effects. Nucl. Phys., A **689**, 75 (2001)
56. Gabrielse, G., et al.: Precision mass spectroscopy of the antiproton and proton using. Phys. Rev. Lett. **82**, 3198 (1999)
57. Korobov, V., Baklanov, D.: Fine and hyperfine structure of the (37, 35) state of the $^4\mathrm{He}^+\bar{p}$ atom. J. Phys., B At. Mol. Opt. Phys. **34**, L519 (2001)
58. Korobov, V., Baklanov, D.: Hyperfine structure of antiprotonic helium energy levels. Hyperfine structure of antiprotonic helium energy levels. Phys. Rev., A **57**, 1662 (1998)
59. Widmann, E., et al.: Hyperfine structure of antiprotonic helium revealed by a laser-microwave-laser resonance method. Phys. Rev. Lett. **89**, 243402 (2002)
60. Hori, M., et al.: Direct measurement of transition frequencies in isolated p-bar He+ atoms, and new CPT-violation limits on the antiproton charge and mass. Phys. Rev. Lett. **91**, 123401 (2003)
61. Hori, M., et al.: Determination of the antiproton-to-electron mass ratio by precision laser spectroscopy of $\bar{p}\mathrm{He}^+$. Phys. Rev. Lett. **96**, 243401 (2006)
62. Berg, G.P., et al.: Dual magnetic separator for TRIμP. Nucl. Instrum. Methods, A **560**, 169 (2006)
63. Jungmann, K., et al.: TRIμP - trapped radiactive atoms - μicrolaboratories for fundamental physics. Phys. Scr., T **104**, 178 (2003)
64. Traykov, E., et al.: Production of radioactive nuclides in inverse reaction kinematics. nucl-ex/0608016 (2006)
65. Pederson, S.G., et al.: β-decay studies of states in ^{12}C. Proc. of Science, NIC-IX, 244 (2006)

Hyperfine Interact (2006) 171:57–67
DOI 10.1007/s10751-006-9500-5

Experimental test of special relativity by laser spectroscopy

C. Novotny · B. Bernhardt · G. Ewald · C. Geppert ·
G. Gwinner · T. W. Hänsch · R. Holzwarth ·
G. Huber · S. Karpuk · H.-J. Kluge · T. Kühl ·
W. Nörtershäuser · S. Reinhardt · G. Saathoff · D. Schwalm ·
T. Udem · A. Wolf

Published online: 25 January 2007

Abstract The Doppler-free laser-spectroscopic frequency measurement of Doppler-shifted optical lines in forward and backward direction of a fast ion beam permits a sensitive test of the relativistic Doppler-formula and, hence, the relativistic time dilation factor $\gamma_{SR} = (1 - v^2/c^2)^{-1/2}$. An experiment on metastable $^7\mathrm{Li}^+$, stored at a velocity of $v=0.064c$ in the Heidelberg heavy-ion storage ring TSR, has confirmed time dilation with unprecedented accuracy. Latest tests at two different ion-velocities ($v=0.03c$ and $v=0.064c$) will enhance these measurements. An improved version of this experiment will be carried out at the experimental storage ring (ESR) at the Gesellschaft für Schwerionenforschung (GSI) in Darmstadt. The ESR permits $^7\mathrm{Li}^+$ to be stored at $v=0.33c$ which promises an improvement of the sensitivity to deviations from γ_{SR} by an order of magnitude. A first test at the ESR has shown the feasibility for this kind of experiment.

Key words precision spectroscopy · test of Lorentz invariance · storage ring · generation of metastabile lithium ions

C. Novotny (✉) · G. Huber · S. Karpuk · W. Nörtershäuser
Johannes Gutenberg-Universität Mainz, D-55128 Mainz, Germany
e-mail: christian.novotny@uni-mainz.de

B. Bernhardt · T. W. Hänsch · R. Holzwarth · T. Udem
Max-Planck-Institut für Quantenoptik, D-85748 Garching, Germany

G. Ewald · C. Geppert · H.-J. Kluge · T. Kühl
Gesellschaft für Schwerionenforschung, D-64291 Darmstadt, Germany

G. Gwinner
University of Manitoba, Winnipeg, MB R3T 2N2, Canada

S. Reinhardt · G. Saathoff · D. Schwalm · A. Wolf
Max-Planck-Institut für Kernphysik, D-69029 Heidelberg, Germany

1 Lorentz invariance and time dilation

Lorentz invariance is a basic principle underlying all currently accepted theories describing the fundamental interactions of nature – the Quantum Field Theories of the strong and electroweak forces as well as General Relativity. This is a strong motivation to test the theory of Special Relativity (SR) experimentally with ever higher precision, as any deviation from its predictions would have profound consequences for our understanding of nature. In particular, many unifying theories, such as string theory, allow for violations of Lorentz invariance at some level [1]. While kinematic test theories describe possible violations of the observer Lorentz transformation (e.g., the rest frequency of an atomic transition is assumed to be the same in all inertial frames, while the transformation between the frames is tested), dynamical test theories like the Standard-Model Extension [1] deal with possible breakdowns of the particle Lorentz transformation. Time dilation experiments are sensitive to several parameters of kinematic as well as dynamic test theories [2, 3].

In order to quantify deviations from the relativity principle, Robertson developed a kinematical test theory [4] that was later modified by Mansouri and Sexl [5]. They consider generalized Lorentz transformations between a hypothetical preferred frame $\Sigma(T, X)$ and a frame $S(t, \mathbf{x})$ moving relative to Σ at a velocity V along the X axis, and the speed of light is assumed to be isotropic in Σ only. Using Einstein synchronization, these transformations read

$$T = \Gamma\left(\frac{t}{\widehat{a}} + \frac{Vx}{\widehat{b}c_0^2}\right); \; X = \Gamma\left(\frac{x}{\widehat{b}} + \frac{Vt}{\widehat{a}}\right); Y = \frac{y}{\widehat{d}}; Z = \frac{z}{\widehat{d}} \tag{1}$$

with $\Gamma = \left(1 - V^2/c_0^2\right)^{-\frac{1}{2}}$ and c_0 being the speed of light in Σ. Note that due to the abolition of the relativity principle these transformations are in general not valid between two arbitrary, constantly moving reference frames but only with respect to Σ. This model contains three velocity-dependent test functions $\widehat{a}(V^2)$, $\widehat{b}(V^2)$, and $\widehat{d}(V^2)$, which modify time dilation as well as Lorentz contraction in longitudinal and transverse direction. They reduce to $\widehat{a}(V^2) = \widehat{b}(V^2) = \widehat{d}(V^2) = 1$ in the case SR holds. In the low-velocity limit, these functions can be expanded in powers of V^2/c^2, i.e., $\widehat{a}(V^2) = \left[1 + \alpha V^2/c_0^2 + O(c_0^{-4})\right]$, $\widehat{b}(V^2) = \left[1 + \beta^* V^2/c_0^2 + O(c_0^{-4})\right]$ and $\widehat{d}(V^2) = \left[1 + \delta V^2/c_0^2 + O(c_0^{-4})\right]$. In this regime one is therefore left with three test parameters α, β^* and δ.

In this framework, the speed of light $c(\theta, V)$ in the moving frame S,

$$\frac{c(\theta, V)}{c_0} = 1 + (\beta^* - \delta)\frac{V^2}{c_0}\sin^2(\theta) + (\alpha - \beta^*)\frac{V^2}{c_0} \tag{2}$$

is in general not constant, but is dependent on the angle θ between the direction of $c(\theta, V)$ and the motion of the moving frame S as well as on the velocity V between Σ and S. Michelson–Morley type experiments are sensitive to an anisotropy of the speed of light thus determining the parameter combination $|\beta^* - \delta|$ and so called Kennedy–Thorndike tests the velocity-dependence of c described by $|\alpha - \beta^*|$. Ives–Stilwell type experiments independently measure the parameter α that describes time dilation.

No deviation from SR has been found to date in any experiment. The most sensitive interferometric experiments have yielded limits of $|\beta^* - \delta| < (0.5 \pm 3 \pm 0.7) \times 10^{-10}$ [6] and $< (-0.9 \pm 2.0) \times 10^{-9}$ [7], respectively, for an anisotropy and of $|\alpha - \beta^*| < (1.6 \pm 3.0) \times 10^{-7}$ [8] for a velocity-dependence of c. The best limit on deviations from time dilation prior to the storage ring experiment was $|\alpha| < 1.4 \times 10^{-6}$ [9].

 Springer

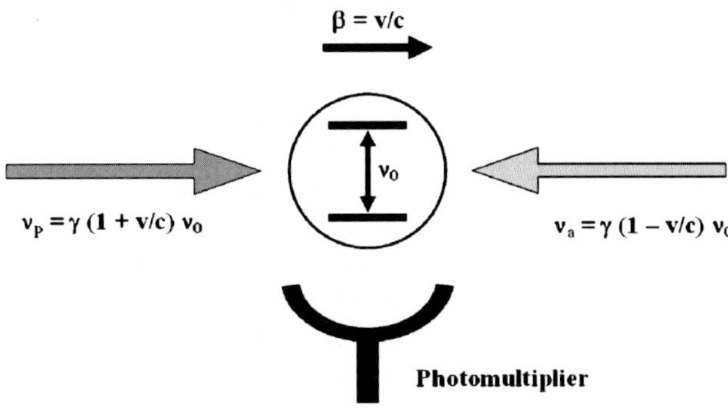

Fig. 1 Principle of Ives–Stilwell experiments with spectroscopy of fast ions

2 Ives and Stilwell experiments

Time dilation can be measured using the optical Doppler Effect, for example via parallel and anti-parallel excitation of a clock moving at a velocity $v=\beta c$, see Fig. 1.

In SR the Doppler-shifted frequencies are given by the relativistic Doppler-formula $v_{p,a} = \gamma_{\mathrm{SR}}(1 \pm \beta)v_0$, where the Lorentz factor γ_{SR} appears as a direct consequence of time dilation. Multiplication of these equations yields the β-independent relation $v_a v_p = v_0^2$ if SR holds. The Mansouri–Sexl test theory parameterizes possible deviations of time dilation from SR with the test parameter α by $\gamma = (1 - \beta^2)^{-1/2+\alpha} = \gamma_{\mathrm{SR}}(1 + \alpha\beta^2 + \ldots)$ [5]. A non-vanishing test parameter α would indicate a violation of SR and modify the outcome of the Ives–Stilwell experiment as

$$\frac{v_p \cdot v_a}{v_0^2} = \left(1 - \beta^2\right)^{-2\alpha} \approx 1 + 2\alpha \cdot \beta^2 \tag{3}$$

The first experiment of this kind was performed by Ives and Stilwell (IS), who measured the Doppler-shifted frequencies v_p and v_a of the H$_\beta$ line (v_0) of a hydrogen beam in parallel and anti-parallel direction. The result of the original IS experiment set an upper bound on α of $|\alpha| < 1 \times 10^{-2}$ [10].

Later, significant improvements have been achieved using laser techniques instead of conventional spectrometers; two-photon spectroscopy on a $\beta=0.0036$ neon atomic beam has set an absolute bound of $|\alpha| < 2.3 \times 10^{-6}$ [11] and, considering the cosmic background frame as the preferred one ($\beta_{lab}\approx350$ km/s), even $|\alpha| < 1.4 \times 10^{-6}$ [9] from limits on sidereal variations.

3 A modern Ives–Stilwell experiment at the TSR

Today, heavy ion storage rings equipped with electron coolers like the TSR in Heidelberg or the ESR in Darmstadt provide low-divergence ion beams at high velocities. The combination of these fast ion beam techniques with high resolution laser spectroscopy allows for a significant improvement of the time dilation measurement. The latest experiment, at the TSR, uses metastable ^{7}Li^{+} ions in a storage ring at a velocity of $\beta=0.064$ and $\beta=0.03$, respectively. The triplet spectrum of the helium-like ^{7}Li^{+} has a strong optical transition $2s\,^{3}S_1 \rightarrow 2p\,^{3}P_2$ at 548.5 nm with a well-resolved and precisely known hyperfine structure multiplet. A first version of this experiment that employed collinear optical-optical double-resonance spectroscopy on a Λ-type three-level system formed by the

 Springer

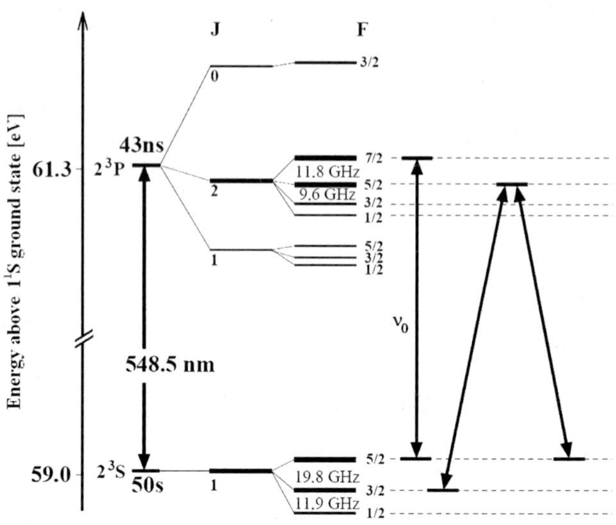

Fig. 2 Level-scheme of ortho-helium-like triplet system of $^7\mathrm{Li}^+$

$2^3S_1(F = 3/2)$, $2^3S_1(F = 5/2)$, and $2^3P_2(F = 5/2)$ states, compare Fig. 2, has set the thitherto best absolute bound of $|\alpha| < 8 \times 10^{-7}$ [12]. The limiting factor was the large observed linewidth of the Λ-resonance of almost 60 MHz, compared to a natural width of 3.8 MHz. This broadening was caused by velocity changes between subsequent excitations of the two transitions of the Λ-system. Changing the experimental scheme to collinear saturation spectroscopy on the $2^3S_1(F = 5/2) \rightarrow 2^3P_2(F = 7/2)$ two-level transition avoids this problem.

The preparation process for the production of metastable $^7\mathrm{Li}^+$ ions starts with negative charged Li ions, which are accelerated and stripped in a Tandem Van de Graaff accelerator. A fraction of about 10% emerges in the metastable $^3S_1$ state from the stripping process and typically an amount of 10^8 ions can be injected into the TSR. There the ions are subjected to electron cooling to narrow the velocity distribution. While the natural lifetime of the metastable ions is 50 s, collisions with the residual gas reduce the beam lifetime to about 13 s. At equilibrium (after a cooling time of about 5–10 s) a longitudinal momentum spread of $\Delta p/p = 3.5 \times 10^{-5}$ leads to a Doppler-width of the transition of about 2.5 GHz (FWHM). A more detailed description of the storage ring can be found in [13] and [14].

For each experiment (viz for each ion-velocity) a frequency-stabilized laser system for the parallel and the anti-parallel excitation of the metastable $^7\mathrm{Li}^+$ ions has been established. All lasers that are used to perform one of the spectroscopy experiments in the storage ring are listed in Table 1. The laser setup is exemplarily described for the case of the faster ion-velocity (β=0.064) below.

To control the frequency of the laser for the parallel excitation, this has been locked directly to a molecular iodine reference line. The dye laser for the anti-parallel excitation has been controlled via a frequency-offset-locking technique relative to a second dye laser that is, again, directly locked to an iodine reference, see Fig. 3. This method allows tuning the first dye laser in the MHz range, while its frequency is, due to the beating with the second dye laser, kept under control on the kHz level. Therefore an accuracy in the frequency determination of $\Delta\nu/\nu = 2 \times 10^{-10}$ has been achieved. The used iodine references were calibrated at the Max Planck Institute for Quantum Optics in Garching in

 Springer

Table 1 Lasers used in the latest experimental series for testing time dilation at the TSR

β [%c]	Excitation direction	Denotation	Laser	Wavelength [nm]	Iodine-Frequency [kHz]
6.4	parallel	$\nu_{p1}=\nu_p$	Ar-Ion	514	582 490 603 430$\pm$3
3	parallel	ν_{p2}	Nd:YAG	532	563 209 278 600$\pm$7
3	anti-parallel	ν_{a2}	Dye	565	530 222 434 291$\pm$73
6.4	anti-parallel	$\nu_{a1}=\nu_a$	Dye	585	512 671 028 075$\pm$73

Iodine frequencies are taken from [14].

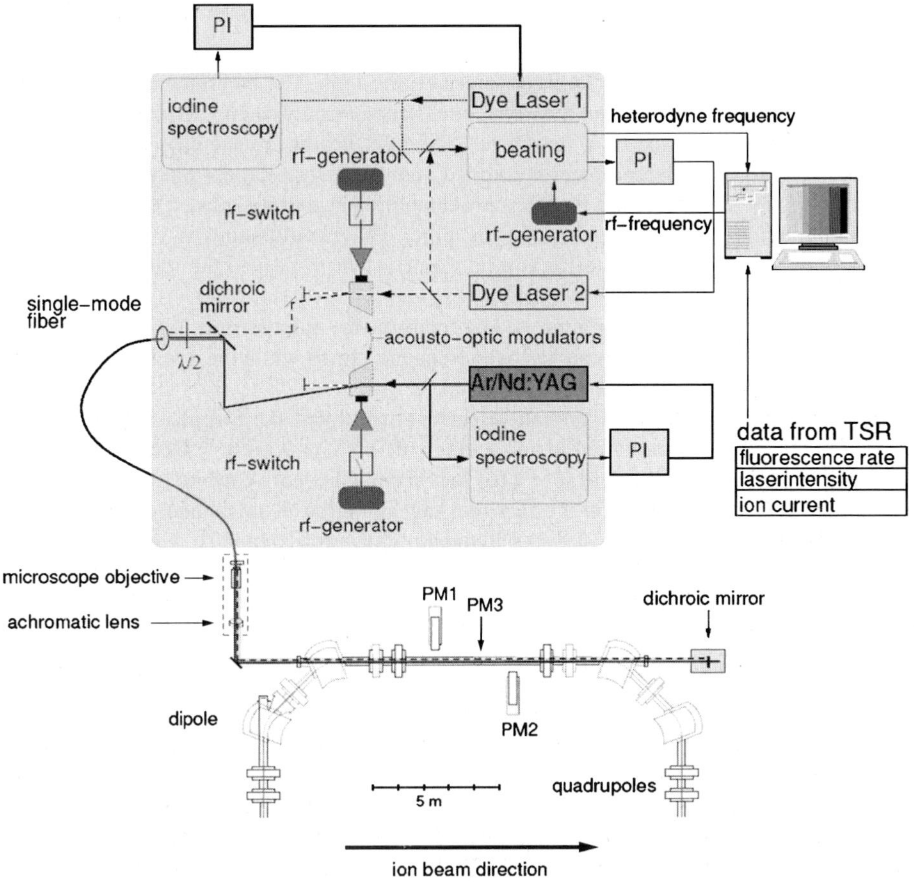

Fig. 3 Setup of the experiment at the ion storage ring TSR. The dye lasers have been used in both experiments (viz for both ion-velocities). Depending on the ion-velocity, the Ar ion laser or the Nd:YAG laser is used

cooperation with the Group of T.W. Hänsch with a frequency comb [15]. Both laser beams going into the TSR are passed through acousto-optic frequency shifters in order to switch the light on and off using fast radiofrequency switches. The beams are then merged with a dichroic mirror and guided to the TSR by a single-mode polarization-maintaining fiber. At the TSR, the laser beams are linearly polarized in the same direction and their intensities are

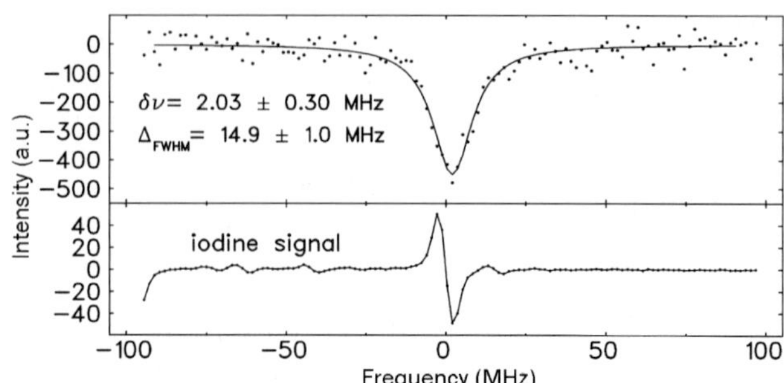

Fig. 4 Upper part: fluorescence signal for a multiple scan of the Lamb-dip at a velocity of β=0.064 (the Doppler-background is subtracted). Lower part: corresponding iodine signal of the tuned dye laser. The zero of the frequency scale corresponds to the position of the iodine reference line

kept equal in order to balance the laser forces on the ions. The bichromatic beam coming out of the fiber is directed via an achromatic telescope through the experimental section of the TSR and retro-reflected by a flat mirror. The laser beams are accurately superimposed with the ion beam with computer-controlled motorized translation/rotation stages. The fluorescence signal is recorded by three photomultipliers that are placed in the ion-laser-beam interaction zone along the beam pipe. By simultaneously maximizing the fluorescence yield, both laser beams can be aligned with respect to the ion beam to better than 70 μrad.

A typical spectrum of the lithium spectroscopy in the storage ring is shown in Fig. 4. The laser for the parallel excitation has been frequency-fixed while the dye laser is tuned by 200 MHz across the resonance.

Applying this technique at an ion-velocity of β=0.064 the Doppler-shifted frequencies ν_p and ν_a could be measured to an accuracy of $\Delta v/v = 1 \times 10^{-9}$. Considering (3) the latest upper limit of $|\alpha| < 2.2 \times 10^{-7}$ [16] has been achieved. A tribute to the quality of the storage ring spectroscopy is the fact that the limitation at this stage comes from the insufficient knowledge of the rest frame frequency Δv_0=400 kHz [17], which enters (3) in quadrature.

To overcome the limitation due to the uncertainty of the rest frequency, a second measurement at a lower velocity of β=0.03 was carried out. Multiplication of the expressions for Doppler-shifted frequencies according (3) for both ion-velocities leads to

$$\frac{v_{p2}v_{a2}}{v_{p1}v_{a1}} = \frac{1 + 2\alpha \cdot \beta_2^2}{1 + 2\alpha \cdot \beta_1^2} \approx 1 + 2\alpha\left(\beta_2^2 - \beta_1^2\right) \tag{4}$$

From the measurements at two different ion velocities it is possible to determine the rest frequency and α independently. Additionally the measurement at high velocity of β=0.064 were repeated with an improved stabilization/frequency scanning scheme for the lasers.

The comparison of the different excitation frequencies at the different velocities cancels out the influence of the rest frame frequency, compare (4), and will lead to an increase of the upper limit for α (the data is currently analyzed and the final value will be presented elsewhere soon).

4 Feasibility-test at the ESR

The next generation of Ives–Stilwell type experiments will be carried out at GSI in the experimental storage ring ESR. Here storage velocities up to β=0.4 are possible and due to the fact that α scales with the square of the ion-velocity, an increase of the upper limit of α

 Springer

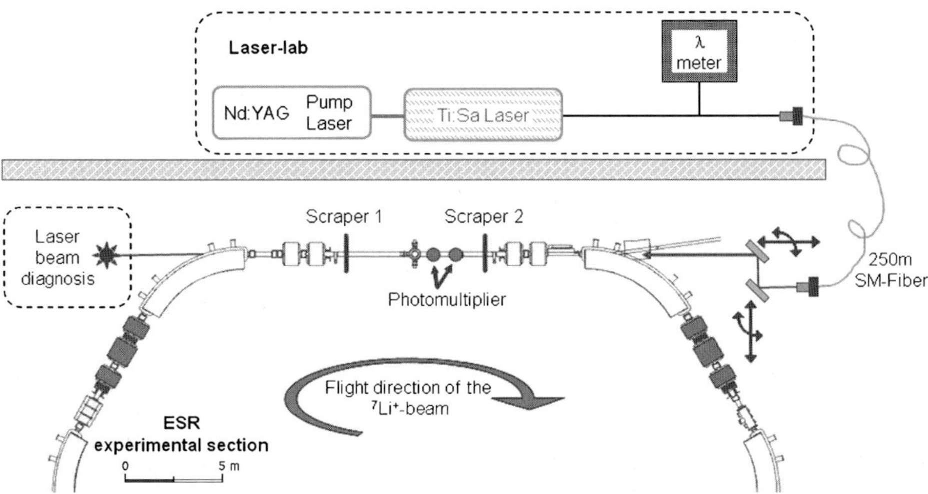

Fig. 5 Setup of the test experiment at the ion storage ring ESR

by more than one order of magnitude is possible. This high velocity leads, of course, to a larger Doppler-shift of the rest frequency and requires new references on the corresponding frequency ranges. With the development of the frequency comb, such a reference with the needed accuracy has become available.

A first test beam-time at the ESR has been performed in October 2005 to show the feasibility of a test of SR at GSI. Before this beam-time only one experiment at GSI dealt with lithium-ions and it had been located just behind the Unilac accelerator (Tinschert, personal communication.), hence, no experience with lithium ions in the subsequent heavy ion synchrotron (SIS) and in the ESR had so far been made. Furthermore at GSI no ion source is available that works with negatively charged ions. Thus, a new production method for metastable ^{7}Li$^+$ ions, different from the one used at the TSR, has to be established. In principle there are two processes to gain metastable ions from an ion source that produces positively charged ions: electron capture of ^{7}Li^{2+} ions or excitation of ^{7}Li$^+$ from the ground state. Indeed a common process for the metastable production is electron capture at energies in the range of 50 keV [18]. Unfortunately no foil nor gas stripper is available at those ion energies at GSI. Hence, the production has to be performed through excitation processes directly in the source. For energy reasons the electron cyclotron resonance (ECR) ion source was chosen. There the raw material (lithium fluoride) was evaporated and ionized using radio frequency radiation. In this process an equilibrium of all possible ion-charges (^{7}Li$^+$, ^{7}Li^{2+}, ^{7}Li^{3+} and even the metastable ^{7}Li^{+*}) will appear. The ion-cloud has been extracted from the source, pre-accelerated to 15 keV and separated by a mass to charge analyzer to the desired ratio of seven to one. Those ions were accelerated to an energy of 58.8 MeV/u by the Unilac and the SIS. About 10^8 ions, corresponding to a current signal of 20–30 μA, were injected into the ESR. There the ions were subjected to electron cooling to narrow the velocity distribution.

The laser system applied for this first test consisted only of one laser for the excitation from the anti-parallel direction. In contrast to the TSR experiment described above, this test gains no Doppler-free signal but shows the velocity distribution of the metastable ions in the storage ring and gives information about the lifetime and the ion beam behaviour. Due to the high velocity of β=0.33 the Doppler-shifted wavelength is λ_α=780.2 nm. This laser

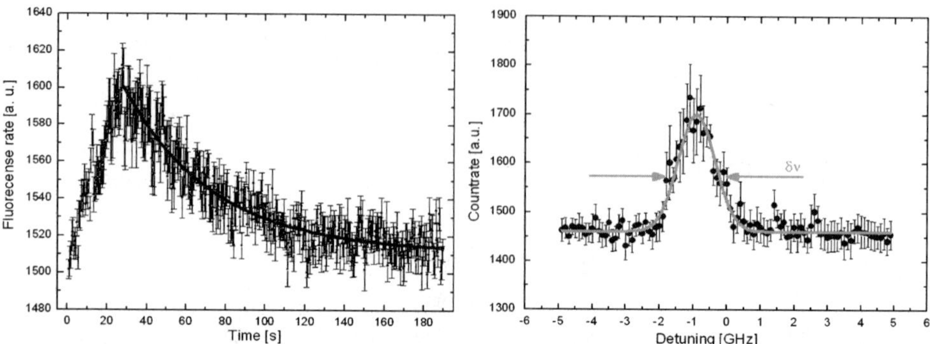

Fig. 6 Left panel: measurement of the lifetime of the metastable ^{7}Li$^+$ ions in the ESR. Right panel: velocity distribution of the ions in terms of frequency in the ESR

light was provided by a cw Ti:Sa laser that was situated in a laboratory ~250 m away from the ESR guided to the storage ring through a single mode fiber. The frequency of this laser was maintained by a λ-meter and tuned in the range of 1 to 10 GHz across the excitation wavelength, to cover the whole velocity distribution of the ions. For the fluorescence detection two photomultipliers where placed at a distance of about 30 cm apart from each other in the experimental section of the ESR, see Fig. 5. To control the overlap of the ion- and the laser-beam, two scrapers were used. These are placed in the middle of the analysis section about 6.5 m apart. At the ESR, the laser beam is guided using fully motorized linear and rotation stages to provide excellent remote control of the laser beam, resulting in an uncertainty of the ion-laser-beam alignment of ±77 µrad.

In contrast to the TSR, where electron cooling takes 5–10 s and the metastable ^{7}Li$^+$ ions have a lifetime of 13 s [13], it turned out that the lifetime at the ESR is close to the natural one of 50 s, see Fig. 6, left panel. The time for the electron cooling in the ESR is in the range of about 30 s. Nevertheless, after reaching equilibrium, the width of the velocity distribution is $\delta\nu$=1.09 GHz±0.08 GHz in the laboratory system (see Fig. 6, right panel), compared to 2.5 GHz at the TSR [16]. Together with the measured centre frequency of $\nu_a = 383\ 621.0$ GHz $\pm\ 0.4$ GHz for the anti-parallel excitation, a velocity ratio of $\Delta\beta/\beta = 3.4 \times 10^{-6}$ is found. This value provides a verification of the velocity of the electrons from the cooler, with an order-of-magnitude better accuracy compared to the electronic measurement setup routinely used at the ESR (Steck, personal communication).

Furthermore it turned out that the amount of stored ions in the metastable state is at most 0.1% of the whole beam (so in the range of <10^5 ions). Comparing this with the efficiency at TSR (~10% of 10^8 ions) it is obvious that the production process has to be significantly improved to increase the fluorescence signal.

5 Recent development

The first test at the ESR has given a positive feedback for the feasibility of the time dilation experiments. Nevertheless some further steps have to be done to perform a new test of special relativity. First of all the production process for the metastable ions has to be optimized (e.g., using a different source material) to a fraction of 1–10% of the whole ion-beam (at a typical ion number of 10^8 stored in the ESR). Tests at an ECR source addressing this point are in progress at the University of Giessen by the group of A. Müller.

 Springer

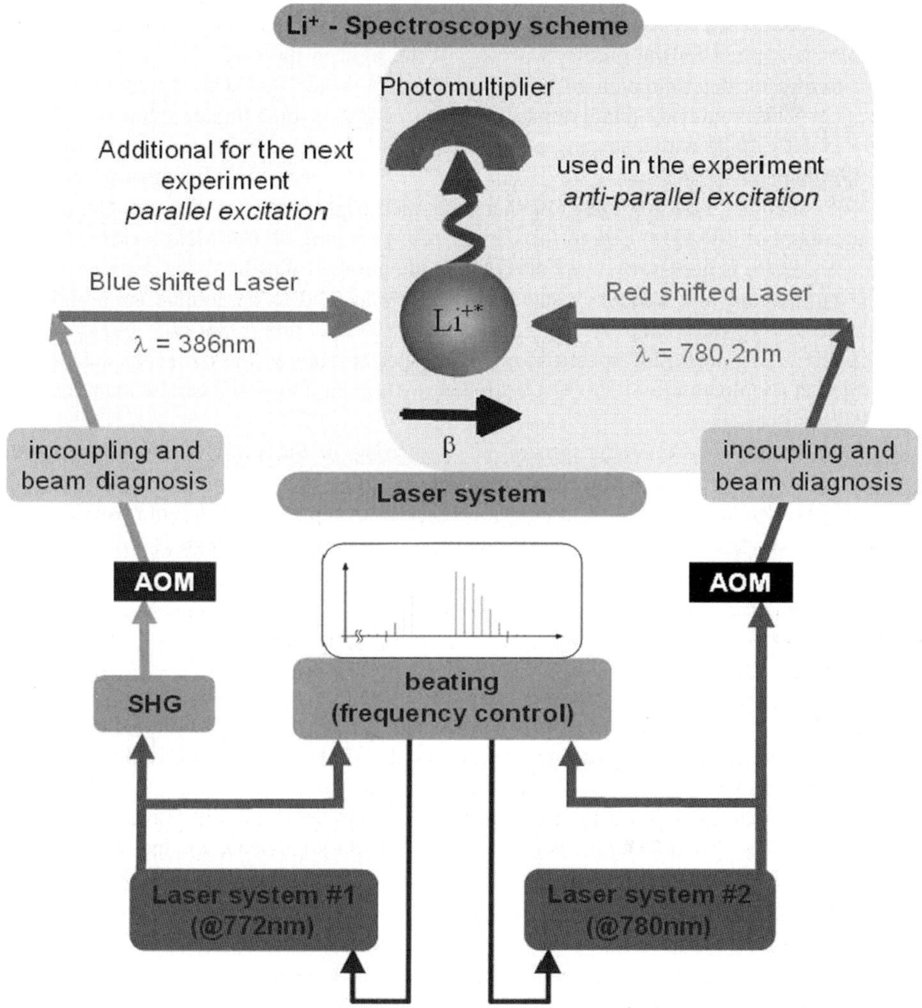

Fig. 7 Scheme of the time dilation experiment at the ESR

Furthermore the laser system for the parallel excitation of the ^{7}Li$^+$ ions has been established. The high ion-velocity induces a Doppler-shift in the parallel excitation scheme to λ_p=386 nm. To reach this wavelength, a laser system has been developed, that consists of a frequency-doubled cw Ti:Sa laser (the fundamental wave is λ_f=771.5 nm) that is pumped by a Nd:YAG laser. The efficiency of the frequency doubler is about 10%. To guide the 386 nm light a photonic crystal fiber, with an efficiency of about 10% and a length of 20 m, is used. To switch the laser light on and off, an acousto-optic modulator is placed right before the fiber, see Fig. 7. With an output power of about 1 W from the Ti:Sa laser, the frequency-doubled light power in the ESR will be in the mW range.

For the anti-parallel excitation, a laser system consisting of a second cw Ti:Sa laser (also pumped with another Nd:YAG laser) at a wavelength of λ_a=780.2 nm is build up. The laser light is guided to the ESR incoupling window via a 50 m fiber. Again, an acousto-optic modulator is applied for the on/off switching of the light in the ring.

The frequency determination of both laser systems is realised with a frequency comb synthesizer. This frequency standard has a comb structure of some hundred thousand modes in the VIS-NIR range with a spacing of 100 MHz. The idea for the time dilation experiment is to simultaneously stabilize both lasers (λ_f=771.5 nm and λ_a=780.2 nm) to different modes of the comb. This provides, together with a Rb-clock reference, the laser frequencies to an accuracy of 200 kHz. Due to the narrow mode spacing of 100 MHz a coarse, online frequency control is necessary to maintain the mode number. This has been realised with an atomic and a molecular frequency standard. λ_a is referenced to the atomic Rubidium line (384 227 995.7 MHz) and λ_f to the molecular iodine line (388 605 093.6 MHz), respectively, using saturation spectroscopy. (The spectroscopy at the molecular iodine was performed at a temperature of 550°C.) With this method the frequency can be maintained in the MHz regime.

For the control of the laser-laser-ion-beam overlap in the ESR, another observation window at a distance of 6 m from the other windows has been installed. Furthermore this window gives another possibility for the fluorescence detection at a different position in the ESR and therefore a chance to gain information about systematic effects in the ring.

6 Conclusions

An upper limit for possible deviations of the time dilation factor from Special Relativity $|\alpha| < 2.2 \times 10^{-7}$ has been found at the TSR in Heidelberg. After finishing the data analysis of the latest experiments at two different velocities at TSR, this upper limit is expected to be further improved.

Furthermore the feasibility of performing time dilation tests at the ESR has been shown in October 2005. The development of the laser excitation systems for the time dilation experiment at an ion-velocity of β=0.33 and their online frequency determination is completed. Experiments to increase the production rate of metastable $^7Li^+$ ions from an ECR ion source are in progress.

References

1. Kostelecky, V.A., Samuel, S.: Phys. Rev., D **39**, 683 (1989)
2. Lane, C.D.: Phys. Rev., D **72**, 016005 (2005)
3. Tobar, M.E., Wolf, P., Fowler, A., Hartnett, J.C.: Phys. Rev., D **71**, 025004 (2005)
4. Robertson, H.P.: Rev. Mod. Phys. **21**, 378 (1949)
5. Mansouri, R.M., Sexl, R.U.: Gen. Relativ. Gravit. **8**, 497 and 515 and 809 (1977)
6. Antonini, P., Okhapkin, M., Göklü, E., Schiller, S.: Phys. Rev., A **71**, 050101 (2005)
7. Stanwix, P.L., Tobar, M.E., Wolf, P., Susli M., Locke, C.R., Ivanov, E.N., Winterflood, J., van Kann, F.: Phys. Rev. Lett. **95**, 040404 (2005)
8. Wolf, P., Tobar, M.E., Bize, S., Clairon, A., Luiten, A.N., Santarelli, G.: Gen. Relat. Gravit. **36**, 2351 (2004)
9. Riis, E., Andersen, L.-U.A., Bjerre, N., Poulsen, O.: Phys. Rev. Lett. **60**, 81 (1988)
10. Ives, H.E., Stilwell, G.R.: J. Opt. Soc. Am. **28**, 215 (1938)
11. McGowan, R.W., Giltner, D.M., Sternberg, S.J., Lee, S.A.: Phys. Rev. Lett. **70**, 251 (1993)
12. Grieser, R., Klein, R., Huber, G., Dickopf, S., Klaft, I., Knobloch, P., Merz, P., Albrecht, F., Grieser M., Habs, D., Schwalm, D., Kühl, T.: Appl. Phys., B **59**, 127 (1994)
13. Saathoff, G., Reinhardt, S., Buhr, H., Carlson, L.A., Schwalm, D., Wolf, A., Karpuk, S., Novotny, C., Huber, G., Gwinner, G.: Can. J. Phys. **83**, 425 (2005)

14. Reinhardt, S.: *PhD. Theses*, 2005, Max Planck Institute for Nuclear Physics, Heidelberg
15. Reinhardt, S., Saathoff, G., Karpuk, S., Novotny, C., Huber, G., Zimmermann, M., Holzwarth, R., Udem, T., Hänsch, T.W., Gwinner, G.: Opt. Commun. **261**, 282 (2006)
16. Saathoff, G., Karpuk, S., Eisenbarth, U., Huber, G., Krohn, S., Munoz Horta, R., Reinhardt, S., Schwalm, D., Gwinner, G.: Phys. Rev. Lett. **91**, 190403 (2003)
17. Riis, E., Sinclair, A.G., Poulsen, O., Drake, G.W.F., Rowley, W.R.C., Levick, A.P.: Phys. Rev., A **49**, 207 (1994)
18. Hvelplund, P.: J. Phys., B **9**, 1555 (1976)

Hyperfine Interact (2006) 171:69–81
DOI 10.1007/s10751-006-9497-9

A vision for laser induced particle acceleration and applications

K. W. D. Ledingham

Published online: 9 February 2007

Abstract High intensity lasers interacting with solid and gas targets can produce energetic beams of electrons, protons, heavy ions, photons and neutrons. There is the potential for producing "table top" sources of radiation for many different applications.

Key words high energetic laser · positron emiting isotopes (^{11}C, ^{18}F) · positron emission tomography · mono-energetic proton beams · proton oncology

1 Preamble

During the last few years, laser induced particle acceleration and applications have become a very exciting new field of research. The reason for this is principally because of the potential that laser accelerators could be very much smaller than conventional accelerators and maybe even cheaper to build. The University of Strathclyde has been at the forefront of what we called Laser Induced Nuclear Physics, a name which was coined in 1999. During the years since then we have been very much instrumental in carrying out a number of "proof of concept" experiments at the Rutherford Appleton Laboratory and a number of laboratories in Europe principally at the University of Jena. These experiments included laser induced fission [1], laser induced PET isotope production [2–4], laser induced transmutation studies [5, 6], laser induced heavy ion reactions [7, 8], laser induced proton production from short pulse lasers [9, 10], laser induced spallation studies [11], laser induced neutron production [12–14] and the use of laser induced isotope production as a high temperature plasma diagnostic [15]. Many of the details of how laser induced nuclear physics operates are described in a recent Science Article [16].

K. W. D. Ledingham (✉)
Department of Physics, University of Strathclyde, Glasgow G4 0NG Scotland, UK
e-mail: K.Ledingham@phys.strath.ac.uk

K. W. D. Ledingham
AWE plc, Aldermaston, Reading RG7 4PR, UK

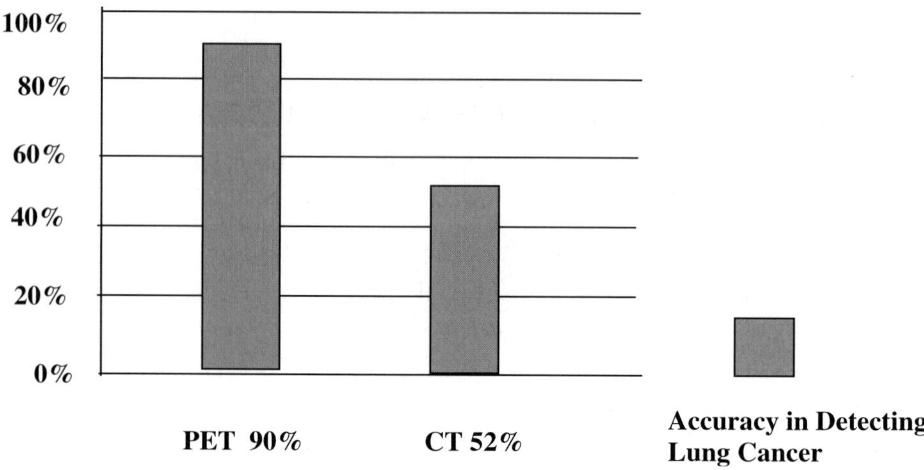

Fig. 1 Accuracy of PET in detecting lung cancer compared with X-ray CT scanning

Finally there are radiation hazards associated with high intensity lasers principally from the production of high energy gamma rays. These are however very much smaller than the hazards associated with conventional accelerators and reactors. The shielding of high intensity lasers has recently been described in detail in a paper by Clarke et al. [17].

Although many of the above nuclear physics applications could have been covered in this presentation only those associated with medical applications will be described in detail; namely PET isotope production and the production of mono-energetic protons which is a prerequisite for one of the potentially most exciting applications of laser induced ion beams that of proton oncology.

2 PET isotope production

One of the exciting applications of the high energy laser induced proton beams is the production of radioactive isotopes for Positron Emission Tomography (PET). PET is a form of medical imaging requiring the production of short lived positron emitting isotopes ^{11}C, ^{13}N, ^{15}O and ^{18}F, by proton irradiation of natural/enriched targets using cyclotrons. PET development has been hampered due to the size and shielding requirements of the nuclear installations but recent results have shown when an intense laser beam interacts with solid targets, tens of MeV protons capable of producing PET isotopes are generated [2–4].

2.1 Positron emission tomography

Positron Emission Tomography (PET) is a powerful medical diagnostic/imaging technique requiring the production of short-lived (2 min–2 h) positron emitting isotopes. The PET process involves the patient receiving an injection of a pharmaceutical labelled with a short-lived β^+ emitting source which collects in 'active' areas of the body such as tumours. The principal tracers used in the PET technique are ^{11}C, ^{13}N, ^{15}O and ^{18}F. Many chemical compounds can be labelled with positron emitting isotopes and their bio-distribution can be determined by PET imaging as a function of time. However the most commonly used radiopharmaceutical is 2-fluoro-2-deoxyglucose 2-[^{18}F]FDG. Over the last few years the

 Springer

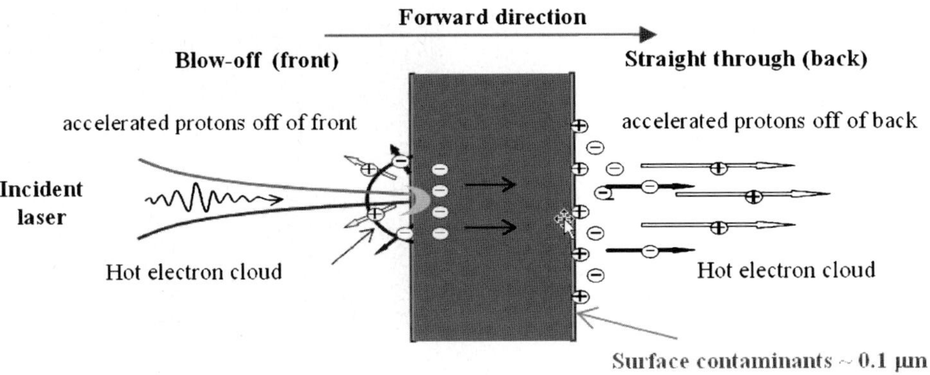

Fig. 2 Pictorial representation of the Target Normal Sheath Acceleration (TNSA) scheme [22]

value of PET FDG in the management of cancer patients has been widely demonstrated. Figure 1 highlights the success rate of PET in diagnosing lung cancer compared with conventional X-ray computed tomography (CT) scanning.

PET isotopes are normally generated using energetic proton beams produced by cyclotrons or Van de Graafs via (p,n) or (p,α) reactions.. Proton induced reactions are favoured since the resultant isotope differs in atomic number from the reactant, thus simplifying the separation process and makes it possible to produce carrier free sources allowing the patient to be injected with the minimum amount of foreign material.

One of the main factors limiting the wider use of FDG PET imaging is the requirement for expensive infrastructure at the heart of which lies the cyclotron and the associated extensive radiation shielding. A more simplified approach to isotope production would be to develop a miniaturised, on site resource with eventual capability similar to that of a cyclotron. As was stated previously, recent results show when an intense laser beam ($I>$ 10^{19} W/cm^2) interacts with solid targets, beams of MeV protons capable of producing PET isotopes are generated. It should be pointed out at this stage that the laser approach to PET isotope production is not intended to be a competitor of cyclotron sources but in the fullness of time as the lasers develop and become smaller this could be a complimentary approach.

2.2 Proton production with a high intensity laser

Recent advances in laser technology with the introduction of Chirped Pulse Amplification [18] (CPA) have led to the development of multi-terawatt pulsed laser systems in many laboratories worldwide. After amplification, these laser pulses are recompressed to deliver 10^{18-20} W/cm^2 on target. Proposed techniques, including optical parametric chirped pulse amplification (OPCPA) [19, 20] promise to extend the boundaries of laser science into the future and also reduce the large lasers used presently to compact table-top varieties.

High intensity laser radiation may now be applied in many traditional areas of nuclear science. As the laser intensity and associated electric field is increased then the electron quiver energy, the energy a free electron has in the laser field, increases dramatically. Thus, when laser radiation is focused onto solid and gaseous targets at intensities $>10^{18}$ W/cm^2, electrons quiver with energies greater than their rest mass (0.511 MeV) creating relativistic plasmas [21]. At these intensities, the Lorentz force $-c(v \times B)$ due to the laser interacting with charged particles produces a ponderomotive force allowing electrons to be accelerated into the target in the direction of laser propagation. The resulting electron energy

Fig. 3 Image of inside the target chamber showing the incident laser beam directed onto varying thicknesses of materials of foils held in a target wheel. The copper stacks for proton energy measurements and Boron samples for C production are shown

distribution can be described by a quasi-Maxwellian distribution yielding temperatures, kT, of a few MeV [15].

The protons emanate from water and from hydrocarbons as contaminant layers on the surfaces of the solid targets. These contamination layers are due to the poor vacuum ($\sim 10^{-5}$ Torr) achievable in the target chambers. The main mechanism thought to be responsible for proton acceleration is the production of electrostatic fields due to the separation of the electrons from the plasma ions. Proton beams are observed both in front of (blow-off direction) and behind (straight through direction) the primary target. In front of the target, ion beams are observed from the expansion of the plasma generated on the target surface, produced either by a pre-pulse or the rising edge of the main pulse itself, also known as "blow-off" plasma, directed normal to the target surface.

Several acceleration mechanisms have been proposed to describe where the protons in the straight through direction originate, either the front surface, back surface or both. One such mechanism is the Target Normal Sheath Acceleration (TNSA) [22]. In this scheme, shown in Fig. 2, the ion acceleration mechanism results from the cloud of hot electrons (generated in the blow-off plasma from the laser pre-pulse interacting with the front surface of the target) travelling through the target and field ionizing the contaminant proton layer on the back surface of the target. The protons are then pulled off the back surface by the cloud of electrons and accelerated normal to the target to tens of MeV's in tens of μm. The initial laser pre-pulse may be of the order of 10^{-6} ($I \sim 10^{12-14}$ W/cm^2) of the main laser pulse, sufficient to ionize the front surface of the target. Thus, at the back of the target where no pre-plasma is formed, the accelerating field is greater, resulting in higher energy ions. Recent studies have reported on direct experimental evidence of back-surface ion acceleration from laser irradiated foils by using sputtering techniques to remove contaminants from both the front and back surfaces [23]. It has also been proposed that the protons are accelerated via an electrostatic sheath formed on the front surface of the target and dragged through the target to produce a proton beam at the rear of the target [24]. Comparative reports on the ion acceleration schemes can be found in references [25, 26].

Proton energies with an exponential distribution up to 58 MeV have been observed [27] for a laser pulse intensity of $3 \cdot 10^{20}$ W/cm^2 and production of greater than 10^{13} protons per pulse has been reported [28]. With the VULCAN laser at the RAL delivering close to petawatt powers, it is now possible to demonstrate the potential for high power lasers to produce intense radioactive sources.

The new petawatt arm of the VULCAN Nd:Glass laser at RAL is employed in this experimental study. The 60 cm beam is focused to ~ 5.5 μm diameter spot using a 1.8 m focal length off-axis parabolic mirror, in a vacuum chamber evacuated to $\sim 10^{-4}$ mbar.

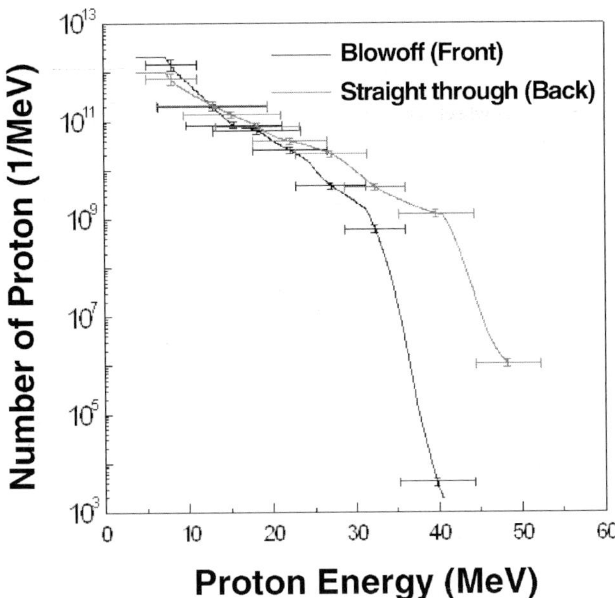

Fig. 4 Typical proton spectra in front of and behind a 10 μ Al target. The number of protons generated per laser shot at about 300 J and $2 \cdot 10^{20}$ W/cm^2 is typically 10^{12}

The energy on target is between 220 and 300 J while the average pulse duration is 750 fs. The peak intensity is of the order of $2 \cdot 10^{20}$ W/cm^2.

2.3 Proton energy measurements

To measure the energy spectra of the accelerated protons, nuclear activation techniques are employed. Copper stacks (5 cm × 5 cm) are positioned along the target normal direction and exposed to the protons accelerated from both the front and back surfaces of the primary target foil. Figure 3 shows an image of the experimental set up inside the chamber.

The activity in the copper foils from the ^{63}Cu(p,n)^{63}Zn reaction with a half-life of 38 min is measured in a 3″×3″ NaI coincidence system. The efficiency of the system is measured using a calibrated ^{22}Na source, thus the absolute activity can be determined. The measured activity in the foils from the ^{63}Cu(p,n)^{63}Zn, convoluted with the reaction cross-section and proton stopping powers is used to produce the energy distributions shown in Fig. 4. From the proton spectrum shown here it can be seen that the shape is typically exponential with a wide energy spread. It will be shown in the next section dealing with mono-energetic protons for oncology that this spectrum shape must be modified.

2.4 ^{18}F and ^{11}C generation

The isotope ^{18}F is generated from a (p,n) reaction on ^{18}O enriched (96.5%) target. The enriched ^{18}O targets were irradiated in the form of 1.5 ml of [^{18}O]H$_2$O placed in a 20 mm diameter stainless steel target holder. For the production of ^{11}C, the copper stacks described above were replaced by boron samples (5 cm diameter and 3 mm thick). After irradiation, the boron targets were removed from the vacuum chamber and the ^{11}C activity produced by the (p,n) reaction on ^{11}B was measured in the coincidence system up to two hours after the laser shot, a safety precaution because of the high activity. The counting rate was determined at time zero and converted to Bq using a calibrated ^{22}Na source.

 Springer

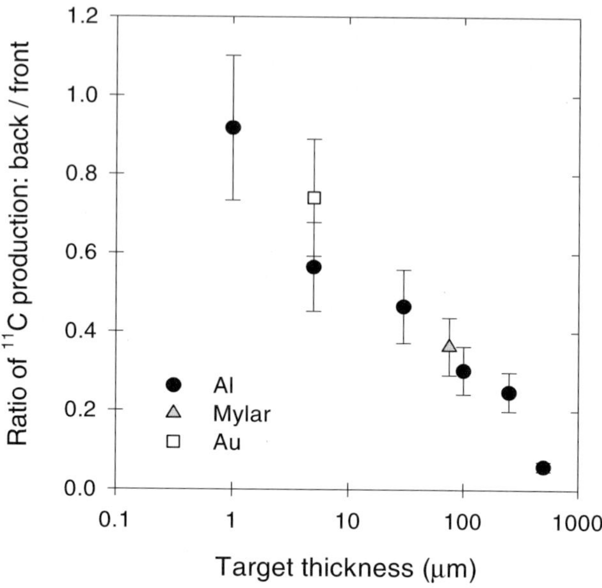

Fig. 5 Back/front ratio of ^{11}C from the (p,n) reactions on ^{11}B as a function of target thickness. At the highest pulse energy on target ~300 J the ^{11}C activity maximally was about $6\cdot10^{6}$ Bq per shot on each side. This is greater than 10^{7} Bq in total

2.5 Target selection

In order to determine the thickness of primary target which generated the highest activity sources, the ^{11}C activity generated in the secondary ^{11}B targets is measured as a function of sample material and thickness. The ratio of the back to front activities is shown in Fig. 5. This was carried out using the production of the PET isotope ^{11}C rather than the more novel ^{18}F because of the cost of carrying out systematic work using the very expensive separated ^{18}O isotope as a target is prohibitive. It is clear from Fig. 5 that very thin targets provide the highest activity sources when the total activity produced per laser shot is the sum of the back and front activities.

Furthermore to first order it does not matter what the target is made of. Aluminium, mylar and gold targets could be used. From further experiments recently carried out at RAL although it is important that for isotope production the intensity must be greater than about 10^{20} W/cm^{2} by far the most important thing is laser pulse energy which in our case was a few hundred joules. If high repetition rate table top lasers were used with just joules in pulse energy then the activities shown in Fig. 5 would have to be integrated over many pulses.

2.6 ^{18}F and ^{11}C production

It was reported earlier that ^{18}F is the most widely used tracer in clinical PET today due to its longer half-life allowing for the synthesis of a number of samples within a half-life decay of the isotope and because fluorine chemistry is readily introduced in many organic and bio-inorganic compounds. It was necessary to determine how much ^{18}F could be produced per laser shot. The isotope is generated from a (p,n) reaction on ^{18}O enriched (96.5%) target. At the highest laser pulse energies (300 J), 10^{5} Bq total activity of ^{18}F was produced (shown later).

The half-life for the ^{18}F source is shown in Fig. 6. The measured half-life of 110±3 min was determined over more than three half-lives and demonstrates the purity of the ^{18}F source and agrees closely with the generally accepted value (109 min). Also shown is the measured half-life of ^{11}C of 20 min and agrees well with the accepted value.

Springer

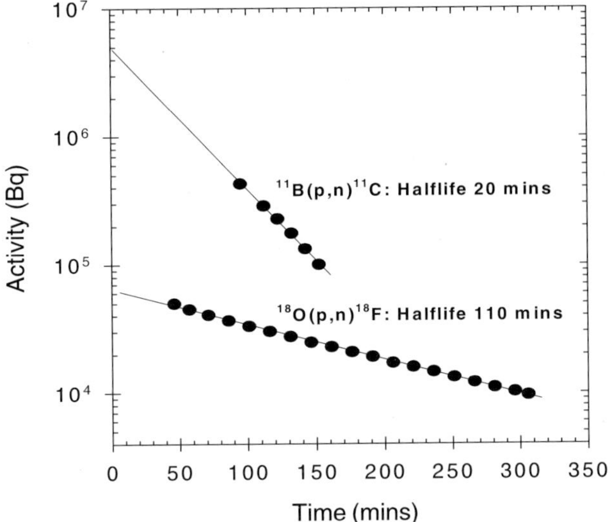

Fig. 6 The measured half-lifes for ^{18}F and ^{11}C. The values are close to the accepted ones indicating the purity of the sources produced

Figure 7 summarises our measurements to date in this programme of research into laser-driven ^{11}C and ^{18}F PET isotope production on VULCAN. The circles (^{11}C) correspond to a number of different laser irradiances and pulse energies up to 300 J with pulse duration of ~750 fs. The single triangular point is the activity from the ^{18}F measurements at the highest laser pulse energy. The hatched areas at the top of the graph provide an indication for the level of required ^{18}F activity (1 GBq), from which an ^{18}F-FDG patient dose would be generated, and the required ^{11}C activity (0.5 GBq), e.g., in the form of [^{11}C]CO.

2.7 Future development and conclusions

2.7.1 How to increase the PET isotope activity to 10^9 Bq?

Although, the results presented here were obtained from a large single shot laser, it is important to highlight the progress made using compact high repetition rate lasers. Fritzler et al. [2] have calculated that 13 MBq of ^{11}C can be generated using the LOA "table-top" laser (1 J, 40 fs) $6 \cdot 10^{19}$ W/cm^2 after 30 min at 10 Hz and that this can be extended to GBq using similar lasers with kHz repetition rates. Alternatively at JanUSP (Livermore) using a single pulse (8.5 J, 100 fs, 800 nm) at $2 \cdot 10^{20}$ W/cm^2, 4.4 kBq of ^{11}C was generated from a single laser shot (Patel, personal communication). Using a compact laser with similar specifications at 100 Hz after 30 min this would amount to close to GBq. A compact "table-top" laser system has recently been designed by Collier and Ross (personal commnication) for this purpose. In addition, the small scale POLARIS [29] all diode pumped petawatt laser currently being built at the Friedrich-Schiller University of Jena has the potential to deliver 10^{21} W/cm^2 (τ=150 fs, E=150 J, $\lambda \sim 1$ µm) with a repetition rate of 0.1 Hz.

3 Mono-energetic proton production

In a recent paper [30] Ledingham stated that treating cancer with beams of high energy protons is just one of the exciting possibilities presented by the advent of laser based

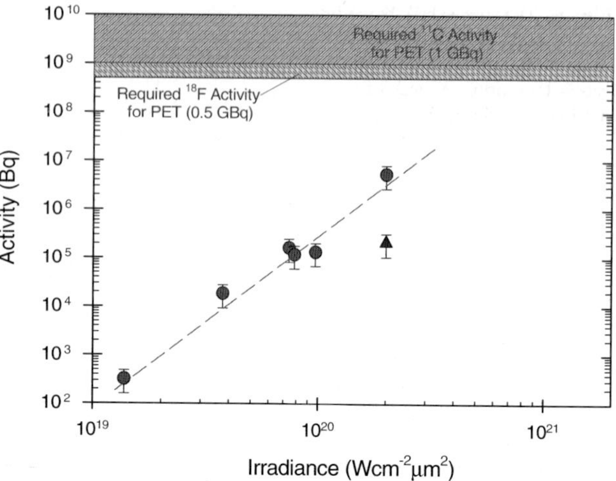

Fig. 7 The total activity (*front* and *back*) generated by a single laser shot for both ^{11}C and ^{18}F as a function of laser irradiance with pulse energies from 15 to 300 J. The circles refer to ^{11}C production and the single triangular point for ^{18}F was measured at the highest energies

particle accelerators. How soon these devices will reach a performance needed for such applications and how soon these improvements will be made are still very contentious issues. However one of the prerequisites for proton oncology is to control the proton beam. At present the proton beams exhibit a quasi exponential energy spectrum which is not suited for the purpose. We need a much more mono-energetic beam for therapy.

The recent publication of three high-profile reports [31–33] on the use of laser-based accelerators to produce high-energy quasi-mono-energetic electron beams of unprecedented quality heralds a new age of high-energy physics. The race is now on to take the logical next step in the development of these devices to generate beams of similar quality from other species such as protons and heavy ions. Such multifunctional desktop accelerators have a wealth of potential applications, from radiographic imaging for peering within the structure of dense objects (for security and non-destructive testing purposes) to proton therapy for treating deep-seated cancerous tissue. There is still a long way to go to increase the energy and improve the quality of the proton and ion beams produced by these devices before such visions become reality. The use of particle accelerators for cancer therapy relies on the fact that when high-energy protons pass through matter – which in this case would be soft-tissue – they create very little damage until they reach the end of their trajectory, where they cause a great deal of damage; and that the stopping distance and distribution of protons of a given energy in a relatively homogenous material is well defined and narrow. It is hoped that by directing an intense, mono-energetic beam of protons, whose energies are tuned so that they stop at the precise location of a deep-seated tumour within a patient, the tumour can be destroyed whilst minimizing any harm to the surrounding healthy tissue. This requires the ability to generate quasi mono-energetic, or at least well-shaped, beams of protons with energies exceeding 200 MeV. The two central questions being asked by groups around the world regarding the development of laser based accelerators for such applications are: how are the energies of the generated protons affected by the character-istics of the lasers used to drive them and in particular the lasers of the future; and can the broad proton-energy distributions generated by the present systems be narrowed sufficiently to the so-called dream beams necessary for targeted cancer treatment?

I now present the first approach to generating mono-energetic proton beams and that is by using micro-structured targets. This was published by Schwoerer et al. in a recent Nature article [34].

 Springer

Fig. 8 A terawatt (TW)-laser pulse is focused onto the front side of the target foil, where it generates a blow-off plasma and subsequently accelerates electrons. The electrons penetrate the foil, ionize hydrogen and other atoms at the back surface and set up a Debye sheath. The inhomogeneous distribution of the hot electron cloud causes a transversely inhomogeneous accelerating field (Target Normal Sheath Acceleration—TNSA [22]). Applying a small hydrogen-rich dot on the back surface enhances the proton yield in the central part of the accelerating field, where it is nearly homogenous. These protons constitute the quasi-monoenergetic bunch

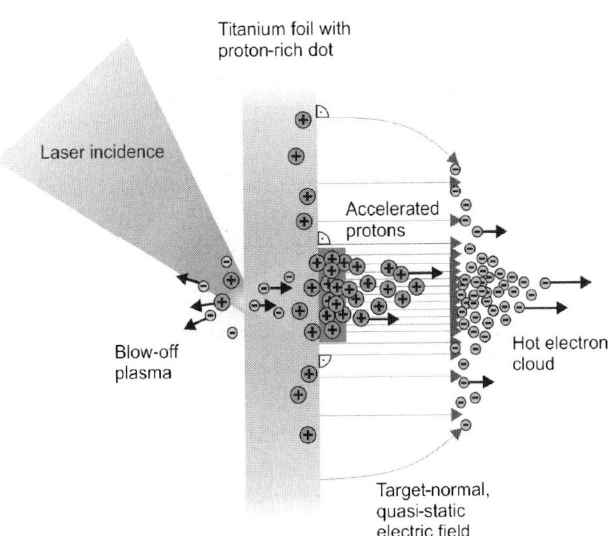

As mentioned previously the interaction of an intense light field with matter yields the generation of hot plasma and the subsequent acceleration of electrons up to relativistic energies. Protons and ions are accelerated by a well controlled mechanism known as 'Target Normal Sheath Acceleration' (TNSA) following the initial electron acceleration, as I described earlier and is shown in Fig. 8. Fast electrons are accelerated by an intense laser pulse (intensity $I > 10^{19}$ W/cm^2) from the surface of a thin metal foil in the forward direction. They penetrate the foil and ionize atoms along their paths. Within about a picosecond those electrons, leaving the target at the rear surface build up a quasi-static electric field. The field acts normally to the target surface, has cylindrical symmetry and decreases in the transverse direction. Owing to the ultra short duration of the electron bunch and its high charge, this field may reach values of several TV (10^{12} V/m) close to the axis and thus the potential can attain several tens of MeV.

Protons and positively charged ions present on the back surface of the foil may be accelerated by this field until they compensate the electron charge. In most cases, the origin of these parasitic protons has been identified to be a hydrocarbon contamination layer on the target surface. As the duration of the acceleration is ultras short and the protons (as well as the ions) are at rest before acceleration, comprising a very small phase space volume, the transverse emittance of the proton beam reaches values as low as a few 10^{-3} mm·mrad for 10 MeV protons. This can be explained by the inhomogeneous distribution of electrons in the sheath causing an accelerating field that is inhomogeneous in the transverse direction. For a plane and unstructured target, the transverse dimension of the electric field and hence the source size of accelerated protons is much larger than the laser's focal spot (Fig. 8). Therefore, different parasitic protons experience a range of potentials, resulting in a broad distribution of energies.

Following this understanding of the mechanism of laser acceleration of protons, Bulanov et al. [35, 36], Esirkepov et al. [37] pointed out that the resulting proton energy spectrum has a strong correlation to the spatial distribution of the protons on the target surface. In order to generate high quality proton beams with mono-energetic features, they proposed a bi-layered, micro-structured target, consisting of a thin high Z metal foil and a small proton-rich dot on the back surface. The transverse dimension of such dots is smaller than that of

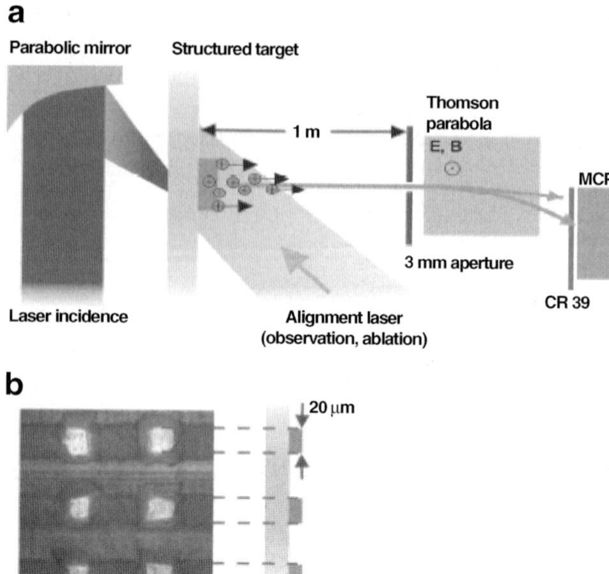

Fig. 9 a A TW-laser pulse is focused by a 45° off-axis parabolic mirror (f/2.5) to an intensity of $3 \cdot 10^{19}$ W/cm. A dot on the back surface of the target foil is positioned opposite to the focus using an alignment laser. A titanium foil of 5 µm thickness carries dots of PMMA with a thickness of 0.5 µm and a transverse size of 20 µ×20 µ. The photograph of the micro-structured back surface in **b** shows the dots as light squares. The protons and ions accelerated from the target are dispersed with respect to energy and charge-to-mass ratio in a Thomson parabola, and then detected by either a microchannel plate (MCP) with phosphor screen and CCD camera or on nuclear track detector plastics (CR39)

the acceleration sheath, and hence the protons will only be subject to the central – that is, homogeneous – part of the acceleration field. In this configuration, the protons all experience the same electric field and are accelerated in the same potential (Fig. 9). The resulting proton beam has a spectrum with a strong mono-energetic peak. The experimental arrangement follows the proposal to use micro-structured targets (Fig. 9). The results are well reproduced by two-dimensional particle in-cell (PIC) simulations. The simulations also show the scalability of the technique.

The high intensity laser pulses (intensity 10^{19} W/cm^2) are generated by the JETI 10TWTi:sapphire laser at the University of Jena. It delivers pulses of 80 fs duration, pulse energy of 600 mJ on target and a maximum repetition rate of 10 Hz. The target is a 5 µm thin titanium foil, coated with a 0.5 mm layer of polymethyl methacrylate (PMMA) on the back surface. In some regions of the sample the PMMA layer was micro-structured, leaving PMMA dots of (20×20) µm^2 on the surface with PMMA-free space around (Fig. 9b). The laser pulse hits the foil on the front surface exactly opposite to one of the dots. Protons and ions, accelerated from the back surface in the normal direction, were analysed by a Thompson spectrometer and detected either by an online system based on microchannel plates (MCPP or by CR) 39 nuclear track detectors.

Figure 10 presents the result of irradiating the micro-structured target foil at the position of a dot in contrast to unstructured material. The data are given as number of protons per energy interval of 0.05 MeV, which corresponds to our MCP resolution, and per 24 msr, which is the solid angle of emission, that results from the simulations described below. The curve indicated by the opened triangles shows the proton spectrum after irradiating a dot. It exhibits a distinct narrow band feature, peaked around $E_{max} = 1.2$ MeV on top of a broad, exponential shaped background. The displayed feature contains about 108 protons per 24 msr, and has a full-width at half-maximum of 25% of its absolute value. For comparison, the black squares data represent an average over six proton spectra recorded if the laser hits a blank position on the same target, where protons can only originate from an unstructured hydrogen contamination layer (parasitic protons). No narrow band feature appears, and the exponential

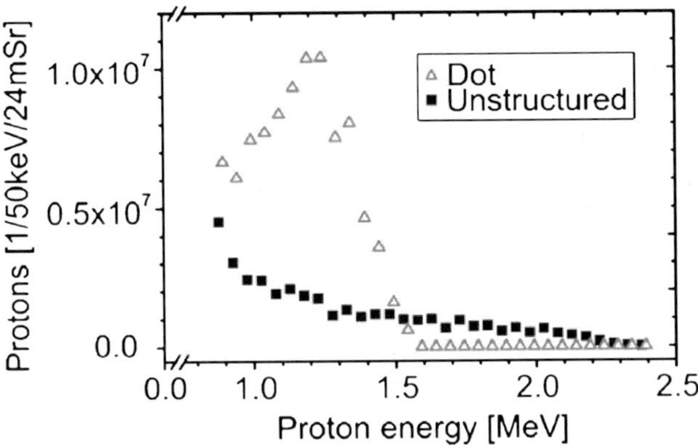

Fig. 10 The triangles graph shows a spectrum obtained from irradiating the foil at the position of a dot using the MCP detection. The spectrum from a dot exhibits a peak at energy of 1.2 MeV as opposed to exponential spectra (*black squares*, average from six shots) in the case of using an unstructured part of the target foil. The peaked structure contains about 10^8 protons per 24 msr

shape of the spectrum can be approximated by a temperature of about 0.5 MeV. The occurrence of the peaked spectrum after irradiating a dot was also observed using CR39 detection. The two curves exhibit the same shape with a peak at 1.2 MeV, which accounts both for the reproducibility of generating peaked spectra as well as the reliability of the MCP detection. The proton yield at spectrum maximum is determined not only by the number of abundant protons in the dot but also by the laser characteristics, which affect the transverse scale of the electrostatic potential, the acceleration time and the spatial distribution of protons.

Owing to the limited size of the multichannel plate, the spectral range of the MCP data is smaller than that one of the CR39 data sets. However, following direct successful comparison of these measurements, the electronic system provides a reliable online observation which allows for systematic investigation of the reproducibility with respect to proton yield, peak position and spectral width as well as for fast and controlled changes of experimental conditions. This experimental accuracy can only be accomplished with table-top lasers with high repetition rate. A maximum in the proton spectrum is reproduced consistently if a micro structure on the rear of the target is irradiated. The position of the peak E_{max}, as well, as its width, varies from shot to shot by about 20%.

From systematic experiments performed earlier on non micro-structured targets, it is known that an optimum target thickness with respect to maximum proton energy and yield follows from a given temporal structure of the laser pulse. Considering our laser and target conditions, the exponential contribution to the proton distribution from the non micro-structured targets is in agreement with previously published results from similar targets. The narrow band spectra of the laser accelerated protons observed in our experiments are due to the small proton-rich area within the centre of the larger quasi-static electric field, set up by the laser accelerated electrons beyond the thin target (Fig. 8). If the scale of the inhomogeneity of the electric field is larger than the proton-rich spot, these protons all experience the same potential. The maximum proton energy is determined by the total charge of electrons constituting the acceleration sheath and occurs on the axis, which in turn depends on the laser intensity and the target thickness. This analysis is supported by two-dimensional PIC multi-parametric simulations based on the code REMP. Figure 11a plots the resulting proton spectrum (solid line) under the conditions laser intensity 10^{19} W/cm^2 and a 5 μm thin titanium target with 20 μm PMMA structure on the back side, which match the experimental parameters. The numerically achieved proton spectrum is dominated by a narrow band structure around 1.2 MeV. Simulation and experiment are in good agreement with respect to both the existence of the narrow structure as well as its

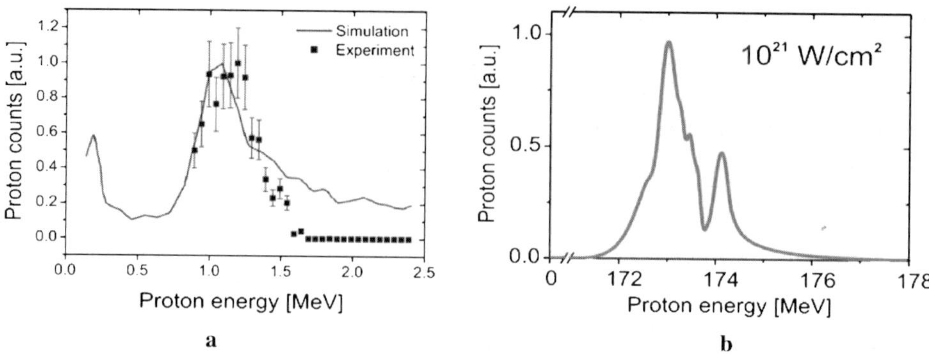

Fig. 11 **a** This figure shows the experimental points taken with the MCP (*black dots*) and the 2D PIC code simulations in continuous line. It can be seen that the agreement is good. **b** the simulation indicates that at laser intensities of 10^{21} Wcm^{-2} the proton spectrum peaks at about 173 MeV which is becoming suitable for proton therapy

position and width. It was estimated that in the case of a 10 μm focal spot and 0.1 μm thick proton layer, a proton energy spread of 1% can be expected. Furthermore, for micro-structured targets the maximum proton energy (E_{max} in MeV) scales as the square root of the laser power (P, in PW) with $E_{max}=230 \times P^{1/2}$. The parameters for the simulation have been changed to smaller dot sizes and higher energies in order to extrapolate the technique to future experiments and investigate the scalability of our results (see Fig. 11b). A high repetition rate table-top laser system with petawatt power (POLARIS) will be available within a few years at Jena. This laser will deliver pulses of 150 J within 150 fs, which leads to an intensity of about 10^{21} W/cm^2 in focus. Also within 2 years the new GEMINI laser at the Rutherford Appleton Laboratory will yield suitable beams for oncology. Simulations performed with these parameters result in a peak proton energy at 173 MeV and relative width dE/E<1% for a dot diameter of 2.5 μm and a reduced layer thickness of 0.1 μm. Under these conditions, the proton yield is no longer limited by the laser energy, but all protons contained in the dot ($8 \cdot 10^8$) are quickly accelerated to an energy of ~173 MeV. Proton beams with such a characteristic might be suitable for treatment of deep sited tumours. More realistic within a few years is the potential for laser induced proton therapy for eye tumours, which only requires 60–70 MeV protons.

We have indicated that this experiment was carried out as "proof of principle." An extensive programme is now underway to reduce the width and increase the energy of our proton peaks by improving the target fabrication procedure. We intend to reduce the dimensions of the dots and change the target material. Our initial targets were dots on titanium, which was chosen on the basis of our preliminary experiments. Gold is expected to be a much better substrate, as it can deliver more electrons. One of the most important aspects of the experimental arrangement is the use of the Nd:YAG laser to ablate the surface of the target before irradiation with the high intensity laser and thus eliminate parasitic protons.

Acknowledgements Although there is a single name on this article it would be amiss of me not to mention the talented groups of people who work with me and who are largely responsible for the work described. My own group at Strathclyde, Paul McKenna, Lynne Robson, Tom McCanny Seiji Shimizu, Jiamin Yang and Ravi Singhal at Glasgow . I also acknowledge my colleagues from RAL and Imperial College and the Laser crew at RAL. Furthermore all of the work on mono-energetic protons was carried by a similar group of talented people at Jena namely H. Schwoerer, S. Pfotenhauer, O. Jackel, K.-U. Amthor, B. Liesfeld, W. Ziegler, R. Sauerbrey. Timur Esirkepov our stalwart theoretician inspired all of this work.

 Springer

References

1. Ledingham, K.W.D., et al.: Phys. Rev. Lett. **84**, 1459 (2000)
2. Fritzler, S., et al.: Appl. Phys. Lett. **83**, 3039 (2003)
3. Spencer, I., et al.: Nucl. Instrum. Methods. **183**, 449 (2001)
4. Ledingham, K.W.D., et al.: J. Phys. D: Appl. Phys. **37**, 2341 (2004)
5. Ledingham, K.W.D., et al.: J. Phys. D Appl. Phys. **36**, L79 (2003)
6. Magill, J., et al.: Appl. Phys., B **77**, 387 (2003)
7. McKenna, P., et al.: Phys. Rev. Lett. **91**, 075006 (2003)
8. McKenna, P., et al.: Phys. Rev., E **70**, 036405 (2004)
9. Spencer, I., et al.: Phys. Rev., E **67**, 046402 (2003)
10. McKenna, P., et al.: Rev. Sci. Instrum. **73**, 4176 (2002)
11. McKenna, P., et al.: Phys. Rev. Lett. **94**, 084801 (2005)
12. Yang, J.N., et al.: Appl. Phys. Lett. **84**, 675 (2004)
13. Lancaster, K.L., et al.: Phys. Plasmas **11**, 3404 (2004)
14. Yang, J.M., et al.: J. Appl. Phys. **96**, 6912 (2004)
15. Spencer, I., et al.: Rev. Sci. Instrum. **73**, 3801 (2002)
16. Ledingham, K.W.D., et al.: Science **300**, 1107 (2003)
17. Clarke, R.J., et al.: J. Radiol. Prot. **26**, 277 (2006)
18. Strickland, D., Mourou G.: Opt. Commun. **56**, 219 (1985)
19. Ross, I.N., et al.: Laser Part. Beams **17**, 331 (1999)
20. Dubietis, A., et al.: Opt. Commun. **88**, 437 (1992)
21. Umstadter, D.: Phys. Plasmas **8**, 1774 (2001)
22. Wilks, S.C., et al.: Phys. Plasmas **8**, 542 (2001)
23. Allen, M., ei al.: Conference proceedings of field ignition high field physics (Kyoto, Japan), pp. 2716 (2004)
24. Clark, E.L., et al.: Phys. Rev. Lett. **84**, 670 (2000)
25. Umstadter, D.J.: Phys. D: Appl. Phys. **36**, R151 (2003)
26. Zepf, M., et al.: Phys. Plasmas **8**, 2323 (2001)
27. Snavely, R.A., et al.: Phys. Rev. Lett. **85**, 2945 (2000)
28. Hatchett, S.P., et al.: Phys. Plasmas **7**, 2076 (2000)
29. http:www.physik.uni-jena.de/qe/Forschung/F-Englisch/Petawatt/Eng-FP-Petawatt.html; 2004
30. Ledingham, K.W.D.: Nature Physics **2**, 11 (2006)
31. Mangles, S.P.D., et al.: Nature **431**, 535 (2004)
32. Geddes, C.G.R., et al.: Nature **431**, 538 (2004)
33. Faure, J., et al.: Nature **431**, 541 (2004)
34. Schwoerer, H., et al.: Nature **439**, 445 (2006)
35. Bulanov, S.V., Khoroshkov V.: Plasma Phys. Rep. **28**, 453 (2002)
36. Bulanov, S.V., et al.: AIP Conf. Proceedings **740**, 414 (2004)
37. Esirkepov, T., et al.: Phys. Rev. Lett. **89**, 175003 (2002)

Hyperfine Interact (2006) 171:83–91
DOI 10.1007/s10751-006-9501-4

Penning trap mass spectrometry for nuclear structure studies

**Klaus Blaum · Dietrich Beck · Martin Breitenfeldt ·
Sebastian George · Frank Herfurth ·
Alexander Herlert · Alban Kellerbauer ·
H.-Jürgen Kluge · David Lunney · Romain Savreux ·
Stefan Schwarz · Lutz Schweikhard ·
Chabouh Yazidjian**

Published online: 15 February 2007
© Springer Science + Business Media B.V. 2007

Abstract High-precision mass measurements as performed at the Penning trap mass spectrometer ISOLTRAP at ISOLDE/CERN are an important contribution to the investigation of nuclear structure. Precise nuclear masses with less than 0.1 ppm relative mass uncertainty allow stringent tests of mass models and formulae that are used to predict mass values of nuclides far from the valley of stability. Furthermore, an investigation of nuclear structure effects like shell or sub-shell closures,

K. Blaum (✉) · S. George
Johannes Gutenberg-Universität, 55099 Mainz, Germany and GSI Darmstadt,
64291 Darmstadt, Germany
e-mail: blaumk@uni-mainz.de

D. Beck · F. Herfurth · R. Savreux · C. Yazidjian
GSI Darmstadt, 64291 Darmstadt, Germany

M. Breitenfeldt · L. Schweikhard
Ernst-Moritz-Arndt-Universität, 17487 Greifswald, Germany

A. Herlert
Physics Department, CERN, 1211 Geneva 23, Switzerland

A. Kellerbauer
MPI für Kernphysik, 69117 Heidelberg, Germany

H.-J. Kluge
GSI Darmstadt, 64291 Darmstadt, Germany and Ruprecht-Karls-Universität,
69120 Heidelberg, Germany

D. Lunney
CSNSM-IN2P3-CNRS, 91405 Orsay-Campus, Orsay, France

S. Schwarz
NSCL, Michigan State University, East Lansing MI-48824-1321, USA

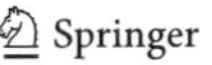

deformations, and halos is possible. In addition to a sophisticated experimental setup for precise mass measurements, a radioactive ion-beam facility that delivers a large variety of short-lived nuclides with sufficient yield is required. An overview of the results from the mass spectrometer ISOLTRAP is given and its limits and possibilities are described.

Key words ISOLTRAP · mass spectrometry · nuclear structure · Penning trap · short-lived nuclides

PACS 7.75.+h Mass spectrometers · 21.10.Dr Binding energies and masses · 32.10.Bi Atomic masses · mass spectra · abundances · isotopes

1 Introduction

Nuclear masses play an important role in a large number of physical systems [1]. Especially for nuclear structure studies and tests of mass models and formulae, nuclear binding energies of radionuclides far from the valley of stability are required [2].

Mass measurements along isotopic and isotonic chains allow one to study the fine structure of the mass surface and clarify discontinuities in order to extract nuclear structure information such as shell and subshell closures and the onset of deformation from binding energies (see e.g. [3–7]). Along this line it was important to demonstrate that nuclear ground and isomeric states can be resolved [5, 8–12] in order to, e.g., confirm the coexistence of nuclear shapes at nearly degenerate energies ($\approx 100\,\mathrm{keV}$), which had been previously deduced from laser and nuclear spectroscopy [13–15].

Different mass models are available which either make estimates on unknown masses with a global approach, or try to predict locally the mass values [2]. The predictive power of these models and formulae has to be tested in regions where masses are accurately known. In most cases a mass uncertainty of 10–100 keV is sufficient. Various experiments located at radioactive ion beam facilities aim at the investigation and precise determination of the masses of short-lived nuclides. The applied techniques range from traditional mass spectrometers to storage rings and Penning trap mass spectrometers [16].

However, the mass uncertainty of a large number of nuclides is above 10 keV and the relative mass uncertainty rapidly rises the farther out the nuclides lie from the valley of stability (see Fig. 1). This is strongly correlated to the half-life and the production rates of the respective nuclides, which makes their investigation an experimental challenge.

In this work, an outline of the ISOLTRAP Penning trap mass spectrometer and its contribution to the field of high-precision mass measurements for nuclear structure studies are presented. The possibilities, challenges, and limitations of ISOLTRAP will be discussed as well.

2 Experimental setup and procedure

The ISOLTRAP mass spectrometer is the prototype setup for Penning trap mass measurements of short-lived nuclides. Various additional facilities are meanwhile

 Springer

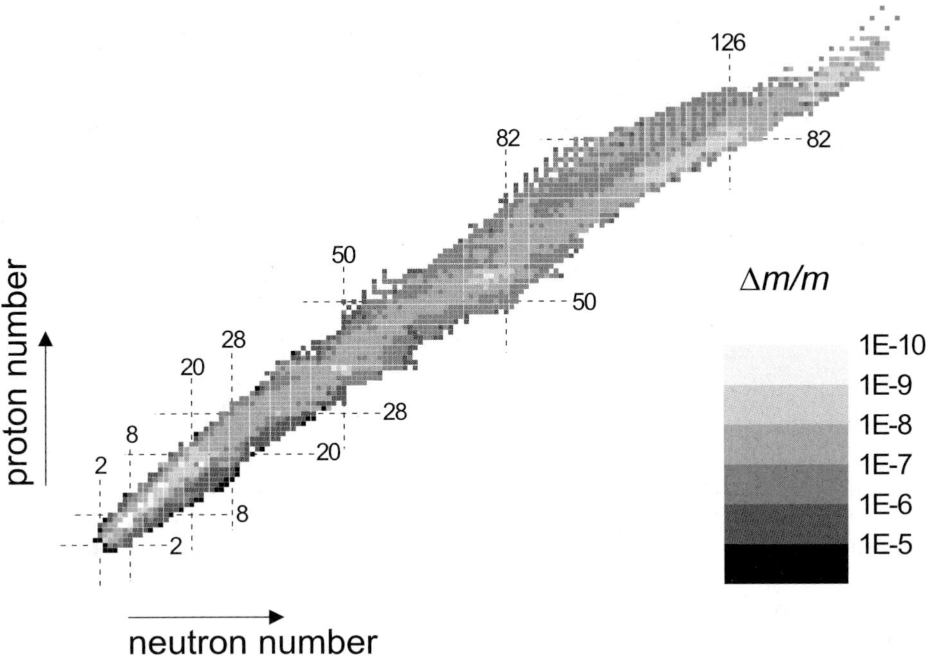

Fig. 1 Experimental relative mass uncertainty $\Delta m/m$ of nuclides; data from [17]

in operation or under construction. For a comprehensive overview see [1, 16]. A schematic layout of the experimental setup is shown in Fig. 2. The radioactive ions are delivered from one of the two target stations at ISOLDE [18]. The short-lived nuclides are produced by bombarding a fixed target with 1.4-GeV protons. Due to spallation, fragmentation, and fission a large number of various radionuclides are formed. The target container is heated and the radioactive particles diffuse into an ion source. Three types of ion sources are available, a surface ion source, a resonant laser ion source, and a (hot or cold) plasma ion source. After passing the separator magnets the usually singly-charged ions are sent as a 60-keV continuous beam to ISOLTRAP.

By use of a linear radiofrequency quadrupole structure [19] mounted inside a 60-keV high voltage cage, the continuous beam is stopped, accumulated, and bunched as well as buffer-gas cooled for the further injection into the first Penning trap. This cylindrically shaped trap also employs a helium buffer gas. A sophisticated cooling technique is applied to mass-selectively center only the radionuclide ions of interest by use of a quadrupolar rf excitation [20]. Thus unwanted isobaric contaminant ions can be removed with a resolving power of $R = 10^4 - 10^5$ depending on the respective mass and excitation duration [21].

The cooled, isobarically pure ion bunch is finally transferred to the hyperbolically-shaped precision Penning trap. There, the cyclotron frequency

$$v_c = \frac{1}{2\pi} \frac{qB}{m} \tag{1}$$

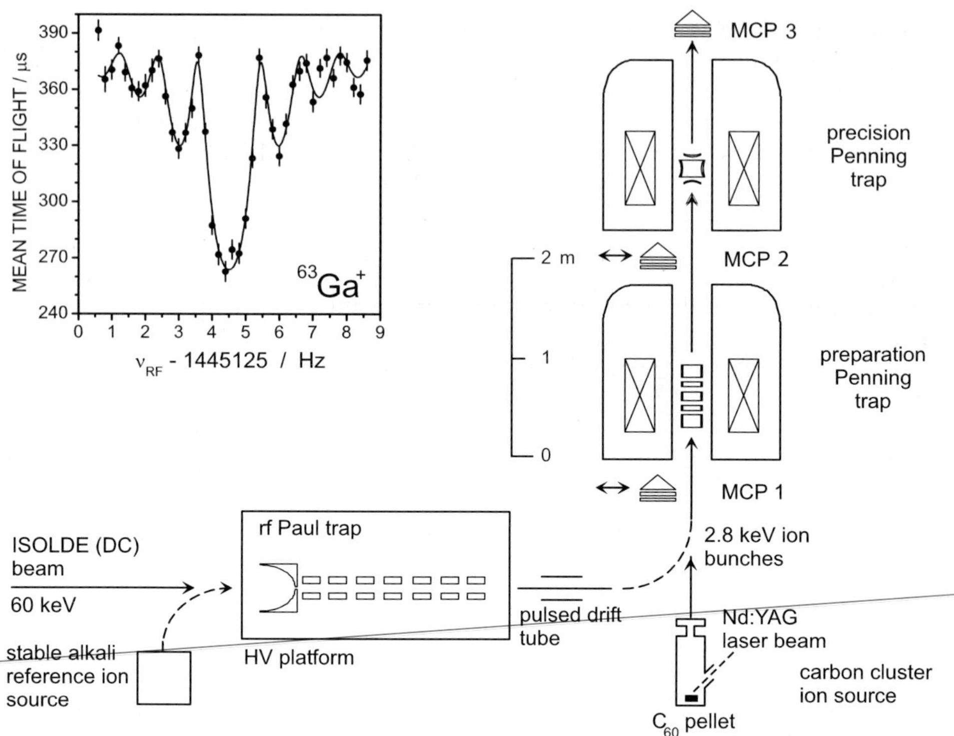

Fig. 2 Schematic layout of ISOLTRAP. The main components are a radiofrequency quadrupole ion beam cooler and buncher, a preparation Penning trap for isobaric cleaning of the radioactive ion ensemble, and a 5.9-T Penning trap mass spectrometer. Micro-channel-plate (MCP) detectors are used to monitor the ion transfer as well as to record the time-of-flight resonance (MCP3) for the determination of the cyclotron frequency. The inset shows a cyclotron resonance curve for the short-lived radionuclide ^{63}Ga$^+$ ($T_{1/2} = 31.4$ s) with 900 ms duration of the quadrupolar rf excitation in the precision Penning trap

and thus the radial energy of the stored ions with mass m and charge q moving in a magnetic field B is probed after some preparatory steps [22] with a time-of-flight cyclotron resonance detection technique [23, 24]. In order to deduce the mass of the investigated radionuclide ions, the magnetic field strength B needs to be determined at the same level of precision. This is performed via a cyclotron-frequency measurement of a well-known stable nuclide ion either delivered from the ISOLTRAP alkali ion source or the ISOLDE target. By calculating the frequency ratio, the mass ratio is obtained.

3 Recent results at ISOLTRAP

3.1 Nuclear structure studies

The contribution of mass measurements to nuclear structure studies is based on the possibility to determine the nuclear binding energy

$$B(N, Z) = (Nm_n + Zm_p - m(N, Z))c^2 , \qquad (2)$$

which represents the sum of all the nucleonic interactions that give rise to correlations in many-body systems. Since the binding energy depends on the detailed composition of protons and neutrons, the mass of each of the more than 3000 nuclides as observed to date [17] is highly specific and a key property of a nuclear system. By taking a closer look at this ensemble of mass data covering the whole nuclear chart, one can examine the hills and valleys that form this mass surface and study the effects of certain nuclear configurations. To unveil such effects, mass measurements with a precision of $\delta m/m < 10^{-6}$ are required.

Mass differences lead to separation energies, i.e., the energy needed to separate some nucleons from the nucleus, providing information on the shell structure and phase transitions. The most striking way in which the shell structure manifests itself in mass systematics is through differences of masses, i.e., the two-neutron separation energy

$$S_{2n}(N, Z) = B(N, Z) - B(N - 2, Z) \qquad (3)$$

in the case of the neutron shells, and the two-proton separation energy

$$S_{2p}(N, Z) = B(N, Z) - B(N, Z - 2) \qquad (4)$$

in the case of proton shells. Here, specific classes of interactions can be isolated [2]. The single-nucleon separation energy is a less clear-cut indication because of the pairing effect.

In general, S_{2n} decreases as neutrons are added and a deviation from this behavior points to manifestations of microscopic nuclear structure effects, as it has been observed with ISOLTRAP in e.g. the Hg and Pt region [5]. Similar investigations have been performed recently in the neutron-rich Ni-Cu-Ga mass region [25, 26].

3.2 Resolution and weighing of isomeric states

An important issue in direct mass measurements with respect to nuclear structure studies is to resolve isomeric and ground states. Nearly one third of the nuclides in the nuclear chart have long-lived isomeric states with, in many cases, unknown excitation energies. In the context of nuclear physics, isomers are excited states of atomic nuclei at excitation energies typically from 100 keV up to a few MeV. Their half-lives range from nanoseconds to beyond the age of the Universe, e.g. 10^{15} years in the case of ^{180m}Ta, the only naturally occurring isomer on Earth [27].

Direct high-precision mass measurements of isomeric states with ion traps have already been performed several times. The first isolation of an isomeric state in a Penning trap was demonstrated by Bollen et al. in 1992 [8]. In that work, the ^{84m}Rb isomer could be eliminated due to its shorter half-life as compared to the ground state by simply storing the ensemble of nuclide ions for a sufficiently long period in the trap. In the case of ^{78}Rb, for which this approach was not possible, the large resolving power of the Penning trap enabled a direct resolution of the two states ^{78}Rb and ^{78m}Rb, separated by an energy $E = 111.2$ keV. In a series of measurements, several isomeric excitation energies were determined for some odd-mass mercury isotopes ^{185m}Hg, ^{187m}Hg, ^{191m}Hg, ^{193m}Hg, and ^{197m}Hg [5]. In these cases an extremely high resolving power of up to $3.7 \cdot 10^{6}$ was applied, corresponding to excitation times

of up to 8 s. Recently, the energy difference of ^{187}Pb and ^{187m}Pb was determined with Penning trap mass spectrometry to be $E = 33(13)$ keV [12]. This is the lowest isomeric excitation energy ever determined by weighing nuclei. The most intricate example is the one of ^{70}Cu, for which a simultaneous combination of laser ionization, decay spectroscopy, and mass spectrometry succeeded in assigning the three low-lying states, the ground state and the two excited isomers ^{70m}Cu and ^{70n}Cu, with excitation energies $E_m = 100.7(2.6)$ keV and $E_n = 242.0(2.7)$ keV, respectively, to the correct spin values [11].

3.3 Test of nuclear mass models and mass formulae

The nucleus is a self-organized, many-body quantum system that interacts through the strong, weak and electromagnetic forces. Due to the lack of an exact description of the strong interaction and the complexity of the many-body nucleonic system, the binding energy can not be described by *ab-initio* theories. Instead, one has to rely on mass predictions by models (with the aim of a quantitative prediction of the total binding energy of a nucleus) and formulae (with the aim of a numerical calculation of masses on a physical basis) [2]. For decisive tests of the predictive power of the different models, large-scale mass measurements of exotic short-lived nuclei have been performed in recent years, especially at the experimental storage ring ESR at GSI-Darmstadt [9, 28, 29] but also at ISOLTRAP [3–5, 25, 30] and recently at JYFLTRAP [31]. High-precision mass data will contribute to further improvement and refinement of mass models and formulae and to more a detailed understanding of the nuclear forces.

Aside from global mass formulae, there are also a number of local mass formulae, which address the problem to obtain the required mass of a so far unmeasured nucleus that lies fairly close to a considerable number of nuclei with known masses. Stringent tests of one of the most powerful local mass formulae, the isobaric multiplet mass equation (IMME) [32], have been performed at ISOLTRAP in recent years. In light nuclei, isobaric analog states (IAS) have nearly identical wave functions. The charge dependent energy difference of these states can be calculated using first-order perturbation theory assuming only two-body Coulomb forces. This leads to the simple equation, noted first by Wigner [33] and Weinberg and Treiman [34], that gives the mass m of a member of an isospin multiplet as a function of its isospin projection $T_Z = (N - Z)/2$:

$$m(T_Z) = c_0 + c_1 T_Z + c_2 T_Z^2 . \tag{5}$$

This quadratic relation of the IMME is of fundamental importance in isospin symmetry in nuclear physics [35]. Looking at the isospin quartets [32] it was found that the IMME worked very well for 21 out of 22 cases. Due to its success and lack of newer experimental data, the IMME is widely used to predict masses as well as level energies, for example for the mapping of the proton drip line over a wide mass range, which is important for determining the *rp*-process path.

In 2001 the ISOLTRAP mass value for ^{33}Ar resulted in a breakdown of the quadratic IMME for the $A = 33$, $T = 3/2$ quartet [36]. These surprising results triggered new reaction spectroscopy experiments. The outcome was that one level energy in ^{33}Cl was found to be inaccurate [37], which shows the importance of refined measurements. Note that a direct mass measurement can only determine the mass

 Springer

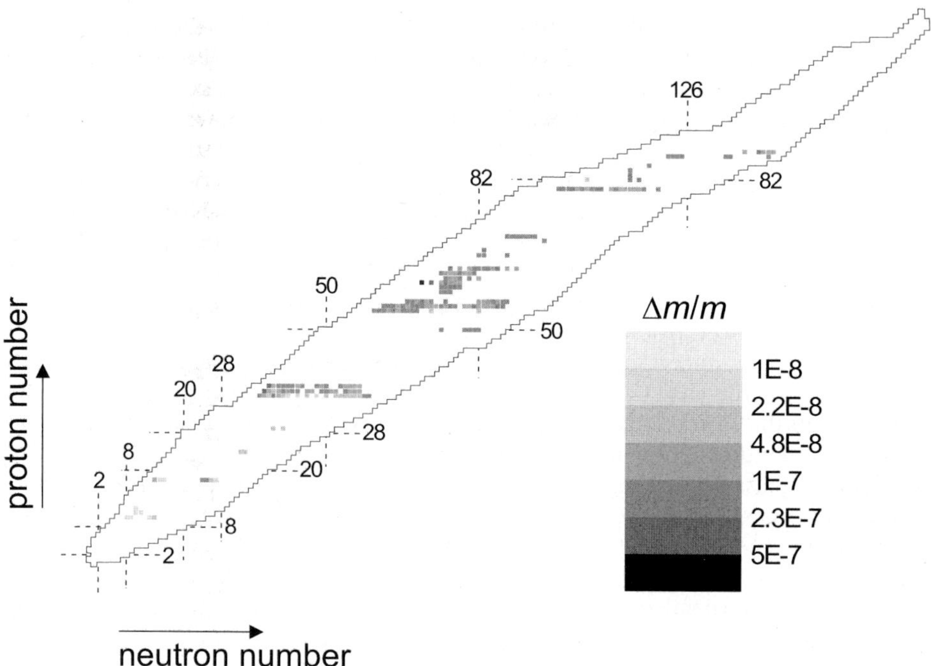

Fig. 3 Relative mass uncertainty of published mass values obtained at ISOLTRAP

of ground state multiplet members. But these nuclei are often also the most exotic members of an isospin multiplet with a rather large mass uncertainty [35].

To test the IMME with a precision never obtained before, ISOLTRAP performed mass measurements on the very short-lived nuclides ^{32}Ar ($T_{1/2} = 98\,$ms) and ^{33}Ar ($T_{1/2} = 173\,$ms) and used the mass excess values of the other states of the $T = 2$ quintet for $A = 32$ and the $T = 3/2$ quartet for $A = 33$ [38]. For the test of the IMME, an additional cubic term $c_3 T_Z^3$ is assumed in Eq. (5) that should fit with $c_3 = 0$ if the quadratic form of the IMME is correct. The experiment yielded a coefficient $c_3 = -0.11(30)$ in the case of $T = 2$, $A = 32$ and $c_3 = -0.13(45)$ in the case of $T = 3/2$, $A = 33$. Thus, in both cases the c_3 coefficient is consistent with zero within the uncertainty. These two multiplets now represent the most stringent test of the IMME.

4 Present status and technical developments at ISOLTRAP

In the past, ISOLTRAP has measured the mass of about 300 nuclides. The relative mass uncertainties resulting from so far published frequency ratio values measured at ISOLTRAP are shown in Fig. 3. While in the beginning the systematic deviations of the spectrometer were not known and therefore a conservative limit of the relative mass uncertainty of 1×10^{-7} was given, the systematic investigation of the setup with the mass measurement of carbon clusters was a major improvement [39]. Carbon clusters consisting of a multiple of ^{12}C atoms are the ideal reference masses since the

atomic mass unit is defined with respect to the mass of ^{12}C [40]. A systematic study showed a limitation in the accuracy of ISOLTRAP of $\delta m/m = 8 \times 10^{-9}$ [41], which is added as a systematic uncertainty to the statistical uncertainty from the frequency determination. Nowadays results regularly reach down to relative mass uncertainties in the order of 1×10^{-8} (see Fig. 3).

Recent technical improvements at the ISOLTRAP mass spectrometer include a sensitivity improvement by a new ion detector. The existing micro channel plate (MCP) detector was replaced by a channeltron detector with a conversion dynode, where secondary electrons are monitored with an efficiency close to 100%. With this setup an increase of the total efficiency by a factor of about 3 was obtained and nuclides with a lower production yield could be addressed [42].

The systematic uncertainty of the mass measurements can be improved with a stabilization of the magnetic field. As has been observed recently, the cyclotron frequency of the stored ions depends strongly on the temperature in the warm bore of the superconducting magnet [43]. A new stabilization system has been installed which aims at a stabilization of this temperature with variations of less than ± 20 mK.

Finally, a new technique to obtain radionuclides that are not delivered from ISOLDE has been demonstrated: the production of radionuclides by in-trap decay [44]. To this end, short-lived nuclides are first stored in the preparation Penning trap and after a respective storage period, enough daughter nuclides from the decay will be collected in the trap. After the usual mass-selective cooling and centering procedure, they are transferred to the precision trap for mass determination. This technique has been successfully applied to produce and to measure the mass of ^{61,62,63}Fe isotopes, which are not available as an ISOLDE beam. The results are currently under investigation.

5 Conclusion and outlook

The ISOLTRAP mass spectrometer at ISOLDE is a source for precise mass values of short-lived radionuclides, where the relative mass uncertainty has already been pushed down to $\delta m/m = 8 \times 10^{-9}$ [41]. With the possibility to reach nuclides with half-lives well below 100 ms [45] and production yields down to 100 ions per second [38], a large number of radionuclides have been and can be investigated at ISOLDE.

Besides the target development at ISOLDE, which aims at higher yields, access to further elements, and less contamination, the experimental setup of ISOLTRAP is constantly subject of technical development. A new ion detector for a better overall efficiency, a stabilization of the magnetic field, and new ion sources that deliver a variety of so far not applicable reference masses are recent improvements of the setup. Further developments are directed to a higher precision in the frequency determination by use of new rf-excitations schemes [46] and by highly-charged ions [47].

Acknowledgements This work was supported by the German Ministry for Education and Research (BMBF) under contract 06GF151 and 06MZ215, by the European Commission under contracts HPMT-CT-2000-00197 (Marie Curie Fellowship), HPRI-CT-2001-50034 (NIPNET), and RII3-CT-2004-506065 (TRAPSPEC), and by the Helmholtz association of national research centres (HGF) under contract VH-NG-037. We also thank the ISOLDE Collaboration as well as the ISOLDE technical group for their assistance.

 Springer

References

1. Blaum, K.: Phys. Rep. **425**, 1 (2006)
2. Lunney, D., Pearson, J.M., Thibault, C.: Rev. Mod. Phys. **75**, 1021 (2003)
3. Ames, F., et al.: Nucl. Phys., A **651**, 3 (1999)
4. Beck, D., et al.: Eur. Phys. J., A **8**, 307 (2000)
5. Schwarz, S., et al.: Nucl. Phys., A **693**, 533 (2001)
6. Rinta-Antila, S., et al.: Phys. Rev., C **70**, 011301(R) (2004)
7. Litvinov, Yu. A., et al.: Nucl. Phys., A **756**, 3 (2005)
8. Bollen, G., et al.: Phys. Rev., C **46**, R2140 (1992)
9. Litvinov, Yu. A., et al.: Nucl. Phys., A **734**, 473 (2004)
10. Blaum, K., et al.: Europhys. Lett. **67**, 586 (2004)
11. Van Roosbroeck, J., et al.: Phys. Rev. Lett. **92**, 112501 (2004)
12. Weber, C., et al.: Phys. Lett., A **347**, 81 (2005)
13. Otten, E.W.: In: Allan Bromley, D. (ed.) Treatise on heavy-ion science, 8, vol. 515. Plenum Press, New York (1989)
14. Andreyev, A.N., et al.: Nature **405**, 430 (2000)
15. Kluge, H.-J., Nörtershäuser, W.: Spectrochim. Acta, B **58**, 1031 (2003)
16. Schweikhard, L., Bollen, G.: A special Issue on Ultra-accurate mass determination and related topics. Int. J. Mass Spectrom. **251**, (2006)
17. Audi, G., Wapstra, A.H., Thibault, C.: Nucl. Phys., A **729**, 337 (2003)
18. Kugler, E.: Hyperfine Interact. **129**, 23 (2000)
19. Herfurth, F., et al.: Nucl. Instrum. Meth., A **469**, 254 (2001)
20. Savard, G., et al.: Phys. Lett., A **158** 247, (1991)
21. Raimbault-Hartmann, H., et al.: Nucl. Instrum. Meth., B **126**, 378 (1997)
22. Blaum, K., et al.: J. Phys. B: At. Mol. Opt. Phys. **36**, 921 (2003)
23. Gräff, G., Kalinowsky, H., Traut, J.: Z. Phys., A **297**, 35 (1980)
24. Bollen, G., et al.: Nucl. Instrum. Meth., A **368**, 675 (1996)
25. Guénaut, C., et al.: J. Phys. G: Nucl. Part. Phys. **31**, S1765 (2005)
26. Guénaut, C., et al.: Eur. Phys. J., A **25**, 33 (2005)
27. Audi, G., et al.: Nucl. Phys., A **729**, 3 (2003)
28. Litvinov, Yu. A., et al.: Hyperfine Interact. **132**, 283 (2001)
29. Novikov, Yu. N., et al.: Nucl. Phys., A **697**, 92 (2002)
30. Delahaye, P., et al.: Phys. Rev., C **74**, 034331 (2006)
31. Hager, U., et al.: Phys. Rev. Lett. **96**, 042504(2006)
32. Benenson, W., Kashy, E.: Rev. Mod. Phys. **51**, 527 (1979)
33. Wigner, E.P.: In: Millikan, W.O. (ed.) Proceedings of the Robert A. Welch Foundation Conference on Chemical Research, Houston, vol. 1. Robert A. Welch Foundation, Houston, (1957)
34. Weinberg, S., Treiman, S.B.: Phys. Rev. **116**, 465 (1959)
35. Britz, J., Pape, A., Antony, M.: At. Data Nucl. Data Tab. **69** 125, (1998)
36. Herfurth, F., et al.: Phys. Rev. Lett. **87**, 142501 (2001)
37. Pyle, M.C., et al.: Phys. Rev. Lett. **88**, 122501 (2002)
38. Blaum, K., et al.: Phys. Rev. Lett. **91**, 260801 (2003)
39. Blaum, K., et al.: Eur. Phys. J., A **15**, 245 (2002)
40. Blaum, K., et al.: Anal. Bioanal. Chem. **377**, 1133 (2003)
41. Kellerbauer, A., et al.: Eur. Phys. J., D **22**, 53 (2003)
42. Yazidjian, C., et al.: Hyperfine Interactions (2006) (in press)
43. Blaum, K., et al.: J. Phys. G: Nucl. Part. Phys. **31**, S1775 (2005)
44. Herlert, A., et al.: New J. Phys. **7**, 44 (2005)
45. Kellerbauer, A., et al.: Phys. Rev. Lett. **93**, 072502 (2004)
46. George, S., et al.: Int. J. Mass Spectrom. (2006) (submitted)
47. Herlert, A., et al.: Int. J. Mass Spectrom. **251**, 131 (2006)

Hyperfine Interact (2006) 171:93–107
DOI 10.1007/s10751-007-9509-4

Precision spectroscopy at heavy ion ring accelerator SIS300

Hartmut Backe

Published online: 16 February 2007

Abstract Unique spectroscopic possibilities open up if a laser beam interacts with relativistic lithium-like ions stored in the heavy ion ring accelerator SIS300 at the future Facility for Antiproton and Ion Research FAIR in Darmstadt, Germany. At a relativistic factor $\gamma = 36$ the $^2P_{1/2}$ level can be excited from the $^2S_{1/2}$ ground state for any element with frequency doubled dye-lasers in collinear geometry. Precise transition energy measurements can be performed if the fluorescence photons, boosted in forward direction into the X-ray region, are energetically analyzed with a single crystal monochromator. The hyperfine structure can be investigated at the $^2P_{1/2} - {}^2S_{1/2}$ transition for all elements and at the $^2P_{3/2} - {}^2S_{1/2}$ transition for elements with $Z \leq 50$. Isotope shifts and nuclear moments can be measured with unprecedented precision, in principle even for only a few stored radioactive species with known nuclear spin. A superior relative line width in the order of $5 \cdot 10^{-7}$ may be feasible after laser cooling, and even polarized external beams may be prepared by optical pumping.

Key words laser spectroscopy · relativistic lithium-like ions, laser cooling · hyperfine spectroscopy · nuclear polarization · SIS300

Abbreviations
FAIR   Facility for Antiproton and Ion Research
HI   Heavy Ion
QED   Quantum Electrodynamics
CCD   Charge Coupled Device
EBIT Electron Beam Ion Trap

H. Backe (✉)
Institut für Kernphysik, Johannes Gutenberg-Universität Mainz, D-55099 Mainz, Germany
e-mail: backe@kph.uni-mainz.de

1 Introduction

The central part of the planned FAIR project in Darmstadt is a heavy ion synchrotron called SIS300 with a magnetic rigidity $B\rho = 300$ Tm and a circumference of 1,100 m [8].[1] The magnetic field will be produced by superconducting magnets with a maximum induction of 6 Tesla which can be ramped with a rate of 1 T/s. With this synchrotron bare uranium can be accelerated up to a maximum energy of 34 GeV/u, corresponding to a relativistic factor $\gamma = 1/\sqrt{1 - \beta^2} = 37.5$ and a reduced velocity $\beta = v/c = 0.9996444$, with c the speed of light. A fascinating possibility of this accelerator is the excitation of few electron systems by the interaction with the light of conventional lasers in a collinear geometry. If the laser beam counter-propagates lithium-like uranium with $\gamma = 36$, the laser light at the blue edge of the visible spectral range with an energy of $\hbar\omega_L = 3.898$ eV is Doppler-shifted and appears in the rest frame of the Li-like system with an energy of $\hbar\omega_0 \cong 2\gamma\hbar\omega_L = 280.6$ eV. As will be outlined in more detail in Section 2 this is just the $^2P_{1/2} - {}^2S_{1/2}$ transition energy in lithium-like uranium. A variety of spectroscopic possibilities exist if this transition can be induced. As will be pointed out in Section 3 the combination with a precise single crystal X-ray spectrometer, which detects the fluorescence photons boosted in forward direction up to an energy of $\hbar\omega_X = 2\gamma\hbar\omega_0 = 20.2$ keV allows both, very accurate measurements of the transition energy $\hbar\omega_0$ and of the relativistic factor γ. If radioactive Li-like ions could be injected into the SIS300, a hyperfine spectroscopy would be possible for radioactive species with nuclear spin $I > 0$, see Section 4. A few remarks on laser cooling will be made in Section 5. In Section 6 the possibility of a nuclear polarization by optical pumping with circularly polarized laser light will briefly be touched on. The paper closes in Section 7 with a conclusion.

The essential ideas of this paper were for the first time presented by the author in the year 2000 at GSI in Darmstadt [1], see also [7].

2 The $^2P_{1/2,3/2} - {}^2S_{1/2}$ transitions in lithium-like uranium

The three electron, lithium-like level scheme of uranium is shown in Fig. 1. The third electron outside the closed $1s^2$ shell is a $2s$ electron, consequently the ground state is a $^2S_{1/2}$ term. The lowest excited states belong to the $1s^2 2p$ configuration and form $^2P_{1/2}$ and $^2P_{3/2}$ terms. The large fine-structure splitting of about 4.3 keV originates from relativistic effects.

A first precision measurement of the $^2P_{1/2} - {}^2S_{1/2}$ transition energy of (280.59 ± 0.09) eV was reported by Schweppe et al. [16]. The aim of this experiment was to test QED in few electron uranium. In a number of publications by Lindgren et al. [19], and by Persson et al. [11] calculated (280.52 ± 0.28) eV which was in good agreement with the experiment. In the meantime better experiments by Brandau et al. [5], who measured (280.516 ± 0.099) eV, and by Beiersdorfer et al. [2] were performed. In the latter reference a value of (280.645 ± 0.015) eV is reported which was obtained with SuperEBIT. The accuracy of previous experiments was improved by nearly one order of magnitude. The best calculation of Yerokhin et al. [18] without second order

---

[1]The parameters used in this paper do not match exactly with the parameters of this report.

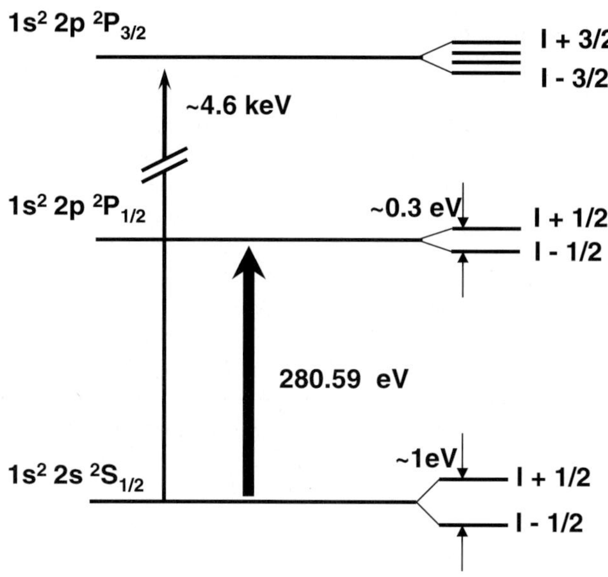

Fig. 1 The lithium-like level scheme of uranium

QED effects is (280.48 ± 0.11) eV, and with inclusion of an estimated value of these effects (280.64 ± 0.21) eV (Yerokhin, private communication), i.e. the precision of the measurement exceeds currently the precision of the calculation by a factor of 14. Despite this situation a scheme is presented in the next section with which in future the experimental precision probably can be further improved by another factor of 4. Alternatively, it may be useful to measure β and γ of the Li-like ions with accuracies in the order of 10^{-8} and 10^{-5}, respectively, see Subsection 3.6.

3 Precision transition energy measurement in Li-like uranium at SIS300

3.1 Proposed experimental setup

The proposed experimental setup is schematically shown in Fig. 2. A laser beam with a photon energy of 5.465 eV, corresponding to a vacuum wavelength $\lambda = 226.87$ nm, counter-propagates in a straight section of SIS300 a Li-like uranium beam with a relativistic factor $\gamma = 25.68$. The photon energy in the rest frame of the Li-like U^{89+}-ion is

$$\hbar\omega_0 = \sqrt{\frac{1+\beta}{1-\beta}}\, \hbar\omega_L = 280.6\ \text{eV} \tag{1}$$

which is with $\beta = \sqrt{1 - 1/\gamma^2} = 0.99924152$ just the $^2P_{1/2} - {}^2S_{1/2}$ transition energy. De-excitation photons emitted in forward direction with respect to the Li-like uranium beam are boosted in the laboratory system to an energy

$$\hbar\omega_X = \sqrt{\frac{1+\beta}{1-\beta}}\, \hbar\omega_0 = 14.41\ \text{keV}. \tag{2}$$

Fig. 2 Proposed schematic experimental setup. The laser beam with photons of energy $\hbar\omega_L$ is reflected by a thin mirror into a *straight section* of SIS300 and interacts in a collinear geometry with lithium-like uranium ions moving in *opposite direction*. De-excitation photons from the $^2P_{1/2} - {}^2S_{1/2}$ transition with an energy $\hbar\omega_0 = 280.6$ eV appear predominantly in forward direction with respect to the HI beam at an X-ray energy $\hbar\omega_X = 14.41$ keV. The photon energy is measured by a single crystal monochromator

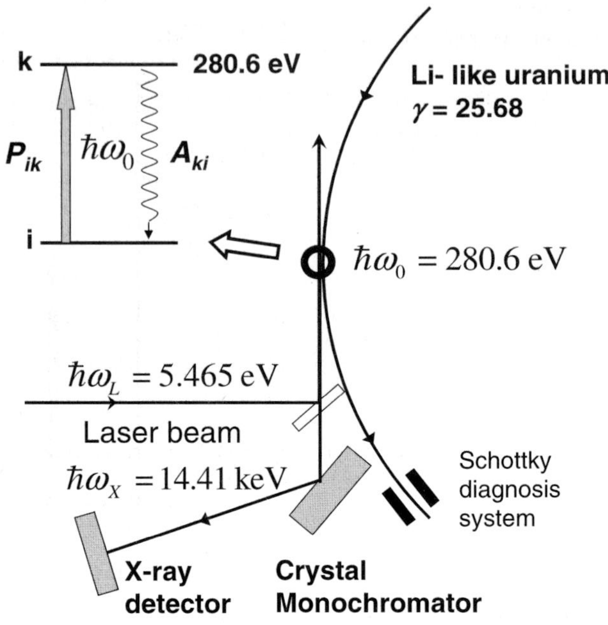

Occurrence of X-rays indicates resonance absorption of laser photons in the Li-like system. However, for an accurate measurement of the transition energy $\hbar\omega_0$ also the reduced velocity β or the relativistic factor γ must be known with high precision. This can be achieved by an energy measurement of the X-ray photons with the aid of a single crystal monochromator. In this manner, two very precise experimental methods are combined. The laser photon energy can be determined easily with a relative accuracy of $5 \cdot 10^{-7}$ and the X-ray energy with $2.8 \cdot 10^{-5}$, see Subsection 3.6. Combining equations (1) and (2), the transition energy $\hbar\omega_0$, the reduced velocity β, and the the relativistic factor γ of the Li-like uranium ion can be determined:

$$\hbar\omega_0 = \sqrt{\hbar\omega_X \cdot \hbar\omega_L} \, , \tag{3}$$

$$\beta = \frac{\hbar\omega_X/\hbar\omega_L - 1}{\hbar\omega_X/\hbar\omega_L + 1} \, , \tag{4}$$

$$\gamma = \frac{1}{2}\frac{\hbar\omega_X/\hbar\omega_L + 1}{\sqrt{\hbar\omega_X/\hbar\omega_L}} \simeq \frac{1}{2}\sqrt{\frac{\hbar\omega_X}{\hbar\omega_L}} . \tag{5}$$

The relative accuracies of a measurement of β and γ are $\Delta\beta/\beta = 2.1 \cdot 10^{-8}$ and $\Delta\gamma/\gamma = 1.4 \cdot 10^{-5}$, respectively, see also Subsection 3.6.[2]

---

[2]It should be mentioned that the precision of the X-ray energy measurement and also the current value of the $^2P_{1/2} - {}^2S_{1/2}$ transition energy $\hbar\omega_0$ is not high enough for an improved test of the time dilatation in special relativity. Such a test is based on equation (3) which can be rewritten with a small additional term as $\hbar\omega_L \cdot \hbar\omega_X/\hbar\omega_0^2 = 1 + 2\hat{\alpha}(\beta^2 + ...)$. The upper limit of the parameter $\hat{\alpha}$ is currently $\hat{\alpha} < 2.2 \cdot 10^{-7}$, see [12]. An improvement of this value would require a measurement of $\hbar\omega_L$, $\hbar\omega_X$, and $\hbar\omega_0$ with at least an accuracy of 10^{-7}.

3.2 Angular distribution and photon energy in the laboratory system

In reality the photons are emitted in the rest frame of the Li-like ion with an angular distribution which will be assumed to be isotropic, i.e. $d\dot{n}_0/d\Omega_0 = \dot{n}_0/(4\pi)$. The angular distribution in the laboratory frame is given by

$$\frac{d\dot{n}}{d\Omega} = \frac{d\dot{n}_0}{d\Omega_0}\frac{d\Omega_0}{d\Omega}\frac{dt_0}{dt} , \tag{6}$$

$$\frac{d\Omega_0}{d\Omega} = \frac{1}{\gamma^2(1 - \beta \cdot \cos\Theta)^2} , \tag{7}$$

$$\frac{dt_0}{dt} = \frac{1}{\gamma} \tag{8}$$

with $d\Omega_0/d\Omega$ the relativistically transformed solid angle ratio which follows from the relation

$$\cos\Theta_0 = \frac{\cos\Theta - \beta}{1 - \beta\cos\Theta} . \tag{9}$$

Here Θ_0 and Θ are the observation angles with respect to the velocity vector $\mathbf{v}$ of an individual Li-like ion in the rest frame of the Li-like ion and the laboratory frame, respectively. Further on, $dt_0/dt = 1/\gamma$ is the relativistic time dilatation.[3] The transition energy in the rest frame of the Li-like system is

$$\hbar\omega_0 = \gamma(1 + \beta\cos\Psi)\hbar\omega_L \tag{10}$$

with Ψ the angle which an individual ion with the velocity vector $\mathbf{v}$ makes with the laser beam axis. In the small angle approximation, with $\vec{\theta}$ the observation angle with respect to the nominal velocity vector $\mathbf{v}_0$ of the Li-like ion beam and $\vec{\psi}$ the angular deviation of an individual ion from $\mathbf{v}_0$, the photon energy is

$$\hbar\omega_X = \frac{\hbar\omega_0}{\gamma(1 - \beta \cdot \cos|\vec{\theta} - \vec{\psi}|)} \simeq \frac{2\gamma}{1 + (|\vec{\theta} - \vec{\psi}|\gamma)^2}\hbar\omega_0 , \tag{11}$$

with $\hbar\omega_0$ the photon energy in the Li-like system. From equations (10) and (11) the relation

$$\hbar\omega_X = \frac{1 + \beta\cos\psi}{1 - \beta \cdot \cos|\vec{\theta} - \vec{\psi}|}\hbar\omega_L \simeq \frac{4\gamma^2}{1 + (|\vec{\theta} - \vec{\psi}|\gamma)^2}\hbar\omega_L . \tag{12}$$

follows which directly relates the laser photon energy $\hbar\omega_L$ to the X-ray energy $\hbar\omega_X$.

Angular distribution $dN/d\Omega$, integrated intensity $\int_0^\Theta (dN/d\Omega)d\Omega$ and X-ray energy $\hbar\omega_X$ are shown in Fig. 3a and b as function of the observation angle Θ. It is worthwhile to notice that in a cone with a polar angle $\Theta = 50$ mrad, corresponding to 2.86° only, already more than 60% of the intensity is concentrated.

---

[3]For a nice survey of relevant formulas of Lorentz transformation in storage rings see [9], and [14]

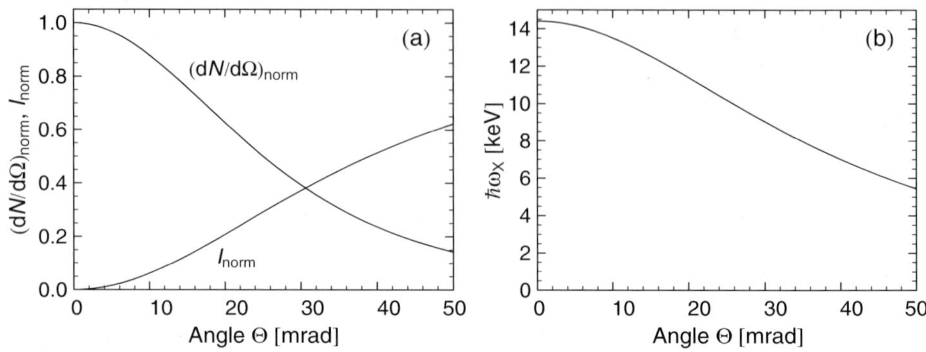

Fig. 3 **a** Angular distribution $(dN/d\Omega)_{norm}$ and integrated intensity $I_{norm} = \int_0^{\Theta}(dN/d\Omega)d\Omega/(4\pi)$, and **b** X-ray energy $\hbar\omega_X$, both as function of the observation angle Θ in the laboratory system for $\Psi = 0$. The angular distribution $(dN/d\Omega)_{norm}$ is normalized to its value at $\Theta = 0$, I_{norm} approaches 1 for $\Theta \to \pi$

3.3 The laser system and fluorescence rate estimate

Laser light with a wavelength of $\lambda = 226.9$ nm can be produced by an Excimer laser running on XeF (351/353 nm) which pumps a dye laser, for example. The output of the latter must be frequency doubled by a BBO crystal. Assuming a mean output power of the Excimer laser $\overline{P} = 250$ W at a repetition rate $f_{rep} = 10$ kHz and a pulse width of $\Delta t_{pulse} = 10$ ns, the instantaneous pulse power output of the frequency doubled laser may amount to 50 kW. A bandwidth of the dye laser radiation $\Delta\nu_D \simeq 1$ GHz can be reached by means of an intracavity etalon which may further be reduced in the frequency doubling unit to $\Delta\nu_L \simeq 0.7$ GHz or $\Delta\nu_L/\nu_L \simeq 5.3 \cdot 10^{-7}$. With these numbers the spectral photon flux within a single laser pulse amounts to $\Delta\dot{N}_{pulse}/(\Delta\nu_L/\nu_L) \simeq 1.1 \cdot 10^{29}$/s.

A lifetime of $61.8 \pm 1.2\text{(stat)} \pm 1.3\text{(syst)}$ ps of the $2p\,^2P_{1/2}$ has been measured [16] which corresponds to a transition rate $A_{ki} = 1.62 \cdot 10^{10}$/s or a relative level width $\Gamma/\hbar\omega_0 = 0.38 \cdot 10^{-7}$. This width is small in comparison to the relative band width of the photon flux in the rest frame of the Li-like ion $\Delta\hbar\omega_0/\hbar\omega_0 = \Delta\nu_L/\nu_L = 5.3 \cdot 10^{-7}$. Under these circumstances the induced transition rate is given by the equation

$$P_{ik} = \frac{d^2\dot{N}^0_{pulse}}{dA \cdot d\hbar\omega_0/\hbar\omega_0} A_{ki} \frac{g_k}{g_i} \frac{\pi^2}{c} \left(\frac{\hbar c}{\hbar\omega_0}\right)^3 , \tag{13}$$

with g_k and g_i the statistical weights of the levels. At a cross-section $A = 10$ mm^2 of the laser beam in the interaction region, the photon flux in the rest frame of the Li-like ion is $d^2\dot{N}^0_{pulse}/(dA \cdot d\hbar\omega_0/\hbar\omega_0) = \gamma \cdot d^2\dot{N}_{pulse}/(dA \cdot d\nu_L/\nu_L) = \gamma \cdot 1.1 \cdot 10^{28}/(\text{mm}^2\text{s})$. The number of induced transitions in a laser pulse of duration $\Delta t_{pulse} = 10$ ns is $P_{ik}\Delta t^0_{pulse} = P_{ik}\Delta t_{pulse}/\gamma = 20.4$ with P_{ik} of equation (13) and $\Delta t^0_{pulse} = P_{ik}\Delta t_{pulse}/\gamma = 20.4$, with Δt^0_{pulse} the pulse length in the rest frame of the Li-like system. This number is much larger as the spontaneous transition rate $A_{ki}\Delta t_{pulse}/\gamma = 6.3$, meaning that the effect of saturation must be taken into account. In the following all estimations of the spontaneous transition rates are performed in the saturation limit, since the experimental conditions are close to saturation. The number of emitted photons per laser pulse and per lithium-like uranium ion is then

$$N_{0,sat} = \frac{1}{2}\left(A_{ki}\frac{\Delta t_{pulse}}{\gamma} + 1\right) = 3.7 \; . \tag{14}$$

The additional summand 1 in the brackets originates from the fact that at saturation the Li-like ion is left with 50% probability in the excited state after the laser pulse has been passed.

3.4 The single crystal monochromator

The monochromator is shown in Fig. 4. A bent silicon single crystal with its surface cut parallel to the (220) crystal planes acts as a cylindrical mirror for the X-rays. It is energy dispersive in the horizontal direction. The deviation ε of the photon energy from the nominal Bragg energy, defined by the equation $\hbar\omega_X = \hbar\omega_B(1 + \varepsilon)$, with

$$\hbar\omega_B = \frac{2\pi\sqrt{h^2 + k^2 + l^2}}{a_0}\frac{\hbar c}{2\sin\Theta_B} \; , \tag{15}$$

is approximately given by the expression [6]

$$\varepsilon = \frac{\chi_0'}{2\sin^2\Theta_B} - \frac{\theta_x}{\tan\Theta_B} \; . \tag{16}$$

Here $\theta_x = \Theta - \Theta_B$ is the horizontal deviation from the nominal Bragg angle Θ_B. The integers h, k, l are the Miller indices, $a_0 = 5.4309$ Å the lattice constant, and χ_0' the real part of the mean dielectric susceptibility $\chi_0 = \chi_0' + i\chi_0''$. The Bragg angle for $\hbar\omega_B = 14.413$ keV amounts for the (220) reflection to $\Theta_B = 12.944°$.

The finite energy width of the Bragg reflection can be calculated from the reflecting power ratio $|R^P|^2$ with the amplitude ratio given by [6, Eq. (3.2)]

$$R^P(\theta_x, \varepsilon) = -y_P(u) + \text{sign}[\Re(y_P(u))]\sqrt{y_P^2(u) - 1} \tag{17}$$

$$y_P(u) = \frac{u + i\chi_0''}{P\chi_H} \tag{18}$$

$$u = 2\sin^2\Theta_B\left[\left(1 - \frac{a}{R\cdot\sin\Theta_B}\right)\frac{\theta_x}{\tan\Theta_B} + \varepsilon\right] + \chi_0' \; . \tag{19}$$

The quantity $\chi_H = \chi_H' + i\chi_H''$ is the Fourier component of the dielectric susceptibility of the analyzer crystal for the reciprocal lattice vector $\mathbf{H}$, R the bending radius of the crystal, and P the polarization factor. The latter is $P = \cos 2\theta_B$ for π polarization, with the polarization vector in the reflection plane, and $P = 1$ for σ polarization, with the polarization vector perpendicular to the reflection plane. Corresponding reflecting power ratios are shown in Fig. 4. Parameters for χ_0 and χ_H were taken from [17].

The energy width of the Bragg reflection is in a good approximation given by the solution of (19) for $\theta_x = 0$ with $u - \chi_0' = \pm|P||\chi_H|$, i.e. the relative energy width is

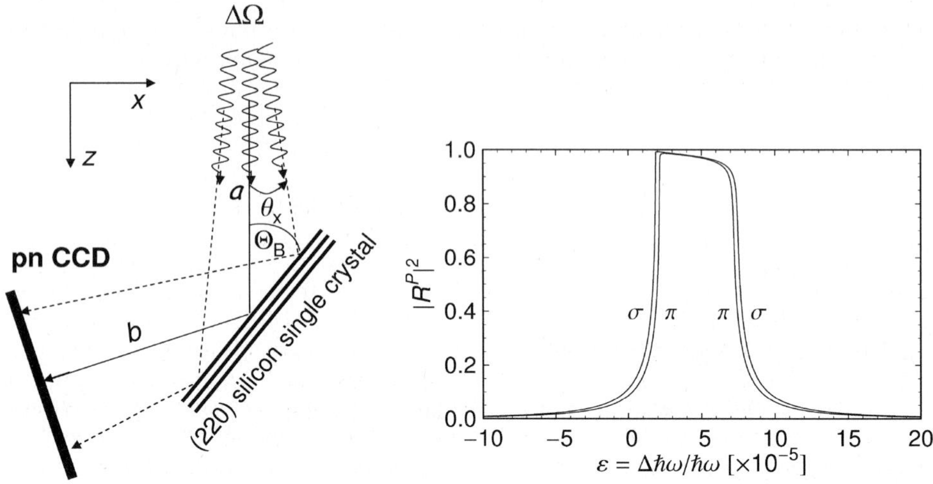

Fig. 4 Scheme of the monochromator (*left*), and reflecting power ratios for π and σ polarized X-rays at $\theta_x = 0$ (*right*). Dielectric susceptibilities $\chi_0' = -0.4677610^{-5}$, $\chi_0'' = -0.3477710^{-7}$, $\chi_H' = -0.2819110^{-5}$, $\chi_H'' = -0.3336410^{-7}$ were used in the calculation. A pn-CCD detector serves as a position sensitive and energy dispersive detector

$\Delta\varepsilon = \Delta\hbar\omega_X/\hbar\omega_X = |P||\chi_h|/\sin^2\Theta_B$. With the energy deviation $\varepsilon = -(\theta\gamma)^2$ of the emitted X-rays as function of the emission angle $\vec{\theta} = \mathbf{e}_x\theta_x + \mathbf{e}_y\theta_y$, which follows from (12) for $\psi = 0$, one obtains from (19) a quadratic equation

$$\theta_x^2 + \theta_y^2 \pm \left(1 - \frac{a}{R \cdot \sin\Theta_B}\right)\frac{\theta_x}{\gamma^2\tan\Theta_B} - \frac{|P||\chi_h|}{2\gamma^2\sin^2\Theta_B} = 0 \tag{20}$$

the solution of which describes the accepted angular region. For a bending radius R chosen such that $1 - a/(R\sin\Theta_B) = 0$ is fulfilled, the accepted angles θ_x and θ_y are located within a circle of radius $\sqrt{|P||\chi_H|/(2\gamma^2\sin^2\Theta_B)}$ and the accepted solid angle is just $\Delta\Omega^P = \pi|P||\chi_H|/(2\gamma^2\sin^2\Theta_B)$. The corresponding solid angle in the rest frame of the Li-like system is $\Delta\Omega_0^P = 4\gamma^2\Delta\Omega^P$. The sum of both polarization states, normalized to 4π, is

$$\frac{\Delta\Omega_0}{4\pi} = \frac{(1 + |\cos 2\Theta_B|)|\chi_H|}{4\sin^2\Theta_B}. \tag{21}$$

With $|\chi_H| = 0.282 \cdot 10^{-5}$ [17] the result is $\Delta\Omega_0/4\pi = 2.67 \cdot 10^{-5}$. The expected count rate at the pn-CCD detector is at saturation with $\dot{n}_0 = A_{ki}/2$

$$\dot{n} = \frac{1}{2}\frac{A_{ki}}{\gamma}\frac{(1 + |\cos 2\Theta_B|)|\chi_H|}{4\sin^2\Theta_B}. \tag{22}$$

It should be mentioned that the focal length of the cylindrical monochromator crystal amounts to $f = (R/2)\sin\Theta_B$ which reduces with $1 - a/(R\sin\Theta_B) = 0$ to $f = a/2$. It follows from the image equation $1/a + 1/b = 1/f$ that at $a = b$ the focus at the pn-CCD detector is just a vertical line.

 Springer

3.5 Count rate estimate

The detected X-ray rate at the pn-CCD detector is given by the equation

$$\dot{N}_X = \frac{\Delta \nu_L / \nu_L}{\Delta \gamma_b / \gamma} f_{rep} \cdot N_{Li} \cdot N_{0,sat} \frac{\Delta \Omega_0}{4\pi} \varepsilon_X \,. \tag{23}$$

The first factor is the fraction $(\Delta \gamma_L / \gamma)/(\Delta \gamma_b / \gamma)$ of Li-like ions in a bunch which can be pumped. With $\Delta \gamma_L / \gamma = \Delta \nu_L / \nu_L$ which follows from (10) for $\Delta \hbar \omega_0 = 0$, it is given by the overlap of the relative laser bandwidth $\Delta \nu_L / \nu_L = 5.3 \cdot 10^{-7}$ with the relative energy spread $\Delta \gamma_b / \gamma = 10^{-4}$ of the Li-like ions in a bunch. A reduction due to the angular spread σ'_{HI} of the Li-like beam can be neglected as long as $\sigma'_{HI} \ll 2\sqrt{\Delta \nu_L / \nu_L} = 1.46$ mrad holds. This latter relation can also be derived from the Doppler-shift formula (10) for which a first order expansion in $\hbar \omega_L$ and a second order expansion in Ψ results in $\Delta \Psi = 2\sqrt{\Delta \hbar \omega_L / \hbar \omega_L} = 2\sqrt{\Delta \nu_L / \nu_L}$ for $\Delta \hbar \omega_0 = 0$. In addition are $f_{rep} = 10^4$/s the laser repetition rate, which must be synchronized with a circulating bunch, $N_{Li} = 10^5$ the number of Li-like ions in a bunch, $N_{0,sat} = 3.7$, and $\Delta \Omega_0 / 4\pi = 2.67 \cdot 10^{-5}$. The overall efficiency $\varepsilon_X = 0.2$ takes into account the photon detection efficiency of the pn CCD as well as photon absorption in the window of the SIS300 vacuum chamber and the mirror for the laser light. The result for the count rate according to (23) is $\dot{N}_X = 104$/s which looks quite reasonable.

3.6 Precision of energy measurement

The relative accuracy of a measurement of the $^2P_{1/2} - {}^2S_{1/2}$ transition energy is, according to (3), given by

$$\frac{\Delta \hbar \omega_0}{\hbar \omega_0} = \frac{1}{2}\sqrt{\left(\frac{\Delta \hbar \omega_L}{\hbar \omega_L}\right)^2 + \left(\frac{\Delta \hbar \omega_X}{\hbar \omega_X}\right)^2}. \tag{24}$$

The precision of the X-ray energy measurement has two contributions. One is a sort of statistical error which is assumed to be 30% of the half width of the Bragg reflex, i.e., $\delta \hbar \omega_X / \hbar \omega_X \simeq 0.3 |\chi_H| / \sin^2 \Theta_B = 1.7 \cdot 10^{-5}$, with $|\chi_H| / \sin^2 \Theta_B = 5.6 \cdot 10^{-5}$. The other one is a systematical error which originates from the energy calibration. Let us assume that the calibration is performed with the 14.41302(32) keV line of a ^{57}Co ($t_{1/2} = 271$ d) source which for this purpose must be placed temporarily in the interaction region of laser and Li-like ion beam. The relative precision $\delta \hbar \omega_{14.4} / \hbar \omega_{14.4} = 2.2 \cdot 10^{-5}$ is of the same order of magnitude as the X-ray energy measurement. Since the error of the laser frequency measurement can be neglected, the total expected relative error of the $^2P_{1/2} - {}^2S_{1/2}$ transition energy is $\Delta \hbar \omega_0 / \hbar \omega_0 = 1.4 \cdot 10^{-5}$ or $\Delta \hbar \omega_0 = 0.0039$ eV. The latter would be a factor of about 4 better as the above quoted value of [2].

According to (5) the relativistic factor γ can be measured simultaneously with the same relative precision $\Delta \gamma / \gamma = 1.4 \cdot 10^{-5}$. It should be mentioned that the relative accuracy of β is $\Delta \beta / \beta = 2.1 \cdot 10^{-8}$ because of $\Delta \beta / \beta = (\Delta \gamma / \gamma)/(\beta \gamma)^2$ which follows from $\gamma = 1/\sqrt{1 - \beta^2}$.

4 Hyperfine spectroscopy

If the nuclear spin I is not zero, ground and excited states of Li-like ions exhibit a hyperfine splitting. The $^2S_{1/2}$ and the $^2P_{1/2}$ states are split only by the magnetic hyperfine interaction while for the $^2P_{3/2}$ also the quadrupole interaction contributes. The hyperfine splitting of the $^2S_{1/2}$ ground-state is typically in the order of 0.5 eV [see, e.g., [13], or [4]]. It will be shown in the following that a hyperfine structure of the $^2P_{1/2} - {}^2S_{1/2}$ or even the $^2P_{3/2} - {}^2S_{1/2}$ transition can be investigated by laser spectroscopy as well.

The experimental setup is depicted in Fig. 5. It might be advantageous to use two laser beams, one pumping, e.g., the $F_g = I - 1/2 \rightarrow F_e = I + 1/2$ transition, and the other the $F_g = I + 1/2 \rightarrow F_e = I + 1/2$ one. Otherwise the signal may cease rapidly because of depopulation pumping. The relative line width $\Delta\hbar\omega_0/\hbar\omega_0 = \Delta\gamma_b/\gamma$ of a transition is entirely determined by the energy distribution of the Li-like ions in a bunch which is in the order of $\Delta\gamma_b/\gamma = 10^{-4}$. Therefore, a line width $\Delta\hbar\omega_0 \simeq$ 0.03 eV is expected which may be small enough to resolve the hyperfine pattern.

Since in such an experiment only small relative energy changes of the hyperfine components must be measured, a high resolution X-ray detector is not required. A large area detector placed in forward direction can be used. This has the advantage that the accepted solid angle can be increased by a large factor, see Fig. 3. For an accepted polar angle of $\Theta = 4$ mrad only, the relative accepted solid angle in the rest frame of the Li-like system is $\Delta\Omega_0/(4\pi) = 0.0104$, i.e., in the order of 1%. Consequently, the count rate is expected to be rather high. Indeed, with $(\Delta\nu_L/\nu_L)/(\Delta\gamma_b/\gamma) = 5.3 \cdot 10^{-3}$, $f_{rep} = 10^4$/s, $N_{Li} = 1$, $N_{0,sat} = 3.7$, $\Delta\Omega_0/(4\pi) = 0.0104$, and $\varepsilon_X = 0.5$ the count rate is according to (23) $\dot{N}_X = 1.0$/s. Notice, that in principle only one stored ion is required for the envisaged hyperfine spectroscopy! However, it is not at all clear that the X-ray detector may be located in forward direction because of expected excess background count rates. But even if a deflection of the X-ray beam out of the forward direction is necessary, e.g., by a pyrolytic graphite crystal which accepts a much larger bandwidth as a single crystal monochromator, the count rate may be sufficiently large with a few tens of Li-like ions in a bunch. Alternatively, the frequency of the pulse laser may be increased. Thereby the event rate increases linearly until the circulation frequency of the ions in the ring of about 275 kHz has been reached. It might also be sufficient to pump only with one laser beam and allow for depopulation pumping if the count rate is high enough.

This kind of hyperfine spectroscopy at Li-like ions has a number of advantages. First of all, hyperfine fields can be calculated with a very high accuracy, at least with a much better accuracy as for neutral atoms which is typically 10% for the isotope shift and 3% for hyperfine fields. This fact is of great importance since relative measurements can be avoided allowing a direct access to the Bohr–Weisskopf effect. Secondly, isotope shifts and magnetic hyperfine splittings can be measured for any element via the $^2P_{1/2} - {}^2S_{1/2}$ transition. At the highest relativistic factor $\gamma = 36$ even a quadrupole interaction can be studied via the $^2P_{3/2} - {}^2S_{1/2}$ transition for all elements with $Z \leq 50$. (For the $2s\,^2S_{1/2} - 2p\,^2P_{1/2,3/2}$ transition energies see [3], or [10].) Finally, it should be mentioned once more that the sensitivity is very high and only a very few or even only one radioactive ion may already be sufficient for the spectroscopy. However, such experiments would require a re-injection of radioactive species, produced by fragmentation reactions, into SIS300 as Li-like ions.

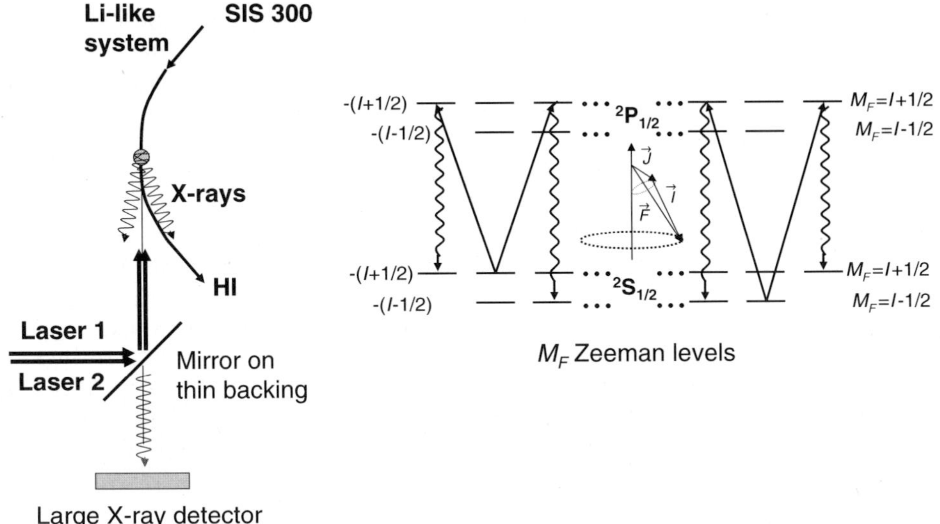

Fig. 5 Scheme of the experimental setup for a measurement of the hyperfine structure (*left*), and pumping scheme (*right*). Shown are the Zeeman levels of the ground and excited state with $F = I \pm 1/2$ the total angular momentum quantum numbers, and I the nuclear spin. While induced laser transitions can populate at resonance only one excited hyperfine level, radiative (downward) transitions can populate both ground state hyperfine levels. For clarity, not all possible transitions are drawn in

5 Laser cooling

The line width of the hyperfine components can be improved by several orders of magnitude by laser cooling as will be demonstrated in the following.[4] Momentum can be transferred to the Li-like ion by absorption of a laser photon as well as by re-emission of the X-ray. While the former momentum transfer is only $\hbar\omega_L/c$, the latter is according to (12) in the order $\gamma^2\hbar\omega_L/c$ since the angular distribution of the X-rays is strongly peaked into forward direction, see Subsection 3.2. The mean longitudinal momentum transfer in the laboratory system due to absorption and emission of a photon of energy $\hbar\omega_0$ in the rest frame of the Li-like system is in the laboratory frame given by

$$\overline{\delta p_{\parallel}} = \frac{\hbar\omega_L}{c} + \int \frac{(\hbar\omega_0/c)\cos\Theta}{\gamma(1-\beta\cos\Theta)} \cdot \frac{f(\cos\Theta_0)}{4\pi} \cdot d\Omega_0 =$$

$$= \frac{\hbar\omega_L}{c} + \int_{-1}^{1} \frac{(\hbar\omega_0/c)\cos\Theta}{\gamma(1-\beta\cos\Theta)} \cdot \frac{d(\cos\Theta)}{2\gamma^2(1-\beta\cos\Theta)^2} = \tag{25}$$

$$= \frac{\hbar\omega_L}{c} + \beta\gamma\frac{\hbar\omega_0}{c} = \gamma\frac{\hbar\omega_0}{c} . \tag{26}$$

---

[4]The subject of laser cooling for relativistic beams has been discussed in [9] and for SIS300 recently in [14], and [15].

Here $(\hbar\omega_0/c)\cos\Theta/(\gamma(1-\beta\cos\Theta))$ is the longitudinal momentum transfer to the Li-like ion in the laboratory system by emission of a photon of momentum $\hbar\omega_0/c$ in the rest frame of the ion. This relation follows from (11) for $\vec{\psi}=0$ and $\theta=\Theta$ after division by c, which transforms the photon energy into the photon momentum, and the projection of the momentum on the velocity axis of the Li-like ion. The function $f(\cos\Theta_0)=1$ represents the angular distribution of the photon in the rest frame of the Li-like ion which is assumed to be isotropic. In (25) the solid angle is transformed by means of (7) into the laboratory system in which the integration is carried out. The integral of (25) can be solved analytically. The final result on the right hand side of (26) has been obtained with (10) for $\Psi=0$ after appropriate reshapings.[5] The instantaneous cooling force is in the laboratory system at saturation $F_{||}=dp_{||}/dt=\overline{\delta p_{||}}\cdot(1/2)\cdot A_{ki}/\gamma=(1/2)\cdot A_{ki}\cdot\hbar\omega_0/c$.

The energy transfer to the Li-like ion of rest mass M_0 is[6] $\delta E_{||}=\delta(\gamma M_0 c^2)=\beta\cdot\delta(\beta\gamma M_0 c)c=\beta\cdot\overline{\delta p_{||}}=\beta\gamma\hbar\omega_0/c$, and the mean relative change of the relativistic factor at resonance absorption and re-emission of a single photon is given by

$$\frac{\delta\gamma}{\gamma}=\beta\frac{\hbar\omega_0}{M_0 c^2}=(1+\beta)\beta\gamma\frac{\hbar\omega_L}{M_0 c^2}\cong 2\gamma\frac{\hbar\omega_L}{M_0 c^2}\;. \tag{27}$$

The numerical result for Li-like uranium at $\gamma=25.68$ is $\delta\gamma/\gamma=1.27\cdot 10^{-9}$. A total energy shift $\Delta\gamma_b/\gamma=10^{-4}$ requires a small de-tuning of the laser frequency in such a manner that first only the ions with largest energies are optically pumped. These are shifted to lower energies by a gradual increase of the laser frequency. In this manner successively more and more ions of the bunch are included in the pumping process. For a complete cooling cycle at least a number $(\Delta\gamma_b/\gamma)/(\delta\gamma/\gamma)$ of transitions is required.

The rate at which γ varies at saturation follows for a pulse laser system with pulse width Δt_{pulse} and repetition rate f_{rep} from (27) and (14) as

$$\left.\frac{\dot\gamma}{\gamma}\right|_{sat}=\left.\frac{d\gamma/\gamma}{dt}\right|_{sat}=\beta\frac{\hbar\omega_0}{A\cdot m_u c^2}\frac{1}{2}\left(A_{ki}\frac{\Delta t_{pulse}}{\gamma}+1\right)\cdot f_{rep}\;. \tag{28}$$

Here A is the atomic number of the Li-like ion and m_u the atomic mass unit. With the numbers of our example the cooling rate is $\dot\gamma/\gamma|_{sat}=0.46\cdot 10^{-4}$/s for ^{238}U. A lower limit for the corresponding cooling time is for $\Delta\gamma_b/\gamma=10^{-4}$

$$\tau_c=\frac{\Delta\gamma_b/\gamma}{\dot\gamma/\gamma|_{sat}}=2.2\text{ s}\;. \tag{29}$$

It might be of interest to know how the cooling rate varies as function of the charge number Z. For ^{120}Sn ($Z=50$), as an example, one obtains with $\hbar\omega_0=107.95$ eV and $A_{ki}=4.606\cdot 10^9$/s, both taken from [10], $\gamma=25.68$, and $f_{rep}=10$ kHz a $\dot\gamma/\gamma|_{sat}=0.135\cdot 10^{-4}$/s. However, cooling of Li-like Sn via the $^2P_{3/2}-{}^2S_{1/2}$ transition at $\hbar\omega_0=374.70$ eV, $A_{ki}=2.057\cdot 10^{11}$/s, and an increased $\gamma=36$ results in a nearly two orders of magnitude faster rate $\dot\gamma/\gamma|_{sat}=9.74\cdot 10^{-4}$/s.

---

[5]This result can also directly be obtained by a Lorentz transformation of the transferred momentum $\hbar\omega_0/c$ in the rest frame of the Li-like ion into the laboratory system.

[6]Also this relation follows directly from the Lorentz transformation formulas.

While in longitudinal direction the ions are cooled, they are heated in transverse direction. The variance of the transverse momentum transfer is given by

$$\overline{\delta p_\perp^2} = \int_{-1}^{1} \frac{(\hbar\omega_0/c)^2 \sin^2\Theta}{\gamma^2(1-\beta\cos\Theta)^2} \cdot \frac{1}{2}\frac{d(\cos\Theta)}{\gamma^2(1-\beta\cos\Theta)^2} = \frac{2}{3}\left(\frac{\hbar\omega_0}{c}\right)^2 . \tag{30}$$

Again, the integral has been solved analytically with the result given at the right hand side of (30).[7] The angular spread is

$$\frac{\sqrt{\overline{\delta p_\perp^2}}}{\beta\gamma M_0 c} = \sqrt{\frac{2}{3}} \cdot \frac{\hbar\omega_0}{\beta\gamma M_0 c^2} \tag{31}$$

with $\beta\gamma M_0 c$ the momentum of the Li-like system. The total angular spread after a number of $(\Delta\gamma_b/\gamma)/(\delta\gamma/\gamma)$ uncorrelated emissions of photons is

$$\sigma'_{HI} = \sqrt{\frac{\Delta\gamma_b/\gamma}{\delta\gamma/\gamma}} \sqrt{\frac{2}{3}} \cdot \frac{\hbar\omega_0}{\beta\gamma M_0 c^2} . \tag{32}$$

With the numbers of our example the result is $\sigma'_{HI} = 0.0113$ µrad which probably is negligibly small in comparison to the angular spread of the HI beam with an emittance of about $1\,\pi$ mm mrad.

Since the energy spread after laser cooling is in the order $\Delta v_L/v_L = 5.3 \cdot 10^{-7}$, a superior line width in the hyperfine structure pattern can be expected with laser cooling.

6 Nuclear polarization by optical pumping

A few remarks will be added on the possibility to prepare a polarized Li-like ion beam which exhibits also a nuclear polarization if the nuclear spin is not zero. As schematically shown in Fig. 6, pumping with left-circularly polarized laser light results after many transitions finally in a population of only the $|F = I + 1/2, M_F = -(I + 1/2)\rangle$ Zeeman level, because this level cannot be pumped anymore. In this state the electronic spin as well as the nuclear spin are polarized as sketched schematically in the vector coupling model inset in Fig. 6.

Unfortunately, a polarization may probably not be maintained within the SIS300 ring since in the strong magnetic field of the bending magnets and beam optical elements a polarized ion with total angular momentum $\vec{F}$ precesses and may randomize. However, it might be conceivable that an external Li-like beam may be pumped in a long straight section, for instance by 100 ns long circularly polarized laser beams which would induce, according to (14), $N_{0,sat} = 32$ transitions in an unpolarized Li-like ion. This is a rather large number and the attained polarization may already be high. For Li-like ions with $Z \leq 50$, it could be pumped via the much faster $^2P_{3/2} - {}^2S_{1/2}$ transition and even much larger numbers may be attained at even shorter laser pulse durations. For quantitative numbers, however, detailed calculations are required which were beyond the scope of this work.

---

[7] The integration may as well be performed in the rest frame of the Li-like system with the same result $\overline{\delta p_\perp^2} = \int_{-1}^{1}(\hbar\omega_0/c)^2 \cdot \sin^2\Theta_0 \cdot (1/2) \cdot d(\cos\Theta_0) = (2/3)(\hbar\omega_0/c)^2.$

 Springer

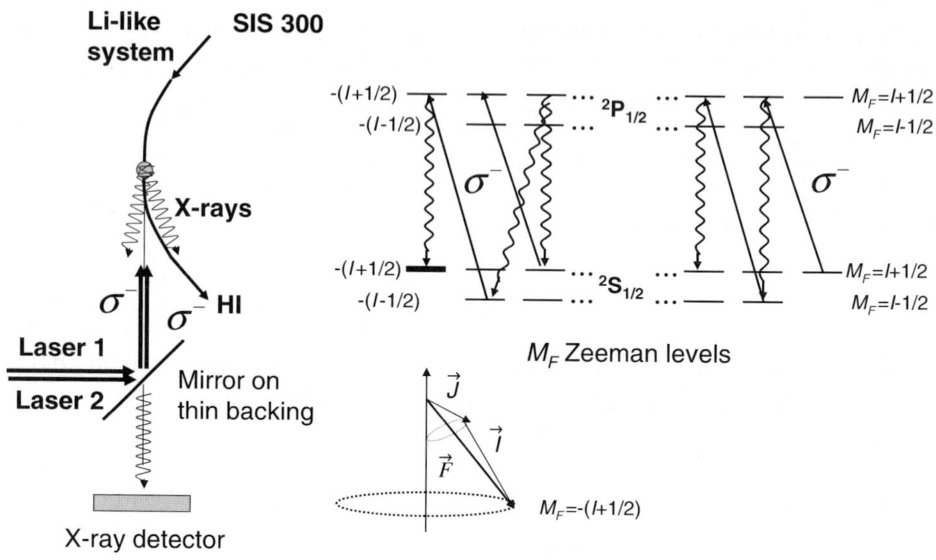

Fig. 6 Scheme of the experimental setup for polarization of the Li-like ion beam (*left*), and pumping scheme with circularly polarized laser light (*right*). Shown are the Zeeman levels for the hyperfine levels of the ground and excited state with $F = I \pm 1/2$, with I the nuclear spin. After a very long pumping time only the $F = -(I + 1/2)$ Zeeman level will be populated

7 Conclusions

Precision transition energies $\hbar\omega_0$ as well as relativistic factors γ can be measured for Li-like ions at the future heavy ion accelerator SIS300 if the inherently very precise laser spectroscopy method is combined with a precision X-ray energy measurement of fluorescence photons by means of a single crystal monochromator. Isotope shifts and magnetic moments can be measured at the $^2P_{1/2} - {}^2S_{1/2}$ transition for all elements, and quadrupole moments for $Z \leq 50$ at the $^2P_{3/2} - {}^2S_{1/2}$ transition, in principle even at a single stored radioactive species far off stability if their nuclear spin is known. Heavy ion beams in SIS300 may be laser cooled, and polarized external heavy ion beams may be prepared by optical pumping with polarized laser beams.

Acknowledgements　I thank Dr W. Lauth, Dr A. Wolf for fruitful discussions, Dr P. Kunz for critical comments on the manuscript, and Dr F. Hagenbuck for valuable information on SIS300. This work has been supported by Bundesministerium für Bildung und Forschung under contract 06 MZ 169 I.

References

1. Backe, H.: Laser spectroscopy and cooling of relativistic ions. Talk presented at the GSI workshop on its future facility, Darmstadt, Germany, 18–20 October 2000
2. Beiersdorfer, P., Chen, H., Thorn, D.B., Träbert, E.: Measurement of the two-loop lamb shift in lithiumlike U^{89+}. Phys. Rev. Lett. **95**, 233003-1–233003-4 (2005)
3. Bosselmann, Ph., Staude, U., Horn, D., Schartner, K.-H., Folkmann, F., Livingston, A.E., Mokler, P.H.: Measurements of $2s^2S_{1/2} - 2p^2P_{1/2,3/2}$ transition energies in lithiumlike heavy

ions. II. Experimental results for Ag^{44+} and discussion along the isoelectronic series. Phys. Rev., A **59**, 1874–1883 (1999)

4. Boucard, S., Indelicato, P.: Relativistic many-body and QED effects on the hyperfine structure of lithium-like ions. Eur. Phys. J., D At. Mol. Opt. Phys. **8**, 59–73 (2000)

5. Brandau, C., Kozhuharov, C., Müller, A., Shi, W., Schippers, S., Bartsch, T., Böhm, S., Böhme, C., Hoffknecht, A., Knopp, H., Grün, N., Scheid, W., Steih, T., Bosch, F., Franzke, B., Mokler, P.H., Nolden, F., Steck, M., Stöhlker, T., Stachura, Z.: Precise determination of the $2s_{1/2} - 2p_{1/2}$ splitting in very heavy lithiumlike ions utilizing dielectronic recombination. Phys. Rev. Lett. **91**, 073202-1–073202-4 (2003)

6. Caticha, A.: Transition-diffracted radiation and Cerenkov emission of x rays. Phys. Rev., A **40**, 4322–4329 (1989)

7. Gutbrod, H.H., Groß, K.-D., Henning, W.F., Metag, V.: An international accelerator facility for beams of ions and antiprotons. In: Henning, W.F. (ed.) Conceptual Design Report, p. 430. GSI, Darmstadt (2001)

8. Gutbrod, H.H., Augustin, I., Eickhoff, H., Groß, K.-D., Henning, W.F., Krämer, D., Walter, G.: FAIR baseline technical report. In: Gutbrod, H.H. (ed.) Accelerator and Science Infrastructure, vol. 2, p. 43. http://www.gsi.de/fair/reports/btr.html

9. Habs, D., Balykin, V., Grieser, M., Grimm, R., Jaeschke, E., Music, M., Petrich, W., Schwalm, D., Wolf, A., Huber, G., Neumann, R.: Relativistic laser cooling. In: Calabrese, R., Tecchio, L. (eds.) Proceedings of the Workshop, Electron Cooling and New Cooling Techniques, Legnaro, Padova, Italy, May 1990, p. 122. World Scientific, Singapore, New Jersey, London, Hong Kong

10. Johnson, W.R., Liu, Z.W., Sapirstein, J.: Transition rates for lithium-like ions, sodium-like ions, and neutral alkali-metal atoms. At. Data Nucl. Data Tables **64**, 279–300 (1996)

11. Persson, H., Lindgren, I., Salomonson, S., Sunnergren, P.: Accurate vacuum-polarization calcualtions. Phys. Rev., A **48**, 2772–2778 (1993)

12. Saathoff, G., Karpuk, S., Eisenbarth, U., Huber, G., Krohn, S., Muñoz Horta, R., Reinhardt, S., Schwalm, D., Wolf, A., Gwinner, G.: Improved test of time dilation in special relativity. Phys. Rev. Lett. **91**, 190403-1–190403-4 (2003)

13. Shabaev, V.M., Shabaeva, M.B., Tupitsyn, I.I., Yerokhin V.A.: Hyperfine structure of highly charged ions. Hyperfine Interact. **114**, 129–133 (1998)

14. Schramm, U., Bussmann, M., Habs, D.: From laser cooling of non-relativistic to relativistic ion beams. Nucl. Instrum. Methods Phys. Res., A **532**, 348–356 (2004)

15. Schramm, U., Bussmann, M., Habs, D., Kühl, T., Beller, P., Franzke, B., Nolden, F., Steck, M., Saathoff, G., Reinhardt, S., Karpuk, S.: Combined laser and electron cooling of bunched C^{3+} ion beams at the storage ring ESR. In: Proceedings of the International Conference 2005, Galena, USA, September 2005. AIP Conf. Proc. **821**, 501–509 (2006)

16. Schweppe, J., Belkacem, A., Blumenfeld, L., Claytor, N., Feinberg, B., Gould, H., Kostroun, V.E., Levy, L., Misawa, S., Mowat, J.R., Prior, M.H.: Measurement of the lamb shift in lithiumlike uranium (U^{89+}). Phys. Rev. Lett. **66**, 1434–1437 (1991)

17. Stepanov, S.: X-ray server. www, http://sergey.gmca.aps.anl.gov/

18. Yerokhin, V.A., Artemyev, A.N., Shabaev, V.M., Sysak, M.M., Zherebtsov, O.M., Soff, G.: Evaluation of the two-photon exchange graphs for the $2p_{1/2} - 2s$ transition in Li-like ions. Phys. Rev., A **64**, 032109-1–032109-15 (2001)

19. Ynnermann, A., James, J., Lindgren, I., Persson, H., Salomonson, S.: Many-body calcualtions of the $2p_{1/2,3/2} - 2s_{1/2}$ transition energies in Li-like ^{238}U. Phys. Rev., A **50**, 4671–4678 (1994)

Hyperfine Interact (2006) 171:109–116
DOI 10.1007/s10751-006-9498-8

Development of a RILIS ionisation scheme for gold at ISOLDE, CERN

B. A. Marsh · V. N. Fedosseev · P. Kosuri · ISOLDE Collaboration

Published online: 15 February 2007
© Springer Science + Business Media B.V. 2007

Abstract At the ISOLDE on-line isotope separation facility, the resonance ionisation laser ion source (RILIS) can be used to ionise reaction products as they effuse from the target. The RILIS process of laser step-wise resonance ionisation of atoms in a hot metal cavity provides a highly element selective stage in the preparation of the radioactive ion beam. As a result, the ISOLDE mass separators can provide beams of a chosen isotope with greatly reduced isobaric contamination. With the addition of a new three-step ionisation scheme for gold, the RILIS is now capable of ionising 26 of the elements. The optimal scheme was determined during an extensive study of the atomic energy levels and auto-ionising states of gold, carried out by means of in-source resonance ionisation spectroscopy. Details of the ionisation scheme and a summary of the spectroscopy study are presented.

Key words radioactive ion beams · resonance ionisation laser ion source · Au

1 Introduction

ISOLDE is an isotope separator on-line (ISOL) type radioactive ion beam facility. Radionuclide production occurs when a pulsed proton beam of 1 or 1.4 GeV is incident upon a thick target. A multitude of reaction channels (e.g. spallation, fragmentation and fission) contribute to the production of a wide spectrum of isotopes.

B. A. Marsh (✉) · V. N. Fedosseev
CERN, CH-1211, Geneva-23, Switzerland
e-mail: bruce.marsh@cern.ch

P. Kosuri
KTH, Stockholm SE-100 44, Sweden

B. A. Marsh
The University of Manchester, Manchester M13 9PL, United Kingdom

Recoiling reaction products are stopped within a short distance inside the heated target material and so their transit to the ion source is by means of diffusion and effusion processes. Unambiguous isotope separation relies on the availability of the reaction products in a single charge state and with a low velocity spread. At ISOLDE, 1^+ ions are most commonly produced and are subject to a 60 keV accelerating voltage. There are two target stations, each connected to its own magnetic dipole mass separator: the General Purpose Separator (GPS) or the High Resolution Separator (HRS). The respective resolving powers of $M/\Delta M \approx 2,500$ and $M/\Delta M \approx 10,000$ are sufficient for the selection of a single atomic mass for transmission. For the chosen atomic mass, often many isobars are produced and isotope selection is not possible by mass separation alone. The somewhat conventional techniques of temperature optimization and beam gating are at hand to give preferential release of the desired element however, such measures are not universally applicable and are often limited in success. The Resonance Ionisation Laser Ion Source (RILIS) combines an unsurpassed level of Z-selectivity with a rapid and efficient ionisation process [1, 2]. For RILIS, ionisation is achieved through a step-wise resonance photon absorption process, providing a high degree of selectivity by exploiting the unique electronic structure of different atomic species. In principle, the RILIS can be used for the ionisation of almost all metallic elements that are released from the ISOLDE target. An important aspect of ongoing development is the extension of its range with the study of resonance ionisation schemes. Recently a request for the development of gold ion beams was addressed to the ISOLDE and Neutron Time of Flight Experiments Committee (INTC) by the letter of intent 'Study of the $\pi h_{11/2}^{-1}$ isomeric states in ^{201,203,205}Au. Following an endorsement of this intent by the committee, the task of finding an efficient RILIS ionisation scheme for gold was put forward.

Resonance ionisation spectroscopy has been applied for gold by several groups. The data obtained as well as the ionisation schemes used are compiled by Saloman [3]. In particular, an efficient ionisation scheme with a three-colour, three-step excitation to an autoionising state has been used for on-line laser spectroscopy of laser-desorbed gold isotopes [4, 5]. Auto-ionising (AIS) states of Au were also studied by resonance ionisation spectroscopy in [6, 7]. Since the RILIS laser system and the conditions in the ionisation region are different from those used in [4, 6, 7], investigating the performance of these known schemes at the RILIS setup was necessary. In addition, the search for transitions to the auto-ionising states was carried out in these works for only a few excited states in a limited spectral range, which is not readily available at RILIS. Therefore, a new, more extensive study of gold resonance ionisation at the RILIS setup was required for defining the optimal ionisation scheme. In this article we present results of the resonance ionisation spectroscopy study of gold atomic transitions performed at the RILIS setup. Based on the experimental results, the optimal ionisation scheme is defined. A summary is presented along with details of this new ionisation scheme.

2 The resonance ionisation laser ion source

References [2, 8, 9] give a thorough description of the ISOLDE RILIS. A master oscillator power amplifier system of copper vapor lasers (CVL) operating at the pulse repetition rate of 11 kHz provides two output laser beams, each with an average power of typically 30–40 W. The RILIS set-up includes three dye lasers and therefore ionisation schemes employing up to three resonant transitions can be used. The wavelength range of the dye lasers is 530–850 nm. Frequency doubling and tripling of dye laser beams are carried out using non-linear BBO (beta-barium borate) crystals to generate 2nd or 3rd harmonics of the fundamental beam, extending the wavelength range to include 214–415 nm. This enables high lying first excited states to be attained and is essential for elements with a high ionisation potential. With this current work included, the RILIS has been used for resonance ionisation of 26 of the elements. Schemes using one, two or three resonant transitions have been used. Most commonly, the ionisation step is a transition to the continuum using an available CVL beam. Alternatively, the final step can be a resonant transition to an auto-ionising state. A transition to an auto-ionising state can have a high cross section, which is favorable for improving the ionisation efficiency. Ionisation takes place in a hot cavity connected to the target. Reaction products enter this cavity as an atomic vapor at a temperature of around 2,300 K. The role of the cavity is to contain the atoms for a certain time within a volume where they can be irradiated by the laser light and to confine the ions during their drift towards the extraction region. The ionisation cavities are refractory metal (W or Nb) tubes with an inner diameter of 3 mm and a length of typically 30 mm. They are resistively heated to a temperature of about 2,300 K with a DC current of 200–350 A. After leaving the source, ions are accelerated to 60 kV, separated in a magnetic field and guided by electrostatic ion-optical elements to the experimental setup.

For this work, a standard ISOLDE target was used with a tungsten surface ioniser cavity. The target/ion source unit was equipped with two ovens. A large gold sample was placed in one of these, for use during the initial spectroscopy study and ionisation scheme search. The second oven was loaded with a precise 3,000 nAh gold sample, for use during the final efficiency measurement. The mass separator was tuned to the mass of ^{197}Au and the transmitted ion current was monitored on a Faraday cup detector.

3 Resonance ionisation spectroscopy of gold

Data on atomic lines and energy levels for gold are taken from the Kurucz atomic line database (http://cfa-www.harvard.edu/amdata/ampdata/kurucz23/sekur.html). The ionisation potential for gold (9.23 eV) is high with respect to the photon energy-range attainable with RILIS (~1.7–5 eV). As a result, ionisation schemes using three excitation steps were investigated. For the first step, the transitions from the ground state $6s^2S_{1/2}$ level to the $6p^2P_{1/2}$ and $6p^2P_{3/2}$ levels at 37,358.991 cm^{-1} and 41,174.613 cm^{-1} respectively are the only known transitions that can be reached with the RILIS laser system. Both transitions are strong and of similar strength ($A = 1.65 \cdot 10^8$ s^{-1} and $1.96 \cdot 10^8$ s^{-1}). At 242.9 nm, the wavelength required to populate the 41,174.61 cm^{-1} level is at the short wavelength end of the RILIS tuning range and can be generated only by tripling the fundamental frequency of the dye laser. The efficiency of 3rd harmonic generation is low (~5% for a 5 W fundamental beam) and so, if this transition is used, a large proportion of the CVL pump beam power must be used to obtain the dye laser power required at the fundamental frequency.

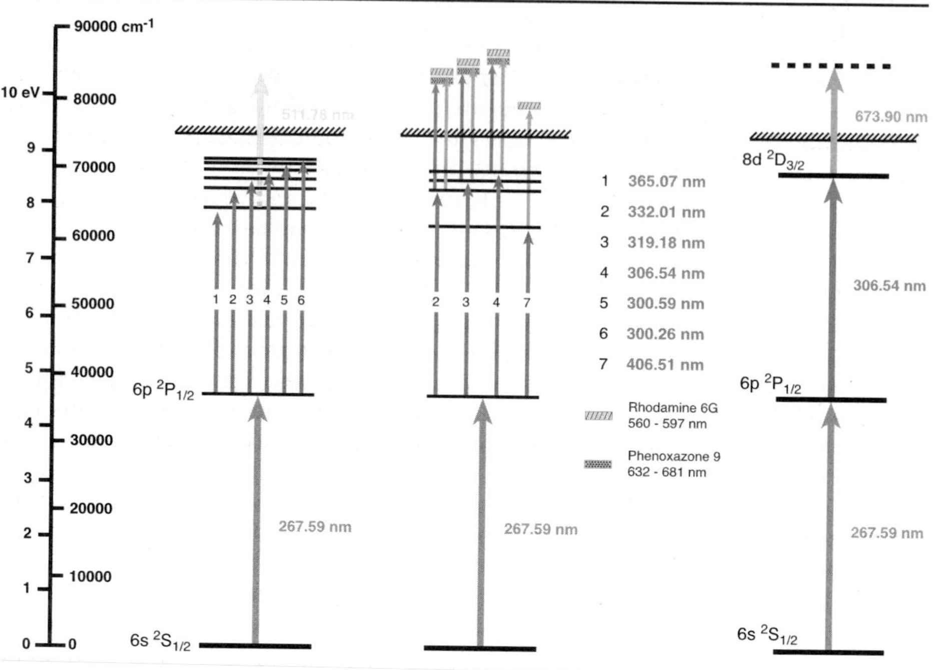

Fig. 1 The scope of the Au RIS study and the optimal ionisation scheme (RHS)

The transition to the lower lying, $^2P_{1/2}$ level is preferable since the transition energy, 37,358.991 cm^{-1}, corresponds to that of a 267.673 nm photon, accessible by 2nd harmonic generation of light at 535.198 nm. This falls within the emission range of the Pyrromethene 546 dye, which is pumped with the green component of the CVL beam. Measurements were carried out using only this transition for the first step in the excitation scheme.

From this first excited state, 11 potential second excited states from 54,485 cm^{-1} to 72,164 cm^{-1} are documented (http://cfa-www.harvard.edu/amdata/ampdata/kurucz23/sekur.html). All but the lowest lying of these excited states exist at a energy of less than 19,581 cm^{-1} below the continuum, meaning that ionisation from these levels is possible via non-resonant absorption of a 511 nm photon, provided by the green CVL beam. Accessing any of these 10 transitions requires 2nd harmonic generation of the fundamental dye laser beam, and hence, a significant proportion of CVL pump power. It is preferable to use all of the CVL pump power at 511 nm for the generation of the frequency for the 1st transition and for the final ionisation stage, leaving only the yellow (578 nm) component for the second step. Therefore, we have limited our study to the group of six second step transitions (Fig. 1), which are accessible with a frequency doubled dye laser pumped with light at 578 nm. The pumping requirements of the dye lasers and amplifiers for the first and second step transitions were met using the total output of the higher power of the two CVL amplifiers after separation of the yellow and green components of the beam. Typical values of the average power for this CVL beam were 22 W and 20 W at 511 nm and 578 nm, respectively. The second CVL beam with a total power (in both components) of 27 W was available for the last excitation step.

 Springer

Table 1 Gold 3 step ionisation schemes

λ_2 (*air*) nm	E_2 cm^{-1}	State II	λ_3 (*air*) nm	E_3 cm^{-1}	Laser Power, mW 2	3	R
365.07	64,742.9	$8s^2\mathrm{S}_{1/2}$	511, 578	Continuum	75	10,000	1
332.01	67,469.68	$7d^2\mathrm{D}_{3/2}$	511, 578	Continuum	75	10,000	9.6
319.18	68,680.63	$9s^2\mathrm{S}_{1/2}$	511, 578	Continuum	60	10,000	2.1
306.54	69,971.42	$8d^2\mathrm{D}_{3/2}$	511, 578	Continuum	75	11,000	7.7
300.59	70,617.73	$10s^2\mathrm{S}_{1/2}$	511, 578	Continuum	65	11,000	1.5
300.26	70,653.25	$J=3/2$	511, 578	Continuum	65	11,000	1.2
406.51	*61,951.89*	*$6d\,^2\mathrm{D}_{3/2}$*	*591.90*	*78,842.0*	*75*	*625*	*55*
319.18	68,680.63	$9s^2\mathrm{S}_{1/2}$	668.65	83,631.9	90	1130	4.4
319.18	68,680.63	$9s^2\mathrm{S}_{1/2}$	644.93	84,181.9	90	1130	3.1
332.01	67,469.68	$7d^2\mathrm{D}_{3/2}$	592.85	84,332.8	90	500	10
332.01	67,469.68	$7d^2\mathrm{D}_{3/2}$	584.84	84,563.8	90	750	6.4
332.01	67,469.68	$7d^2\mathrm{D}_{3/2}$	578.38	84,754.5	90	550	73
306.54	69,971.42	$8d^2\mathrm{D}_{3/2}$	676.27	84,754.3	80	750	150
332.01	67,469.68	$7d^2\mathrm{D}_{3/2}$	576.64	84,806.6	90	950	120
306.54	**69,971.42**	**$8d^2\mathrm{D}_{3/2}$**	**673.9**	**84,806.3**	**80**	**750**	**220**
332.01	67,469.68	$7d^2\mathrm{D}_{3/2}$	574.20	84,880.5	90	750	34
306.54	69,971.42	$8d^2\mathrm{D}_{3/2}$	670.55	84,880.4	80	1000	180
306.54	69,971.42	$8d^2\mathrm{D}_{3/2}$	666.24	84,976.9	80	1130	13
332.01	67,469.68	$7d^2\mathrm{D}_{3/2}$	570.77	84,985.0	90	750	2.6
306.54	69,971.42	$8d^2\mathrm{D}_{3/2}$	665.81	84,986.6	80	1130	38
332.01	67,469.68	$7d^2\mathrm{D}_{3/2}$	569.80	85,014.7	90	975	3.5
332.01	67,469.68	$7d^2\mathrm{D}_{3/2}$	567.24	85,094.1	90	600	4.5
306.54	69,971.42	$8d^2\mathrm{D}_{3/2}$	661.08	85,093.9	80	1250	21
332.01	67,469.68	$7d^2\mathrm{D}_{3/2}$	567.12	85,097.8	90	600	5.4
306.54	69,971.42	$8d^2\mathrm{D}_{3/2}$	660.90	85,098.0	80	1250	20
332.01	67,469.68	$7d^2\mathrm{D}_{3/2}$	564.14	85,190.8	90	450	3.3
306.54	69,971.42	$8d^2\mathrm{D}_{3/2}$	632.80	85,769.9	80	500	4.3
319.18	68,680.63	$9s^2\mathrm{S}_{1/2}$	567.39	86,300.3	80	750	4.5
319.18	68,680.63	$9s^2\mathrm{S}_{1/2}$	565.61	86,355.6	85	625	3.8
319.18	68,680.63	$9s^2\mathrm{S}_{1/2}$	565.36	86,363.5	85	625	3.6
306.54	69,971.42	$8d^2\mathrm{D}_{3/2}$	575.60	87,339.7	85	750	76
306.54	69,971.42	$8d^2\mathrm{D}_{3/2}$	572.38	87,437.6	85	750	59
306.54	69,971.42	$8d^2\mathrm{D}_{3/2}$	570.70	87,488.4	85	750	59
306.54	69,971.42	$8d^2\mathrm{D}_{3/2}$	567.98	87,572.8	85	700	44
306.54	69,971.42	$8d^2\mathrm{D}_{3/2}$	567.14	87,598.7	85	550	7.4
306.54	69,971.42	$8d^2\mathrm{D}_{3/2}$	564.07	87,694.9	85	550	15
306.54	69,971.42	$8d^2\mathrm{D}_{3/2}$	561.71	87,769.1	85	375	4.3

$E_0 = 6s^2\mathrm{S}_{1/2}$, 0 cm^{-1}; $E_1 = 6p^2\mathrm{P}_{1/2}$, 37,358.99 cm^{-1}
The laser power was measured on the laser table, transmission to the ion source is approximately 20% for the first step, 50% for the second step and 50% for the CVL beam (3rd step)

As the first stage of this study, schemes using 3-step ionisation with a non-resonant transition to the continuum (induced by both the green and the yellow components of the CVL beam) were studied. During this study, a ion current of gold was observed in the presence of only the beam for the first step in the excitation scheme. Since the ionisation threshold of gold (74,409.0(2) cm^{-1} [10]) is slightly less than the sum

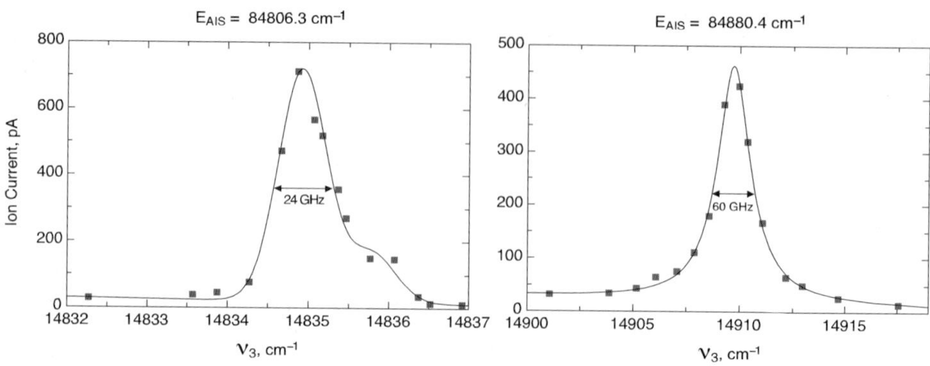

Fig. 2 The two strongest AIS resonances observed

of the energies of two photons at 267.673 nm, this ion current is attributed to the occurrence of two-step ionisation via a transition to the continuum from the first excited state by non-resonant absorption of second photon from the first step beam. The presence of this laser-ionised ion signal facilitated a reliable optimization of the beam focusing and positioning. In addition, the measurements of the ion current produced by the first step transition beam provided a useful reference point for use in the evaluation of the relative efficiencies of ionisation schemes. The ratio of ion current produced with three laser beams versus ion current produced by the first step beam $R = I_{1+2+3}/I_{1+3}$ is taken as an efficiency parameter of the ionisation scheme. In order to keep thermal conditions of laser beam transport optics constant these measurements were carried out whilst blocking only the beam of second step laser, leaving the most powerful laser beam propagating toward the ion source. This beam did not participate in the ionisation process because its photon energy was less than required for transition from the first excited state to the ionisation continuum.

For the second phase of the study a search for auto-ionising states was conducted. The CVL power used for the ionisation stage was diverted to pump a 3rd dye laser and dye amplifier. Ethanol solutions of the laser dyes Rhodamine 6G and Phenoxazone 9 were used in this laser, respectively the wavelength was scanned in the ranges of 560–597 nm and 632–681 nm. When an observable increase of the ion current was detected, scanning was stopped and the wavelength was optimized to the maximal value of the ion current. Its peak value, the corresponding transition frequency, and the values of laser power for each of the three beams were recorded. Measurements of the laser frequency were made with a HighFinesse / Ångstrom WS/7 wavemeter. The accuracy of the transition frequency measurements was defined by the width of resonance and laser line width (~ 0.6 cm^{-1}). For narrow AIS it was possible to define the peak position with the accuracy of about 0.1 cm^{-1}. Energy regions above the ionisation threshold for the transitions from the second excited states corresponding to the four strongest second step transitions were studied.

A summary of these measurements is given in Table 1. Within the energy intervals that were studied, five auto-ionising states had been previously observed [4, 7], but not as resonant transitions from the excited states measured during this study. Three of the known AIS were confirmed and 27 new AIS were observed. Figures 2a and 2b show the profiles of the resonances for the two strongest AIS, at 84,806.3 cm^{-1}

 Springer

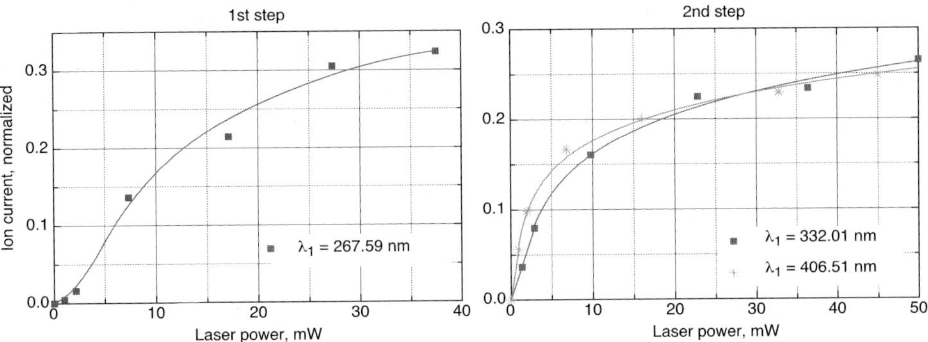

Fig. 3 Saturation curves for the resonant transitions of the most efficient Au schemes

and 84,880.4 cm^{-1}, respectively. Of particular interest was the measurement of a previously applied ionisation scheme for gold [4]. This scheme, shown in italics in Table 1, was one of the more efficient schemes previously measured however the optimum scheme, shown on the right hand side of Fig. 1, gives a factor of 4–5 increase in the efficiency parameter R. The saturation measurements for the 1st and 2nd step in this scheme are presented in Fig. 3, they confirm that the saturation points for the transitions are exceeded with the laser power available.

For the measurement of the absolute value of ionisation efficiency, some experimental difficulties were encountered, limiting the reliability of the assessment of the mass marker evaporation. During the initial search for a resonance ionisation signal, a large quantity of the gold sample was evaporated from the first oven. After detection of the ion signal, the subsequent tests were carried out without the need to heat the mass marker itself, evaporation of the gold that had condensed in the target container provided a sufficient and steady supply of gold vapor for the duration of the spectroscopic work. An attempt was made to extinguish this supply by gradually increasing the target heating from 570–700 A. A small decrease in the ion current was observed after some hours. The evaporation of the 3,000 nAh mass marker was then carried out with the target heating kept at 700 A. For improved laser ionisation stability and experimental ease, only the first step laser was used for ionisation during the efficiency measurement. The ionisation efficiency of the scheme presented in Fig. 1 was estimated to be over 3% by comparing the product of the ion current integration and the parameter R of the most efficient ionisation scheme with the original sample size.

4 Conclusion

With the development of this new ionisation scheme for gold, the RILIS can now provide a means of efficient and selective ionisation of 26 of the elements. In accordance with the requirements of the ISOLDE experimental program, new schemes will continue to be developed for RILIS using procedures similar to those described in this work. Currently, such studies have to fit into the framework of general ISOLDE off-line development and thus, the dedicated mass separator use required is limited to just a few weeks per year. A recent letter of intent addressed to

 Springer

the INTC, 'Development of the RILIS research laboratory at ISOLDE' gives details of a new and dedicated RILIS spectroscopy facility. Following its completion, the laboratory will enable the investigation of new ionisation schemes as a task that is independent of ISOLDE scheduling. Many of the RILIS ionisation schemes rely on an inefficient final step of non-resonant ionisation using a large proportion of the total CVL power. It is hoped that the time consuming search for auto-ionising states for many of these elements could be carried out in this laboratory and significant improvements in RILIS efficiency could be made.

Acknowledgements   The target preparation (assembly, mass marker preparation and initial testing) was carried out by R. Catherall, D. Carminati, B. Crepieux and T. Stora (CERN-AB). This work was supported by the EU 6th programme Integrating Infrastructure Initiative Transnational Access, Contract number: RII3-CT-2004-506065.

References

1. Mishin, V.I., Fedoseyev, V.N., Kluge, H.-J., Letokhov, V.S., Ravn, H.L., Scheerer, F., Shirakabe, Y., Sundell, S., Tengblad, O., ISOLDE Collaboration: Nucl. Instrum. Methods Phys. Res., B **73**, 550 (1993)
2. Fedoseyev, V.N., Huber, G., Köster, U., Lettry, J., Mishin, V.I., Ravn, H.L., Sebastian, V., ISOLDE Collaboration: Hyperfine Interact. **127**, 409 (2000)
3. Saloman, E.B.: Spectrochim. Acta, Part B **45**, 37 (1990)
4. Krönert, U., Beckert, St., Hilberath, Th., Kluge, H.-J., Schulz, C.: Appl. Phys., A **44**, 339 (1987)
5. Krönert, U., Beckert, St., Bollen, G., Gerber, M., Hilberath, Th., Kluge, H.-J., Passler, G., ISOLDE Collaboration: Nucl. Instrum. Methods Phys. Res., A **300**, 522 (1991)
6. Bekov, G.I., Tursunov, A.T., Khasanov, G., Eshkobilov, N.B.: Opt. Spectrosc. (USSR) **62**, 163 (1987)
7. Zhao, W.Z., Xu, X.Y., Ma, W.Y., Cheng, Y., Hui, Q., Wen, K.L., Chen, D.Y.: Appl. Phys., B **52**, 299 (1991)
8. Fedosseev, V.N., Fedorov, D.V., Horn, R., Huber, G., Köster, U., Lassen, J., Mishin, V.I., Seliverstov, M.D., Weissman, L., Wendt, K., ISOLDE Collaboration: Nucl. Instrum. Methods Phys. Res., B **204**, 353 (2003)
9. Catherall, R., Fedosseev, V.N., Köster, U., Lettry, J., Suberlucq, G., Marsh, B.A., Tengborn, E., ISOLDE Collaboration: Rev. Sci. Instrum. **75**, 1614 (2004)
10. Loock, H.-P., Beaty, L.M., Simard, B.: Phys. Rev., A **59**, 873 (1999)

Hyperfine Interact (2006) 171:117–119
DOI 10.1007/s10751-006-9495-y

The ALTO project at IPN-Orsay

S. Franchoo · F. Ibrahim · F. Le Blanc · M. Lebois ·
A. Olivier · C. Phan Viet · B. Roussière · R. Sifi ·
D. Verney

Published online: 9 February 2007
© Springer Science + Business Media B.V. 2007

Abstract The availability of both a tandem deuteron beam and a linac electron beam, the latter converted into Bremsstrahlung, at the new ALTO facility at IPN-Orsay offers a unique opportunity to compare the performance of a laser ion guide under different regimes. The ALTO accelerator has delivered its first electron beam at the end of 2005 and a design for a gas-cell prototype is being studied.

Key words ALTO · IPN-Orsay · Bremsstrahlung

1 The ALTO accelerator

The purpose of the ALTO project at IPN-Orsay is to gather the necessary know-how for the construction of the ISOL-based SPIRAL-2 facility at GANIL. At the same time, it opens a window to study neutron-rich nuclei produced in a thick actinide target by photofission, after conversion of a primary electron beam into Bremsstrahlung [1].

The ALTO accelerator is a 50 MeV electron machine that was recovered from CERN, where it served as the LEP injector linac till the dismantlement of LEP in 2001. The accelerator was moved to IPN-Orsay and reassembled in the main experimental area that also houses the tandem driver and the PARRNE separator. It was put into operation in December 2005 and the first photofission production runs on a uranium carbide target took place in June 2006. For radioprotection issues, the intensity of the accelerated beam was limited to 100 nA instead of its 10 μA nominal current.

The fission fragments were mass separated on-line, deposited on a movable tape and their radiation recorded by a standard germanium detector in close geometry. The dead time-corrected yield of ^{132}Sn from the plasma ion source used so far reached $1.8 \cdot 10^6/\mu C$ [2]. At ISOLDE, this isotope is currently produced with laser ionisation at a rate of $3 \cdot 10^8/$

S. Franchoo (✉) · F. Ibrahim · F. Le Blanc · M. Lebois · A. Olivier · C. Phan Viet · B. Roussière ·
R. Sifi · D. Verney
Institut de Physique Nucléaire (IPN) d'Orsay, 91406 Orsay, Cedex, France
e-mail: franchoo@ipno.in2p3.fr

μC [3]. The ALTO machine is expected to deliver 10 μA, while ISOLDE runs on an average of 1 to 2 μA.

2 The laser ion source

The purity of the extracted beam at an ISOL facility greatly benefits from the use of an element-selective laser ion source. Such a source is currently under development at IPN-Orsay, based on the experience accumulated by teams at ISOLDE [4], Leuven [5], Jyväskylä [6] and TRIUMF [7]. Ionisation schemes for Sn and Cu were tested off-line with a dye-laser set-up pumped at 30 Hz. Efficiencies of 10^{-6} were obtained for both elements, without auto-ionising state [8, 9].

A new laser set-up will be installed early 2007. It will consist of three FL-3002 Lambda Physik dye lasers pumped by a 532 nm Nd:YAG laser at 20 kHz. Together with the use of ionisation schemes that include auto-ionising states, the higher repetition frequency would lift the efficiency of the source to the percentage range.

3 The laser ion guide

Access to neutron-rich isotopes of chemically refractory elements is often difficult because of their slow release from conventional thick targets. A solution is provided by the ion-guide technique, where thin targets are suspended in a buffer-gas cell. At ALTO, we have undertaken preliminary design studies for the implementation of an ion guide. Simulations were carried out that investigate the influence of various parameters such as the thickness of the converter slab, the radius of the impinging electron beam, the target thickness and spacing, and the gas pressure.

It was found that the highest production yield is achieved for a converter of 3.5 mm thickness in close contact to the target foils. The converter was chosen to be tungsten. As it does not stop the electron beam, the buffer gas is still subject to a considerable charge load. The generation of low-energy secondary electrons and X-rays adds to the creation of a local plasma.

Further calculations show the advantage of a single target along the axis of a well-focused electron beam rather than a wide beam that irradiates a multi-layered set-up with additional off-axis foils. For a given fixed volume the number of fissions evidently increases with the number of perpendicular targets, while the volume in which the particles slow down is inversely reduced. Stopping efficiencies in 500 mbar of argon of 30 and 40% were estimated for distances between 15 mg/cm^2 targets of 10 and 20 mm, respectively. If we take the depth of the gas cell as 40 mm, led by practical considerations, then the two configurations can hold three or one foils, so we opt for the former. While a longitudinal shift by 5 mm allows incorporating a fourth perpendicular target, still more of them would stretch resources. The total number of fissions for the proposed geometry amounts to $6.7 \cdot 10^7/\mu C$, out of which an estimated $2.4 \cdot 10^7/\mu C$ would be stopped in the buffer gas.

Coupling of the laser ion guide to the existing PARRNE separator will require a study of the necessary adaptations to the front end, for instance the enlargement of the extraction chamber to accommodate differential pumping and the transit of pumping ducts through the concrete shielding walls to high-capacity Roots blowers. In particular, the current design of the laser ion guide does not foresee for any entrance of the laser light other than the extraction axis.

 Springer

Acknowledgements The authors wish to express their gratitude to the Ion-Source Group of the Accelerator Division at IPN-Orsay for their continuous commitment, in particular M. Cheikh Mhamed, J.-M. Curaudeau, M. Ducour-tieux, S. Essabaa, C. Lau, H. Lefort, J. Lesrel, M. Raynaud, A. Said, and C. Vogel.

References

1. Ibrahim, F., et al.: Eur. Phys. J., **A15**, 357 (2002)
2. Lebois, M., et al.: Proceedings of Exon-2006 (in press)
3. Fedoseyev, V., et al.: Hyp. Int. **127**, 409 (2000)
4. Fedoseyev, V., et al.: Nucl. Instr. Methods, **B 204**, 353 (2003)
5. Kudryavtsev, Yu., et al.: Nucl. Phys., **A 701**, 465c (2002)
6. Moore, I., et al.: J. Phys., **G 31**, S1499 (2005)
7. Lassen, J., et al.: Hyp. Int. **162**, 69 (2005)
8. Hosni, F.: Ph.D. thesis, IPN-Orsay (available on-line at tel.ccsd.cnrs.fr)
9. Sifi, R.: Ph.D. thesis, IPN-Orsay (in press)

Hyperfine Interact (2006) 171:121–126
DOI 10.1007/s10751-006-9496-x

Upgrade to the IGISOL laser ion source towards spectroscopy on Tc

T. Kessler · T. Eronen · C. Mattolat · I. D. Moore ·
K. Peräjärvi · P. Ronkanen · P. Thörle · B. Tordoff ·
N. Trautmann · K. Wendt · K. Wies · J. Äystö

Published online: 1 March 2007
© Springer Science + Business Media B.V. 2007

Abstract A new method of optical pumping has been applied to increase the power of the Ti:Sapphire laser system of the FURIOS laser ion source, Jyväskylä. This upgrade has led to a factor of two improvement in the output power, and has been directly employed in the first off-line laser ionisation tests on the long-lived refractory isotope ^{99}Tc. In the future further studies will be done to determine the efficiency of this ionisation scheme and to employ it for on-line experiments.

Keywords Laser ion source · Technetium · IGISOL

1 Introduction

The IGISOL (Ion-Guide Isotope Separator On-Line) technique has been in use for the past 20 years. During this time there has been a number of technical advances to improve the performance of the technique [1]. Due to the production mechanism the ion guide method is not dependent on target chemistry and therefore

T. Kessler (✉) · T. Eronen · I. D. Moore · K. Peräjärvi · P. Ronkanen · J. Äystö
Department of Physics, University of Jyväskylä, Survontie 9, PL 35 (YFL),
40014 Jyväskylä, Finland
e-mail: thomas.kessler@phys.jyu.fi

C. Mattolat · K. Wendt · K. Wies
AG Larissa/Quantum, Inst. f. Physik, Johannes Gutenberg Universität, Staudinger Weg 7,
55128 Mainz, Germany

P. Thörle · N. Trautmann
Inst. f. Kernchemie, Johannes Gutenberg Universität, Fritz-Straßmann-Weg 2,
55128 Mainz, Germany

B. Tordoff
Nuclear Physics Group, Schuster Laboratory, University of Manchester, Brunswick Street,
Manchester M13 9PL, UK

is the ISOL system of choice for the production of refractory short-lived nuclei [2]. Despite the attractive features of a universal production of exotic nuclei and the fast (submillisecond) release, two drawbacks remain. The IGISOL is rather unselective, a problem in particular for proton-induced fission reactions, and in some cases inefficient. To overcome these deficiencies a laser ion source has been installed at the University of Jyväskylä consisting of both solid-state Ti:Sapphire lasers and dye lasers [3].

One of the elements of interest for study at the IGISOL facility is technetium. From a physics point of view, there is a three-fold motivation: nuclear astrophysics, weak interaction physics and nuclear structure physics. For example, according to the conserved vector current hypothesis (CVC), the matrix elements of the superallowed Fermi transitions between the 0^+ isobaric analog states (IAS) should all be equal, independent of nuclear structure apart from small radiative and isospin-symmetry breaking terms. These latter terms become increasingly important with increasing Z of the nucleus, and uncertainties between different theoretical calculations dominate the uncertainty of the isospin correction [4]. A measurement of the Q-value and half-life of the $N = Z$ superallowed beta emitter, ^{86}Tc, would provide important experimental input to this field.

In addition, nuclei close to the $N = Z$ line play a special role in nuclear astrophysics since the rapid-proton (rp) capture process passes right through them. On the neutron-rich side of the valley of stability, extending the ability to produce even more exotic Tc isotopes will benefit for the studies of nuclear structure in this region, and also provide input for calculations dealing with the astrophysical r-process.

The ability to perform spectroscopy on exotic Tc isotopes relies on the need to produce intense and pure beams of these nuclei. By combining the IGISOL technique with the efficiency and selectivity of a laser ion source this requirement will be achieved.

2 Experimental setup

In order to selectively ionise Tc, an efficient resonance ionisation scheme must first be developed. Due to the refractory nature, particular attention to the production of an atomic vapour must be addressed. High temperatures and a stable operating condition are required, developments which go hand in hand with the continuous improvement of the laser system in use. In this section the experimental setup used for the preparatory experiments for the laser ionisation of Tc is described. After a short introduction to the ion guide and IGISOL technique, the modification to the laser system will be explained.

2.1 The ion guide technique

The ion guide used for the laser ionisation of Tc is described in detail in [5]. It is equipped with a mount for an electrothermally heated filament. The filament was prepared at the nuclear chemistry department of the University of Mainz using an electrodeposition technique as described in [6]. The filament substrate is made from rhenium and is cut into a 20×4 mm^2 foil, with a thickness of 25 μm. A sample of 10^{14} atoms of ^{99}Tc with a half-life, $T_{1/2} = 2.111 \cdot 10^5$ years [7], is deposited onto the foil

 Springer

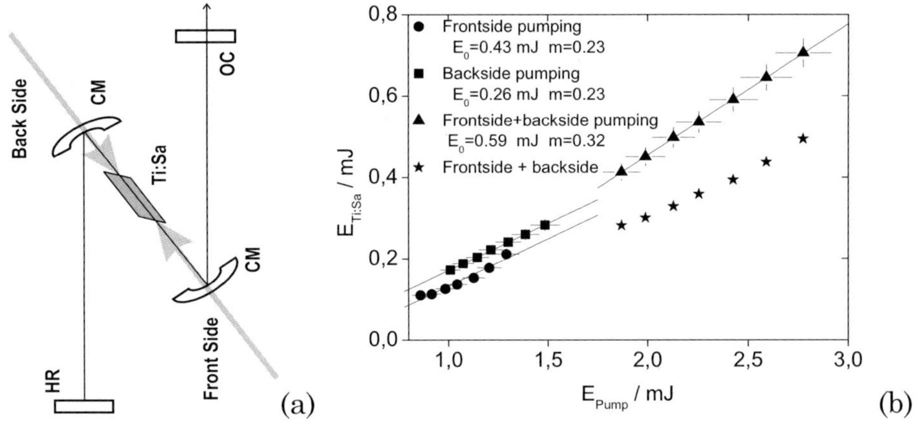

Fig. 1 **a** The double-sided Ti:Sapphire laser. *HR = high reflector, CM = curved mirror*, Ti:Sa = Ti:Sapphire crystal, *OC = output coupler*. **b** Conversion efficiencies for the double-sided Ti:Sapphire laser

and can be evaporated with temperatures of up to 1,500 °C under vacuum. The laser radiation enters the ion guide on-axis via an entrance window in the rear. A helium gas flow at a pressure of 60 mbar extracts the resonant ions through the exit nozzle of the ion guide. The ions are subsequently guided through a radio-frequency sextupole device before being accelerated to 30 kV and mass separated. The final detection is done with a set of microchannel plates (MCP) downstream from the IGISOL focal plane.

2.2 Laser system and new developments

The FURIOS laser system was designed to operate as a twin laser system. One setup consists of two commercial dye lasers pumped by a Copper Vapour Laser (CVL) operating at 12 kHz. In this work an all solid-state system was used consisting of three Ti:Sapphire lasers pumped by a commercial 12 kHz Nd:YAG laser.

Of crucial importance in developing a laser ion source is the ability to efficiently ionise the element of interest. The bottle neck of the ionisation process is often the final step, either non-resonant or resonant, due to the low absorption cross sections. In order to provide the laser power required for achieving an efficient ionisation of Tc, a modified version of the Ti:Sapphire lasers in use has been developed. This involves pumping of the Ti:Sapphire crystal on both sides in order to achieve a higher population inversion in the active medium. Figure 1a shows a schematic of the double-pumping principle. The crystal is positioned between the two curved mirrors (CM) within the z-shaped cavity. The Nd:YAG pump laser is focussed onto the crystal through the rear of these mirrors. Figure 1b illustrates the conversion efficiency for the double-sided pumping cavity compared to that of a standard single-sided laser. The energy of the pump laser was varied by adjusting the repetition rate from 7 to 10 kHz in steps of 0.5 kHz. The circles and squares correspond to the output energy obtained by pumping the crystal from the front and rear side respectively. The

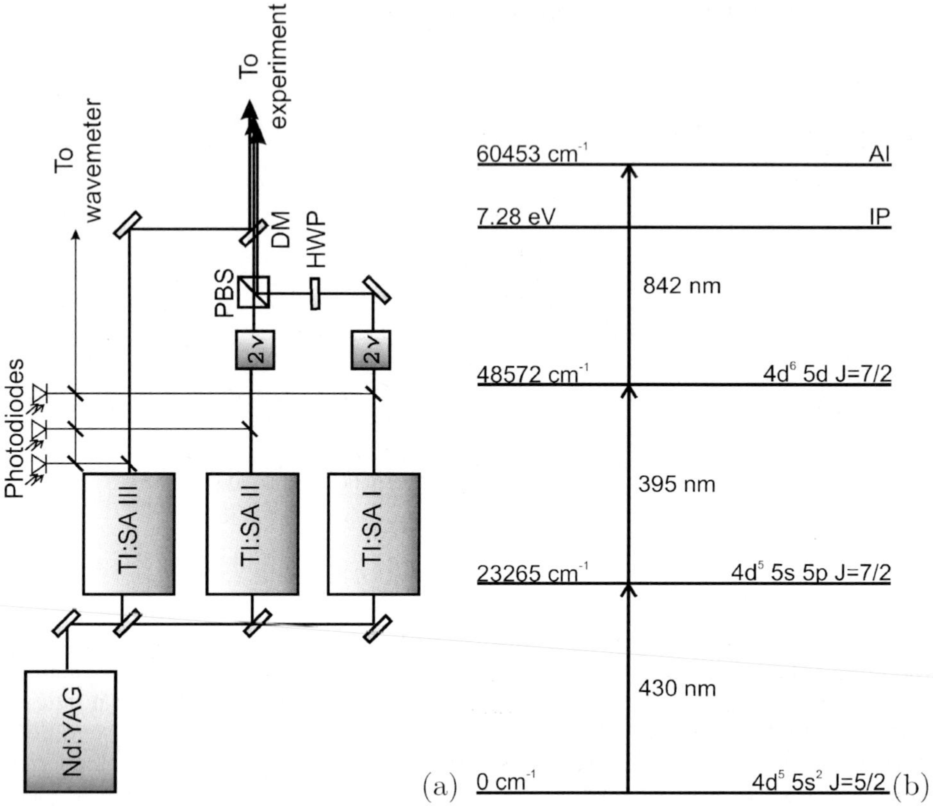

(a) (b)

Fig. 2 **a** Laser system used for resonance ionisation of Tc. **b** Relevant atomic energy levels in Tc and the transitions used in this work

triangles correspond to the output pulse energy obtained when pumping both sides at the same time. A linear fit of the form

$$E_{out} = (E_{in} - E_0) \cdot m \tag{1}$$

was applied to the data, where m is denoted as the slope efficiency and E_0 the lasing threshold. The extracted coefficients are shown in the figure.

The result of the fitting shows an increase in the slope efficiency for the double-sided pumping with a maximum value for the output energy of 0.7 mJ/pulse. This value can not be obtained with single-sided pumping due to the damage threshold of the Ti:Sapphire crystal. The slope-efficiency for the front-side is higher than for the back-side since the overlap of the pump-beam with the laser-mode is optimized for this side.

A comparison between the double-sided pumped cavity with the sum of the two single-sided cavities (indicated by stars in Fig. 1b) suggests that the double-sided Ti:Sapphire laser is more efficient [8]. This can be explained by the fact that the same threshold has to be surpassed for both the single-sided and the double-sided cavities. Consequently the inversion gain due to back-side pumping directly contributes to the

 Springer

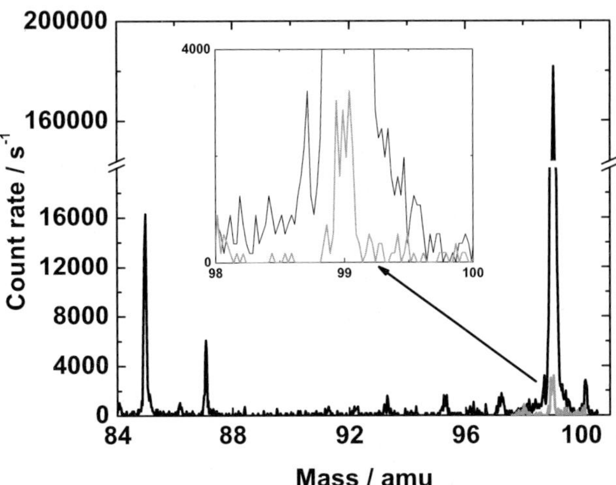

Fig. 3 Mass scan from 84 to 100 amu. The surface ionised isotopes ^{85}Rb and ^{87}Rb as well as the laser ionised ^{99}Tc appear in the spectrum. The insert shows the mass scan in the region of 99 amu with lasers off. Note the *y*-axis break

laser process. During a high power test up to 5 W output power was obtained with a total input power of 28 W at a typical repetition rate of 10 kHz.

2.3 Laser setup for Tc

A schematic of the laser system used for the resonant ionisation of Tc is shown in Fig. 2a, while Fig. 2b shows the ionisation scheme developed by the University of Mainz.

The excitation of the first and second steps at 429.8 and 395.2 nm, respectively, was achieved using two BBO crystals (2ν) to generate the second harmonic frequency. Both laser beams were overlapped spatially using a polarizing beamsplitter cube (PBS) combined with a half-wave plate (HWP). The final ionisation step of 841.7 nm was realised using the fundamental wavelength of the double-sided Ti:Sapphire laser. The infrared light is then combined with the second harmonic light using a dichroic mirror (DM). Temporal synchronisation of all lasers was achieved using Pockels cells serving as q-switches in the laser cavities. The laser powers measured at the entrance window of the ion guide were as follows: 70 mW for the first step; 130 mW for the second step; 1.4 W for the ionisation step. The power available for the first two steps exceeds that needed to saturate the transitions (Wies, personal communication). As no spectroscopic investigations have been carried out for the third step it is unknown as to whether this step is saturated.

3 First laser ions of ^{99}Tc

Following the alignment of the lasers into the ion guide the filament was continuously heated to provide an atomic vapour of Tc. At a current of 18 A corresponding to a filament temperature of $\approx$ 1,400 °C, a first Tc laser ion signal was observed. A maximum count rate of 180,000 counts/s was obtained after optimizing the laser wavelengths, spatial overlap and IGISOL ion beam tuning. Figure 3 shows a mass scan in the range of 84 to 100 amu. A key parameter for a laser ion source is the

selectivity, determined by measuring the ratio of laser ions to non-resonant ions. In this work a selectivity of ≈ 56 was observed. Non-resonant surface ions were observed in the mass scan due to the high temperature needed to evaporate the Tc sample from the Re substrate. As expected, the alkali metal isotopes ^{23}Na, ^{39}K, ^{41}K, ^{85}Rb and ^{87}Rb were all observed. The non-resonant signals observed at positions of 92–101 amu most likely arise from molecular contaminants formed from impurities in the buffer gas.

4 Conclusion and outlook

An ion guide has been built for efficient and selective laser ionisation of exotic refractory radioisotopes at the IGISOL facility, Jyväskylä, using the FURIOS laser system. To provide a more efficient laser ionisation, the solid-state Ti:Sapphire lasers have been upgraded to enable a double-sided pumping technique. This has resulted in a significant increase in the total output power and therefore a more readily achievable ability to saturate resonant transitions. The first off-line laser ion signal of the long-lived ^{99}Tc isotope has been observed at IGISOL using a three step excitation scheme.

In the near future laser ionisation of ^{99}Tc will be attempted within the sextupole ion guide, incorporating the so-called LIST principle [9]. Furthermore an efficiency measurement for the laser ionisation process will be carried out, with a view towards a first on-line test towards the end of the year.

Acknowledgements The work has been supported by the EU within the 6th framework programme Integrating Infrastructure Initiative – Transnational Access, Content Number: 506065 (EURONS). We also acknowledge support from the Academy of Finland under the Finish Centre of Excellence from 2000–2005 (Project no 44875), Nuclear and Condensed Matter Physics Program at JYFL). BT acknowledges the Marie Curie Foundation (EC).

References

1. Penttilä, H., et al.: Eur. Phys. J. **25**, 745 (2005)
2. Äystö, J.: Nucl. Phys. **A693**, 477 (2001)
3. Moore, I.D., et al.: J. Phys. G: Nucl. Part. Phys. **31**, 1499 (2005)
4. Towner, I.S., Hardy, J.C.: Phys. Rev. **C71**, 035501 (2005)
5. Kessler, T., Moore, I.D.: A study of the laser ionization of Yttrium at the IGISOL facility. (in press)
6. Trautmann, N., Folger, H.: NIM B **A282**, 102 (1989)
7. National Nuclear Data Center, Brookhaven National Laboratory: Nuclear Wallet Cards 7th edition. http://www.nndc.bnl.gov/wallet/wccurrent.html. Cited 24 May 2006 (2006)
8. Kessler, T.: Diploma Thesis. University of Mainz, Germany (2004)
9. Blaum, K., et al.: NIM B **204**, 331 (2003)

Hyperfine Interact (2006) 171:127–134
DOI 10.1007/s10751-006-9493-0

TRIUMF resonant ionization laser ion source

Ga, Al and Be radioactive ion beam development

**E. J. Prime · J. Lassen · T. Achtzehn · D. Albers ·
P. Bricault · T. Cocolios · M. Dombsky · F. Labrecque ·
J. P. Lavoie · M. R. Pearson · T. Stubbe · N. Lecesne ·
Ch. Geppert · K. D. A. Wendt**

Published online: 15 February 2007
© Springer Science + Business Media B.V. 2007

Abstract The range of isotopes available at the TRIUMF Isotope Separator Accelerator (ISAC) facility has been greatly enhanced by adding a Resonance Ionization Laser Ion Source (RILIS). A large wavelength range is accessible with the fundamental, second and third harmonic generation of titanium-sapphire laser light. In addition a dedicated laser is available for non-resonant laser ionization. The first on-line beam ^{62}Ga was delivered in Dec. 2004. In general RILIS improves the intensity, purity and emittance of ion beams. ^{62}Ga and ^{26}Al and Be beams have been delivered so far on-line.

Key words laser ion source · radioactive ion beam · resonance ionization · titanium-sapphire laser · TRIUMF-ISAC · Ga · Al · Be

PACS 28.60.+s Isotope separation (on-line) · 42.55.-f Ti:Sa laser · 32.80.Rm Multiphoton ionization and excitation · 29.25.Ni Resonance ionization laser ion source · 42.62.Fi Resonance ionization mass spectroscopy

This work was financed by TRIUMF which is federally funded via a contribution agreement through the National Research Council of Canada.

E. J. Prime · J. Lassen (✉) · T. Achtzehn · D. Albers · P. Bricault · T. Cocolios · M. Dombsky · F. Labrecque · J. P. Lavoie · M. R. Pearson · T. Stubbe
TRIUMF-ISAC, 4004 Wesbrook Mall, Vancouver, BC, V6T 2A3, Canada
e-mail: lassen@triumf.ca

N. Lecesne
GANIL BP 55027, 14076 Caen, Cedex 5, France

Ch. Geppert · K. D. A. Wendt
Institut für Physik, Johannes Gutenberg-Universtät Mainz, 55099 Mainz, Germany

1 Introduction

For the study of isotopes far from stability intense and pure beams are needed for experiments. At the TRIUMF Isotope Separator Accelerator (ISAC) facility, a range of radioactive ion beams have been produced with the ISOL technique and the use of a surface ion source since 1998 [1]. By adding a Resonance Ionization Laser Ion Source (RILIS) to the setup it is possible to efficiently ionize elements with ionization potentials above 6 eV. In resonant laser ionization the valence electron is excited by resonant photon absorption via several intermediate steps into the continuum. The last excitation is either via resonant excitation into a high lying Rydberg state, via an auto-ionizing state, or non-resonant.

The TRIUMF Resonant Ionization Laser Ion Source TRI LIS uses all solid state lasers, with currently up to three titanium-sapphire (Ti:Sa) lasers of the proven Mainz University design [2]. In addition to the fundamental, second and third harmonic laser light has been generated [3]. For optimum ionization efficiency all steps in a scheme should be resonant. However, many of the successful dye laser schemes have employed copper vapour lasers for non-resonant ionization in the last excitation step [4]. The addition of a frequency doubled $Nd:YVO_4$ (Vanadate) based laser unit with a 35 ns pulse length allows access to many of these dye laser schemes. This system at hand produces 6 W at 532 nm and 10 kHz repetition rate. In addition to extending the range of possible ionizable elements, a RILIS results in significantly increased ionization efficiency for many elements already accessible with the surface source. For ^{26}Al beams at ISAC the surface ion source efficiency is 0.6%, whilst with TRILIS 13% total ionization efficiency was achieved. Laser ionization also has the advantage that the ionization volume tends to be tightly defined when compared with other ion sources which can result in a reduced angular distribution of the extracted ion beam [5]. Finally a RILIS moves the complexity of the ion beam production away from the high radiation area of the isotope production. Outlined are the TRILIS setup and the progress made therewith in radioactive ion beam development since the commissioning of the on-line laser ion source at ISAC in Dec. 2004.

2 ISAC setup

TRIUMF's H$^-$ driver cyclotron produces a continuous beam of 500 MeV protons with a beam current up to 100 μA delivered to ISAC [6]. These protons bombard an isotope production target in one of two target stations. The targets are of varied composition depending on the desired ion beam [7] and are either resistively heated (standard targets), or radiatively cooled (high power targets) according to the proton current in order to keep the temperature of the target at around 2,000°C. Radioactive species produced from nuclear reactions diffuse out of the target and are transferred by effusion to the surface ion source. This is a 3 mm diameter and 40 mm long Ta ionizer tube, which is closely coupled to the target for fast extraction. The laser ion source makes use of this hot cavity. Whilst the high temperature reduces sticking times and aids the fast transfer of atoms through the tube, which is particularly beneficial for short lived isotopes, it also provides spatial confinement of the atomic radioisotopes in the laser interaction region sufficiently long enough for the lasers to induce ionization. Atoms are typically in this region for 0.1 ms. Therefore a laser pulse repetition rate of at least 10 kHz is required for efficient ionization. A

 Springer

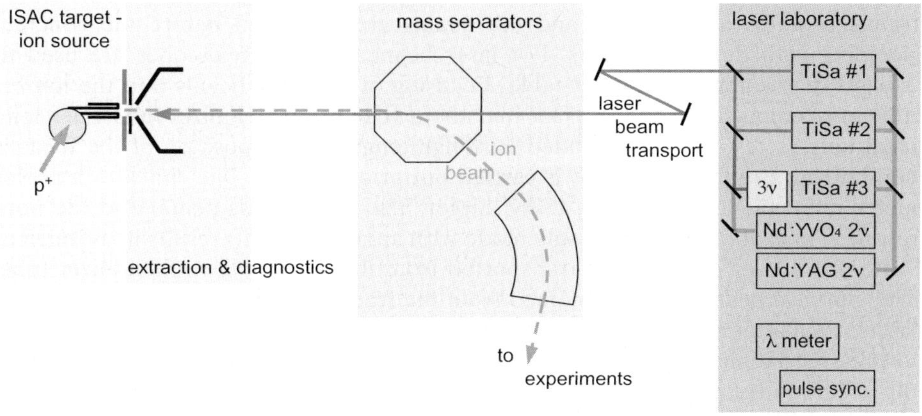

Fig. 1 Diagram of the target ion source system, extraction, mass separator and laser system at TRIUMF

fixed geometry multi-electrode extraction system extracts the ion beam with energies up to 60 keV out of the ionizer tube. In order to extract the isotope of interest from any contaminant background the ion beam enters a magnetic mass separator. Electrostatic ion optics is used for beam transport to the yield station, as well as to the low energy experimental areas or the accelerators. A diagram of the target, ionizer tube and extraction setup is shown in Fig. 1.

3 Laser system

TRI LIS uses an all solid-state laser system in order to achieve reliable operation. It is a pulsed system in order to achieve the power density required to achieve ionization. The pump laser is a frequency doubled, diode pumped, Nd:YAG laser. With up to 50 W power (multimode) all three Ti:Sa lasers can be pumped simultaneously from the same pump laser. Operating at 10 kHz repetition rate, 10–15 W of pump power typically result in 2–3 W output from the single sided pumped Ti:Sa laser and about 4 W from the double sided pumped Ti:Sa (DS-Ti:Sa) laser [8, 9]. Wavelength scanning is achieved with an intra-cavity three plate birefringent filter (BRF) and a motor controlled etalon. Typically the Ti:Sa line width is 6 GHz. The laser wavelengths achieved so far range from 695–990 nm. Second and third harmonic frequency generation is possible with two dedicated units. This adds the wavelength ranges 350–490 and 235–325 nm. Doubling and tripling efficiencies increase with light power and beam quality. The tripling units produce 10 mW from 1 W of fundamental light, increasing to 40 mW when applying 1.7 W of fundamental light. The laser wavelengths are monitored with a 10^{-6} precision wave meter which is calibrated to a wavelength standard, polarization stabilized HeNe laser.

4 TRI LIS setup & beam development

The on-line laser lab is located in a quasi clean-room. The laser beams are transported over a distance of 20 m to the target ion source. Three separate laser beam

transport paths are installed and the beams are combined before the concrete shielding with dichroic mirrors. For laser beam transport telescopes are used to expand the beams on the laser table. Focusing of the laser beams into the ionizer tube is accomplished with lenses located about 10 m from the ionizer tube. Such long focal lengths result in an extended Rayleigh length so the position of the focus is non-critical. Primary alignment is carried out prior to an on-line run with a power meter mounted in place of an ISAC target. The Ti:Sa fundamental and the non-resonant ionization laser are single mode with an M^2=1.3. This results in as much as 90% of the laser power being transported from the laser table to the ionizer tube. The elliptical beam profiles obtained from the frequency doubled and tripled Ti:Sa lasers, even after re-shaping allow for about 60% transport efficiency. It is important to achieve spatial and temporal overlap of the various laser beams. The reflections of the beams off the pre-separator laser port are used as reference spots for maintaining alignment. They are monitored with CCD cameras remotely, as the mass separator area during on-line beam delivery is located in an inaccessible radiation zone. For Ti:Sa laser pulse synchronization two methods have been employed so far: (1) passive, which involves varying the individual laser gain by adjusting the pump power. This, however, can lead to significant reduction in laser output power, and (2) active, by q-switching with Pockel cells. An applied voltage on a Pockel cell within the laser resonator will rotate the polarization of light and thus act as a variable attenuator.

4.1 Gallium two-step resonant ionization

The ISAC physics program priorities as decided on by the TRIUMF Experimental Evaluation Committee determine the radioactive ion beam development. The first task for TRILIS was the enhancement of ^{62}Ga beam produced by the surface ion source. ^{62}Ga is a super allowed β-emitter for which a precision lifetime and branching ratio measurement was needed [10]. The selectivity of resonant laser ionization was important as the conventional high resolution mass separator could not resolve ^{62}Ga and the main contaminant ^{62}Cu. Whilst Ga (IP=5.9993 eV) surface ionizes better than Cu (IP=7.7264 eV), the latter is more abundantly produced in the target and the Ga is masked in the tail of the Cu mass peak. The added selectivity of TRILIS combined with a reduction of surface ionization by the removal of a high work function liner in the ionizer tube resulted in an improved Ga beam purity. With a Ta ionizer tube heated to 2,000°C a 100x Cu background suppression was achieved. The Ga laser excitation scheme involves a two-step resonant excitation from the ground state to a Rydberg-state, followed by electric field ionization as shown in Fig. 2. ^{62}Ga has a fine structure splitting of the ground state and at 2,000°C atoms are thermally distributed between these states. The transition to the first excited state requires frequency tripled light, this was provided with a frequency tripler prototype using LBO and BBO crystals. The specific Ti:Sa fundamental wavelength required for the second step in the scheme was determined by off-line resonance ionization spectroscopy carried out at the University of Mainz [11]. An atomic energy level search was carried out because for efficient ionization it is necessary to excite to a high lying Rydberg state which is easily populated as well as field ionized. For the chosen $n = 40$ state the critical electric field for ionization is 30 V/mm. At the exit of the ionizer tube just before the ground electrode, where the ionization takes place,

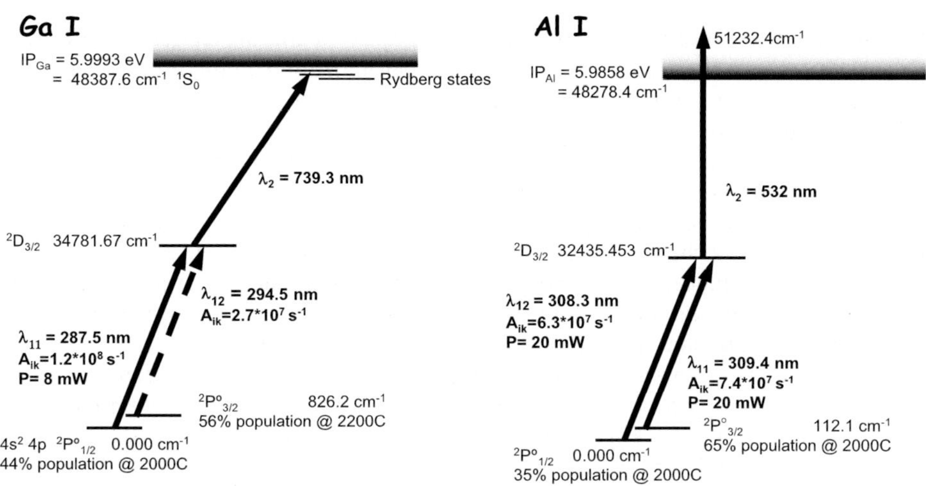

Fig. 2 Ti:Sa laser excitation scheme used for on-line Ga and Al resonant laser ionization

the high voltage from the extraction electrode penetrates and provides the electric field. Additionally in this region the ions are extracted at the full extraction potential so have maximum beam energy. This scheme was tested off-line at ISAC in Nov. 2004 [5] and the first on-line laser ions were achieved at ISAC in Dec. 2004 [11]. A ZrC target was bombarded with 35 μA of proton beam and a short stack of target foils were used for better effusion as ^{62}Ga has a half life of $\tau_{1/2}$=116 ms. A subsequent measurement resulted in a ^{62}Ga yield at the experiment (8π detectors) of 5,000 ions/s for a duration of 1 week. This corresponded to a factor of 28 enhancement over surface ionization and a suppression of Cu background resulting in a ratio of ^{62}Ga to ^{62}Cu of 6.5. The ionization efficiency is measured by knowing the target chemistry for a given target temperature and calculating the expected production yield in target. Comparing this result with the actual yields measured after the mass separator gave a total ionization efficiency of 26% for ^{62}Ga. Ion yield as a function of laser power proves that the first (UV) laser excitation step saturates the transition with the 8 mW available, whereas the second excitation step in the scheme was not saturated with the 1.65 W available. Ion yield is maximized by optimizing the combination of laser intensity and ionization volume.

4.2 Aluminum (1, 1′) resonance ionization

The (p,γ) cross sections for ^{26g}Al ($\tau_{1/2}$= 7.1*10^5 y) as a primordial isotope are important to nuclear astrophysics [12]. In such a cross section measurement 1 particle nA of ^{26}Al beam would result in approximately 10 reaction events per day. Such an intense ^{26}Al beam was requested. A one step resonant, one step non-resonant (1,1′) laser excitation scheme as shown in Fig. 2 was used where the ionization step is provided by a dedicated frequency doubled Nd:YVO$_4$ (Vanadate) laser. Al also has a fine structure splitting of the ground state and the thermal distribution at 2,000°C puts 65% of the electronic population in the $^2P_{3/2}$ state. This scheme was tested off-line and in the first on-line run 70 μA of protons were put on to a newly developed

high power target with SiC as target material [13]. Al beams are difficult to produce, due to the target chemistry where Al release from Si and C material is inhibited [Dombsky, private communication]. A second run resulted in an increase in laser ionized yield by the successful addition of the second first step transition from the ground state (35% population). A new frequency-tripling unit provided the UV light. The beam purity was improved further by using the mass separator to separate off a portion of the abundant ^{26}Na isobar. With the increased Al yield it was possible to employ several techniques to further suppress the surface ionized Na background. Firstly, the mass separator was slightly tuned off the ^{26}Al peak, this obviously resulted in a loss in Al (by a factor of 2), but it also resulted in a greater loss in isobaric Na. The superior beam emittance of the laser ions can be employed to further improve the level of isobar suppression. Despite resulting in lower over all transmission, the slits after the mass separator were narrowed in order to reduce the number of Na surface ions which have a wider ion beam profile. All these measures resulted in the acceleration of 4 particle nA of ^{26}Al, and a yield of $5*10^{10}$ ions/s (as measured at the ISAC yield station) for a duration of 3 weeks. The total ionization efficiency for ^{26}Al was 13%. A 20-fold enhancement over surface ionization was achieved with simultaneous suppression of isobaric Na background resulting in a ratio of ^{26}Al to ^{26}Na of $1*10^4$. Ion yield as a function of laser power shows that both of the first step UV transitions are well saturated with around 10 mW for each excitation. The non-resonant ionization step proved to be the bottleneck.

4.3 Beryllium (2, 1′) resonance ionization

Be is of major interest to theorists and experimenters alike as it allows the extension of precision measurements and theory beyond Li. Although production of Be is high with 35 μA of protons on a Ta target, the IP of 9.32 eV precludes significant levels of surface ionization. Two different (2,1′) excitation schemes were tested, the second one in preparation for Rydberg state spectroscopy. These schemes are shown in Fig. 3. The 234 nm laser light for the first excitation step was derived from frequency tripling a Ti:Sa laser operating at 702 nm, the laser wavelength of 698 nm was provided by the TRIUMF DS-Ti:Sa laser. For non-resonant ionization again the frequency doubled Nd:YVO$_4$ (Vanadate) laser was used. This resulted in an excitation to only 487 cm^{-1} above ionization threshold. Unfortunately this 698 nm transition is weak ($A_{ik}=10^5$ s^{-1}). In the second scheme a transition at 457 nm (frequency doubled Ti:Sa laser) with an $A_{ik}=10^7$ s^{-1} was used. The non-resonant step excites to 8033 cm^{-1} above ionization threshold, which should result in a lower ionization efficiency. However, the orders of magnitude of the various transitions strengths appear to cancel out as these two schemes both resulted in an ionization efficiency of 3%, which is almost identical to the ionization efficiency quoted for this scheme using dye lasers [4]. The UV step is saturated with the 16 mW produced. In order to synchronize the pulses, the pump power to the DS-Ti:Sa, was adjusted. In this case the use of the passive method for timing control resulted in power loss for the DS-Ti:Sa. The two Ti:Sa's in use were at two very different wavelengths represent two very different points on the Ti:Sa gain curve. In order to bring them to a similar gain the pump power to the DS-Ti:Sa was greatly reduced. Combined with power loss due to achieving the correct wavelength by tuning the etalon resulted in 1.3 W at 698 nm and only 160 mW at 457 nm, consequently the second step in either of the

 Springer

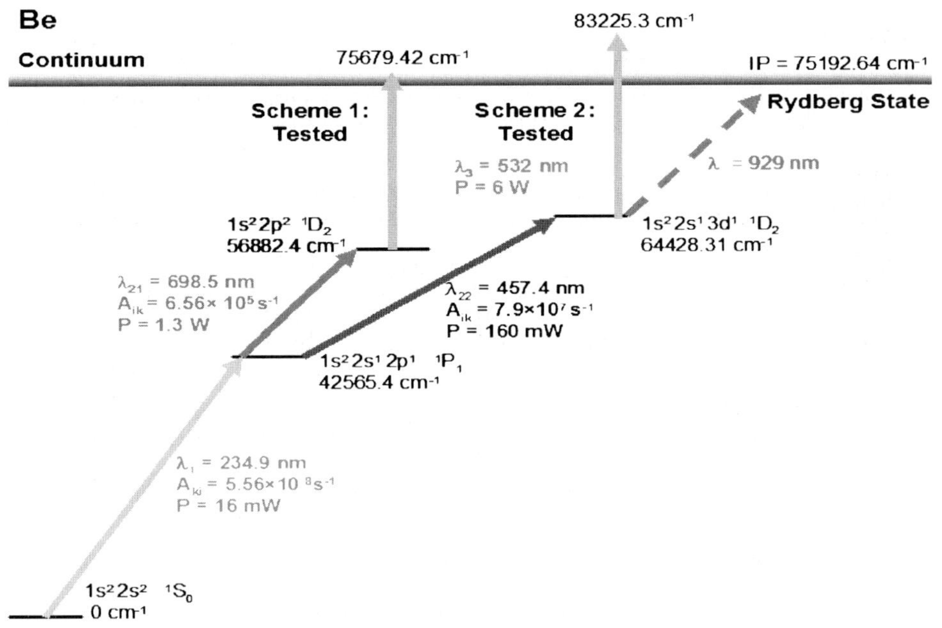

Fig. 3 Ti:Sa laser excitation scheme used for on-line Be resonant laser ionization. The measured ionization efficiencies were identical

schemes was not saturated. Again it was not possible to saturate the non-resonant steps. The laser wavelengths were tuned to compensate for isotope shifts and all the Be isotopes from mass 9 to mass 12 were laser ionized. In order to try and improve on the achieved ionization efficiency an all resonant excitation scheme will be tested. The second scheme will be used and a final resonant Ti:Sa fundamental step will be scanned to find a suitable Rydberg or auto-ionizing state. The Be isotopes are easily separated with the mass separator and its resolution is sufficient to separate out the Li isotopes which are the only major contaminant due to efficient surface ionization. Mass selectivity is rarely an issue with the lightest elements as the binding energy represents a large fraction of the total energy. Therefore no other means of suppression of surface ionization is required.

5 Conclusions and outlook

The resonance ionization laser ion source setup for the west target station at ISAC has been completed and is continually being improved. The subsequent radioactive ion beam development program has lead to the successful on-line production of ^{62}Ga, ^{26}Al, $^{9-12}$Be laser ions. In the second half of the year 2006 laser q-switching will be implemented.

Acknowledgements The contributions from the collaboration with the research group of Dr. K.D.A. Wendt from the University of Mainz are key to the rapid success of Ti:Sa laser based resonance ionization.

 Springer

References

1. Dombsky, M., Bishop, D., Bricault, P., Dale, D., et al.: Rev. Sci. Instrum. **71**, 978 (2000)
2. Horn, R.: Aufbau eines systems gepulster, abstimmbarer Festkörperlaser zum Einsatz in der Resonanzionisations-Massenspektrometrie. Dissertation, Johannes Gutenberg-Universität Mainz (2003) (in German)
3. Sirotzki, S.: Konzeption und Aufbau einer Frequenzverdreifachungseinheit für ein gepulstes Titan-Saphir Lasersystem, Staatsexamensarbeit, Johannes Gutenberg-Universität Mainz (2003) (in German)
4. Köster, U., Fedoseyev, V.N., Mishin, V.I.: Spectrochim. Acta, B **58**, 1047 (2003)
5. Rauth, Ch., Geppert, Ch., Horn, R., Lassen, J., et al.: Nucl. Instrum. Methods Phys. Res., B **215**, 268–277 (2004)
6. Baartman, R., Bricault, P., Bylinsky, I., Dombsky, M., et al.: Proceedings of the Particle Accelerator Conference, 1584 (2003)
7. Dombsky, M., Bricault, P., Schmor, P., Lane, M.: Nucl. Instrum. Methods, B **204**, 191 (2003)
8. Kessler, T.: Optimierung eines Ti:Saphir-Lasersystems für den Einsatz an einer On-line-Ionenquelle, Spektroskopie an Zinn. Diploma Thesis, Johannes Gutenberg-Universität Mainz (2004) (in German)
9. Albers, D.: Double sided pumped Titanium Sapphire laser system. Diploma thesis, Fachhochschule, University of Applied Sciences, Oldenburg, Ostfriesland, Wilhelmshaven (2007)
10. Hyland, B., Svensson, C.E., Ball, G.C., Leslie, J.R., et al.: Phys. Rev. Lett. **97**, 102501 (2006)
11. Geppert, Ch.: Resonanzionisation zum Nachweis und zur Erzeugung radioactiver Ionenstrahlen: Vom hochselektiven Ultraspurennachweis zur selektiven on-line Laserionenquelle. Dissertation, Johannes Gutenberg-Universität Mainz (2005) (in German)
12. Ruiz, C., Parikh, A., José, J., et al.: Phys. Rev. Lett. **96**, 252501 (2006)
13. Bricault, P.G., Dombsky, M., Schmor, P.W., Dowling, A.: Proceedings of the Particle Accelerator Conference, 439 (2003)

Hyperfine Interact (2006) 171:135–141
DOI 10.1007/s10751-006-9494-z

Resonance ionization spectroscopy of bismuth at the IGISOL facility

I. D. Moore · T. Kessler · J. Äystö · J. Billowes ·
P. Campbell · B. Cheal · B. Tordoff · M. L. Bissel ·
G. Tungate

Published online: 13 February 2007
© Springer Science + Business Media B.V. 2007

Abstract A programme of research has commenced at the IGISOL facility, Jyväskylä, combining the technique of resonance ionization spectroscopy with the development of the highly selective laser ion source trap (LIST). The first element of interest is bismuth, which contains three isomers of multi-quasiparticle states in near-spherical nuclei, namely ^{207}Bi $(21/2^+, 182\ \mu s)$, ^{204}Bi $(10^-, 13\ ms)$ and ^{204}Bi $(17^+, 1.07\ ms)$. A measurement of the optical isomer shift provides a direct comparison of the mean–square charge radii between the isomer and the nuclear ground state. Due to the short isomer lifetimes the spectroscopy will be done either within the ion guide or in a sextupole ion beam guide (SPIG), located after the ion guide and used in the development of the LIST. A mixed dye-Ti:Sapphire laser ionization scheme has been successfully tested for bismuth and first off-line results have been obtained.

Key words resonance ionization spectroscopy · bismuth · laser ion source

1 Introduction

In recent years a general feature of multi-quasiparticle (qp) isomers has been revealed through measurements of the nuclear charge radii at the laser-IGISOL facility: all isomeric states are found to have a smaller nuclear mean-square charge radius than their respective ground state, even though they are not less quadrupole-deformed [1]. This property was first observed by Boos et al. [2] for the celebrated 16^{+178}Hf^{m2} isomer. This effect is most

I. D. Moore (✉) · T. Kessler · J. Äystö
Department of Physics, University of Jyväskylä, P.O. Box 35 (YFL), 40014 Jyväskylä, Finland
e-mail: iain.moore@phys.jyu.fi

J. Billowes · P. Campbell · B. Cheal · B. Tordoff
Nuclear Physics Group, Schuster Building, University of Manchester, Brunswick Street, Manchester
M13 9PL, UK

M. L. Bissel · G. Tungate
School of Physics and Astronomy, University of Birmingham, Edgbaston, Birmingham B15 2TT, UK

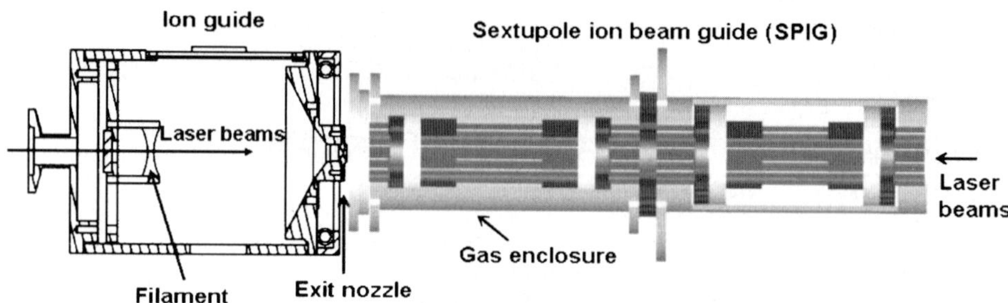

Fig. 1 At the JYFL IGISOL facility, resonance ionization spectroscopy is achieved either within the ion guide, or within the volume of the SPIG. Laser beams either enter from the *left* or through a window on the mass separator, not shown, on the *right*

likely either due to a decrease in the dynamic quadrupole contribution to $<r^2>$ or due to a reduction in the diffuseness of the nuclear surface. The relative importance of the changes in rigidity and pairing are difficult to assess from the current data that only cover quadrupole-deformed states. However both effects have been suggested in explanations of the odd-even staggering of nuclear charge radii, where the orbital blocking by the odd quasiparticle could reduce pairing correlations and increase the rigidity of the nuclear shape.

In order to separate the two effects it is planned to measure the mean-square charge radius variation of multi qp isomers in spherical nuclei, where any change of shape in the isomer can only increase the rms radius. This will be done by measuring the optical isomer shifts which are directly proportional to these variations. Bismuth is an element well-known to the collaboration (see [3] and references therein). The proximity to the doubly-magic ^{208}Pb isotope means the bismuth core will be spherical, and there are a few excellent cases in the system that will be attempted at the IGISOL facility, Jyväskylä. The ^{207}Bi $(21/2^+)$ 3-qp isomer has a lifetime of only 182 μs, however it has been extracted from the IGISOL previously with a usable efficiency of 0.5% [4]. Two other, slightly less challenging, isomers are of interest: ^{204}Bi $(10^-, 17^+)$ isomers with lifetimes of 13 and 1.07 ms, respectively. The former is a 2-qp state and the latter a 4-qp state. With the ability to perform laser spectroscopy effectively at the source of the production of exotic nuclei, these measurements will open up a new field of research of short-lived isotopes and isomers at the IGISOL facility.

2 Experimental development

The recent installation and on-going development of the laser ion source in Jyväskylä [5] in order to improve the efficiency and selectivity of the IGISOL technique has led to the possibility of measuring very short-lived isotopes and isomers. One technique being developed to achieve high selectivity is that of the Laser Ion Source Trap (LIST) method [6], illustrated in Fig. 1. Lasers enter through a window on the mass separator and selectively re-ionize the neutralized nuclear reaction products after they exit the ion guide within the gas flow. The laser-produced ions are captured within the volume of a RF Sextupole Ion Beam Guide (SPIG) and are guided towards the mass separator. By repelling any non-neutral fraction at the entrance to the SPIG, any produced ion transported to the extraction and acceleration stage of the mass separator is guaranteed to be a resonantly produced ion.

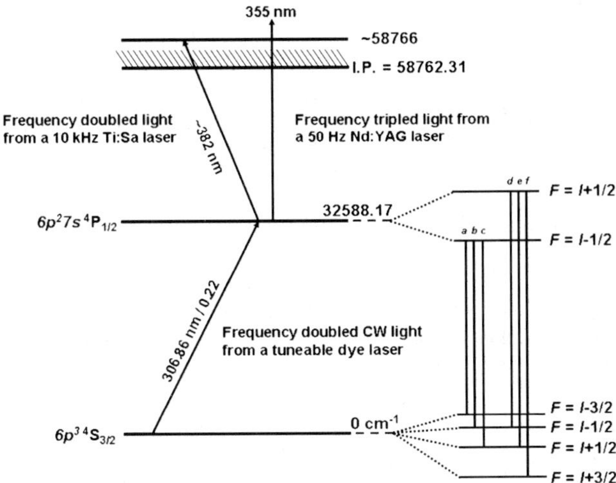

Fig. 2 Relevant atomic energy levels in bismuth and the transitions investigated in this work. The hyperfine splitting of the ground state and intermediate state is shown

The facility has the flexibility to do Resonance Ionization Spectroscopy (RIS) at the production source (within the ion guide) or in the SPIG volume. The application of RIS is not to maximize isotope production but to achieve the sensitivity needed to perform spectroscopic measurements. In addition to the broadband pulsed lasers used for the laser ion source [5], a pulsed dye amplifier (Spectra Physics) seeded by a single mode CW laser is available to produce the narrowest-possible linewidth for the pulsed laser light. It is pumped by a 50 Hz Nd:YAG laser (Spectron Laser Systems Ltd SL801) operating at frequency doubled 532 nm with a pulsewidth of 6 ns. This pump laser also has the option of running in parallel frequency tripled light at 355 nm and has been used successfully at the IGISOL facility in an earlier work [7].

The relevant atomic energy level scheme of bismuth showing the transitions investigated in this work is shown in Fig. 2. The atomic excitation energy of the levels is labeled in units of cm^{-1}, and the levels are identified in the Russel–Saunders–Coupling notation. The log-gf (transition strength) is listed next to the wavelength of the first step transition. Bismuth has a first ionization potential of 7.29 eV. Perhaps the easiest ionization scheme to investigate is a two step $\omega_1+\omega_1$ scheme, using a single laser to produce both steps at 306.8 nm. The advantage is that spatial and temporal overlap of both steps is ensured, however problems arise when the high laser power needed to ensure saturation of the ionization step is also used for the excitation step. Power broadening and smearing of the lineshapes can be extremely detrimental to a high-resolution study of hyperfine structures (shown in Fig. 2 for bismuth isotopes with a non-zero nuclear spin) and isotope shifts. Two different schemes were therefore tested, both using a single mode frequency doubled CW laser for the first excitation step which removed any concerns regarding temporal overlap. The first scheme used a non-resonant ionization step to the continuum with 355 nm from the frequency tripled 50 Hz Nd:YAG laser, and the second to a known auto-ionizing level [8] using a 10 kHz frequency doubled Ti:Sapphire laser operating at ~382 nm.

2.1 Resonance ionization spectroscopy in an atomic beam unit

An atomic beam unit, described fully in [9], was used to test the resonant ionization scheme of bismuth involving the auto-ionization state and the saturation power needed for the final

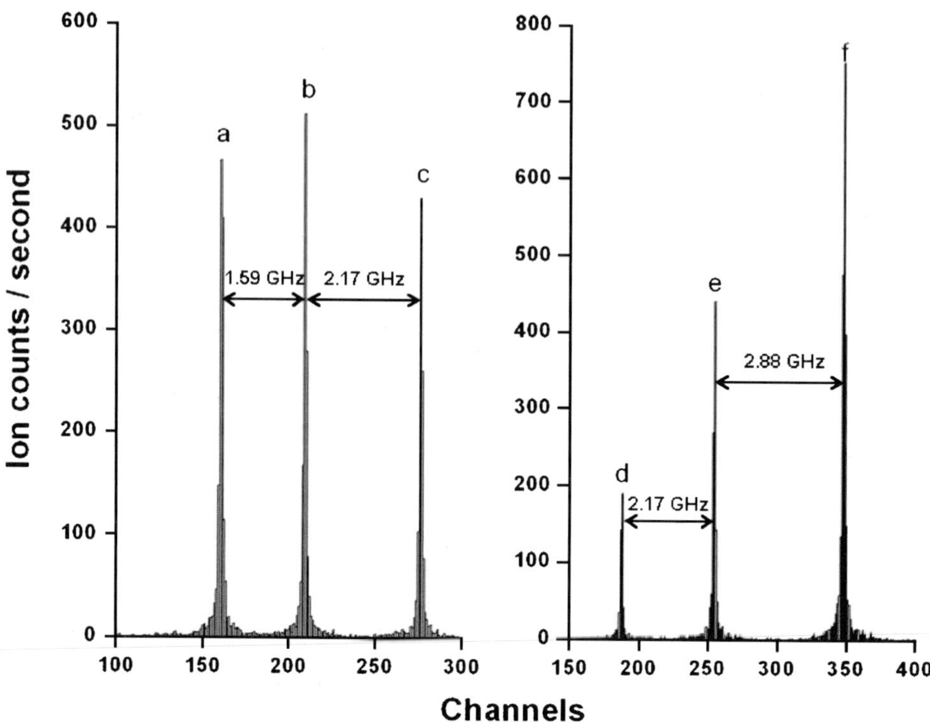

Fig. 3 A laser wavelength scan over the lower frequency triplet of hyperfine states in ^{209}Bi (*left*) and the upper frequency triplet (*right*). The peaks are labeled using the notation shown in Fig. 2

step. In this work ~400 mW of 382 nm laser light was available from the Ti:Sapphire. By measuring the ion count rate as a function of laser power it can be deduced that 400 mW measured over a beam diameter of ~2 mm provides 70% saturation of the final step. The first resonant transition step was realized using 1 mW of CW light from an intracavity doubled Spectra Physics 380D tunable dye laser. The frequency of the fundamental beam was locked to an external reference standard and was monitored by an iodine absorption cell. ^{209}Bi was loaded in powdered form in a joule-heated oven. A low pressure (0.01 mbar) bismuth vapour was produced with an oven temperature of about 670°C.

The hyperfine spectrum of ^{209}Bi is composed of six components, well-separated into two groups of three due to the large splitting of the $J=1/2$ upper atomic level (24.6 GHz). Using the notation in Fig. 2 for the hyperfine transitions, Fig. 3 shows two separate wavelength scans of the resonant step over the lower three components and upper components respectively. The splittings of the individual components are shown in GHz, with 22.4 GHz separating peak "c" from peak "d". The laser excitation probability of a transition between hyperfine substates in a two level system can be written in terms of the electronic transition probability, the lower state weighting $(2F+1)$ and angular momentum coupling coefficients. The relative intensities, called Racah intensities, are shown in Table 1 for ^{209}Bi. It is clear from Fig. 3 that the intensities are not matching the Racah intensities and work is presently being done to understand this mismatch, including the effect of the second step power which has been seen to alter the population within the hyperfine levels.

 Springer

Table 1 Relative hyperfine intensities for ^{209}Bi

Peak	a	b	c	d	e	f
Relative hyperfine intensity	7.0	6.6	4.4	2.4	6.6	13.0

2.2 Resonance ionization spectroscopy in an ion guide

The IGISOL technique is described fully in [10] and will not be discussed here. The ion guide used for the resonance ionization of bismuth is shown with the SPIG in Fig. 1. Atoms of ^{209}Bi were produced from a resistively heated Ta filament (~7 mg/cm^2 thickness), with bismuth sputtered onto the surface. The laser light entered the ion guide longitudinally through a sapphire window, ionizing neutral atoms along the axis of the cell. The diameter of the laser beam was brought to a gradual focus of approximately 2 mm near to the exit hole (1 mm in diameter). In this test up to 100 mbar of He gas was used to evacuate the ions from the cell, where they were subsequently transported through the SPIG and injected into the mass separator via stages of differential pumping. The ions were accelerated to 30 kV, mass separated, and detected on a set of microchannel plates downstream from the separator focal plane.

In order to increase the power of the first resonant step the fundamental light from the 380D CW dye laser was sent into an external frequency doubling cavity, Coherent MBD-200. This was to offset losses experienced during the laser transport to the IGISOL. A factor of three could be gained over the intracavity doubling. Two laser scans were made over the upper three hyperfine peaks within one hour of each other, at a constant filament current of 7 A. The first scan used the frequency tripled light from the 50 Hz Nd:YAG as a non-resonant ionization step. On success, the frequency doubled Ti:Sapphire was then spatially overlapped with the CW beam from the MBD-200 cavity and was used to reach the auto-ionizing level. Results from both these scans can be seen in Fig. 4.

During the scan on the left of Fig. 4 it is possible that the MBD-200 cavity temporarily dropped out of lock with the dye laser at around channels 20 and 45. This could be the reason for the sharp edges of the spectra seen. It is difficult to make a direct comparison between the two spectra as the repetition rates and hence the pulse energies of the ionizing lasers were significantly different. However, the increase in count rate by about a factor of 60 on the largest peak is almost certainly due to the higher probability of ionizing when using an auto-ionizing level compared to a non-resonant transition. In addition, spectra were taken at differing ion guide pressures to provide information about pressure broadening effects. The data are presently being analysed, however a few conclusions can be immediately drawn. The resolution achieved in Fig. 4 is clearly adequate to resolve expected hyperfine structure. The advantage of RIS in the IGISOL gas volume is that the pressure broadening helps when narrow-band laser light is used, and there is almost no Doppler shift of the line profile (only 10 m/s drift velocity inside the cell). From the off-line data, however, no conclusions can be made about the neutral/ionic fraction, or the absolute ionization efficiency. Only on-line conditions will be able to provide the important information on both selectivity and efficiency of the RIS process.

3 Conclusion and outlook

Resonance ionization spectroscopy on ^{209}Bi using a CW laser for the first excitation has been shown to be successful in an ion guide at the IGISOL facility, Jyväskylä. The effect of

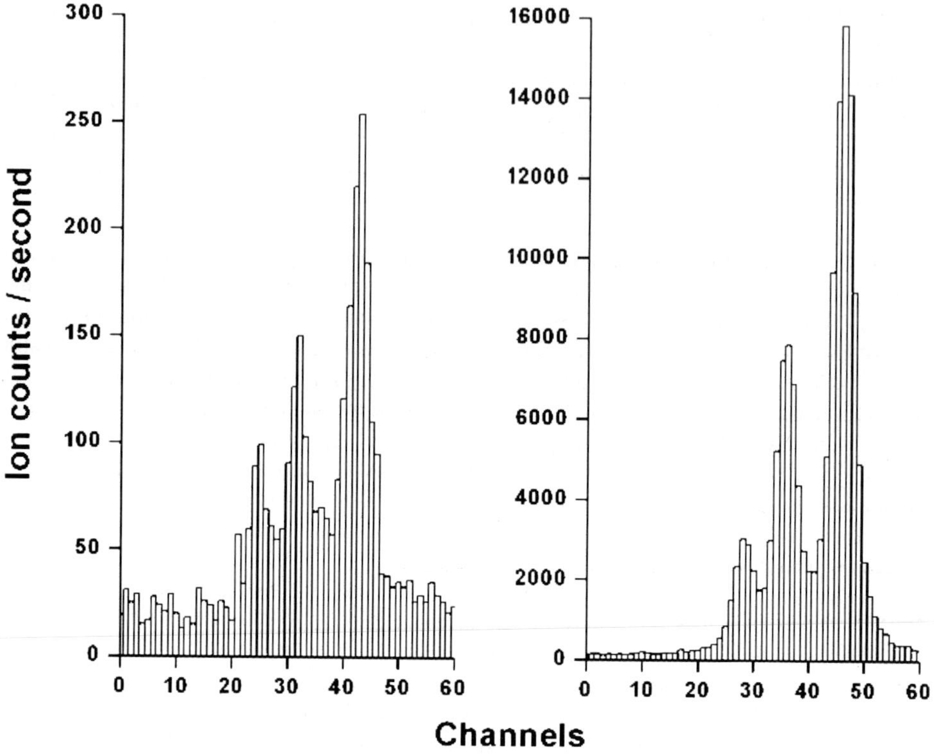

Channels

Fig. 4 Resonance ionization spectroscopy in the ion guide at 50 mbar He pressure. The CW dye laser was scanned over the upper hyperfine peaks. The *left figure* shows results obtained using the non-resonant 355 nm ionization step, while the Ti:Sapphire was used to reach the auto-ionizing state in the *figure shown on the right*

the auto-ionizing state has been clearly confirmed to be an advantage for spectroscopy in this system. The next step will be to perform the high resolution RIS within the volume of the SPIG, the lasers transported through the mass separator dipole magnet in the same manner as that of the LIST technique. The important advantage over RIS in the gas cell volume is that the resultant ion signal is clean, without any non-laser ion related background. However, it is expected that the hyperfine profiles will exhibit both Doppler broadening and Doppler shift, and there will be no pressure broadening to help with the narrow band first step. This option has been previously achieved on the bismuth system earlier this year, and the effect of the repelling voltage to ensure all detected ions are indeed created in the SPIG volume was successful. However, the low repetition rate laser system was used in this test, both for pumping the pulsed dye amplifier in order to achieve the first resonant transition and also for producing the non-resonant 355 nm. This had the disadvantage of hugely power broadening the two sets of hyperfine structures. The possibility of using the CW dye laser and external cavity to produce the first step transition was tried in a LIST mode in the present work, however was unsuccessful – believed to be due to the extremely low laser intensity finally reaching the SPIG.

In the near future the use of a high repetition rate (12 kHz) copper vapour laser (Oxford Lasers LM100X(KE)) to pump the PDA will be tested, which has the advantage of having a significantly lower energy per pulse than the Nd:YAG laser (the repetition rates are 12 kHz and 50 Hz, respectively). This will bring the added advantage of a higher power in the first transition, not provided by the CW laser, and the reduction in power broadening compared with the low repetition rate pumping. In order to study the isomer shifts of ^{207}Bi $(21/2^+)$ and ^{204}Bi $(10^-, 17^+)$ the resonance ionization spectroscopy will be performed on-line. By using a ^{209}Bi target and the reactions (p,p2n) and (p,p5n), not only will the isomers be populated but additional ^{209}Bi will be produced from sputtering of the target material which can be used as a calibration.

A future option not fully optimized at present at the IGISOL facility, is the Collinear Resonance Ionization Spectroscopy (CRIS) method [11]. Longer-lived Bi isotopes/isomers can be extracted from the ion guide in the standard manner, cooled and bunched in the radio-frequency quadrupole and released to the collinear laser spectroscopy line. By neutralizing in a charge-exchange cell, the bunched-atom beam can be re-ionized with narrow band pulsed lasers. High efficiency can be achieved by matching the release time of the ion bunch from the cooler with the low-repetition rate of the laser system (50 Hz), thereby removing any duty cycle losses. The laser resonance is located by the measurement of the ion yield (easy), rather than by photon detection (low solid angle, poor quantum efficiency).

Acknowledgements This work has been supported by the EU within the 6th Framework Program "Integrating Infrastructure Initiative – Transnational Access," Contract No. 506065 (EURONS), by the Academy of Finland under the Finnish Centre of Excellence Program 2006–2011 (Nuclear and Accelerator based physics program at JYFL) and by the UK Engineering and Physical Sciences Research Council (EPSRC).

References

1. Bissel, M.L., et al.: Phys. Lett. B (submitted)
2. Boos, N., et al.: Phys. Rev. Lett. **72**, 2689 (1994)
3. Pearson, M.R., et al.: J. Phys. **G 26**, 1829 (2000)
4. Ärje, J., et al.: Nucl. Instrum. Methods **A 247**, 431 (1984)
5. Moore, I.D., et al.: J. Phys. **G 31**, S1499 (2005)
6. Moore, I.D., et al.: In: Harissopulos, S.V., Demetriou, P., Julin, R. (eds.) Frontiers in Nuclear Structure, Astrophysics, and Reactions, FINUSTAR, AIP Conference Proceedings, vol. 831, p. 511 (2006)
7. Yeandle, G., et al.: Hyperfine Interact. **127**, 91 (2000)
8. Bühler, B., et al.: Z. Phys. **A 320**, 71 (1985)
9. Tordoff, B., Billowes, J., Campbell, P., Cheal, B., Forest, D.H., Kessler, T., Lee, J., Moore, I.D., Popov, A., Tungate, G., Äystö, J.: Investigation of the low-lying isomer in ^{229}Th by collinear laser spectroscopy (this issue)
10. Äystö, J.: Nucl. Phys. **A 693**, 477 (2001)
11. Schulz, Ch., et al.: J. Phys. **B 24**, 4831 (1991)

Hyperfine Interact (2006) 171:143–148
DOI 10.1007/s10751-006-9491-2

Optical pumping in an RF cooler buncher

P. Campbell

Published online: 25 January 2007
© Springer Science + Business Media B.V. 2007

Abstract The use of optical pumping in an RF linear Paul trap is shown to permit efficient collinear laser spectroscopy in systems where atomic ground state resonance lines are ill-suited to the fast beam technique. The case of yttrium, where laser spectroscopy has been used to study $^{86-90,92-102}$Y and isomeric states of $^{87-90,93,96,97,98}$Y, is highlighted.

Key words laser spectroscopy · radiofrequency ion traps

1 Introduction

The change in the nuclear mean-square charge radius, $\delta\langle r^2\rangle^{A,A'}$, between isotopes with mass A and A', is a fundamental quantity which may be extracted from optical isotope shift measurements with little model dependency [1, 2]. In conjunction with measurements of optical hyperfine structures, which provide information on the nuclear multipole structure, it is possible to construct a detailed picture of the nuclear ground (or isomeric) state. The fundamental feature such measurements are sensitive to is the distribution of charge within the nucleus. This distribution of charge is found to display a variety of bulk (macroscopic) and valence (microscopic) phenomena [1, 2].

Laser spectroscopy has made a huge impact in this field since it is the only method available for measuring charge radii of *radioactive* nuclei and, of all methods for stable nuclei, it is the most sensitive to small isotopic variations. Elements with favourable atomic and chemical properties have been studied over long isotope sequences which stretch almost to the edges of the chart of nuclides.

P. Campbell (✉)
Department of Physics and Astronomy, Manchester University, Manchester M13 9PL, UK
e-mail: pc@mags.ph.man.ac.uk

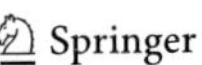

In the most common laser technique, 'collinear fast beams,' an atomic or ionic beam is overlapped with a collinear beam of counter-propagating laser light and, for fluorescence observation, a length of the overlap region is imaged onto a photo-multiplier. At JYFL, Jyväskylä, the laser-IGISOL collaboration has developed and commissioned a gas-filled RF quadrupole 'ion cooler'[3] which is positioned between the IGISOL ion source [4] and the collinear laser station. The cooler consists of a HV platform mounted linear Paul trap similar to smaller devices previously developed at McGill University and at ISOLDE, CERN (where they were primarily used as first stage 'coolers' and 'bunchers' upstream of high precision Penning traps). At the IGISOL the larger scale, general purpose, device services a variety of experimental stations. Unlike conventional isotope separators the gas-cell based IGISOL, while fast and universal, normally produces ion beams with large energy spreads and of large spatial and phase spatial size. Such beam properties are particularly damaging to the efficiency of collinear laser spectroscopy and the development of the JYFL cooler-buncher has been crucial to the success of the laser spectroscopy programme at the facility. The reduction in the energy spread of the cooled ion beams allows spectroscopy to be perform close to the natural linewidth limit in resolution.

The bunching of radioactive beams in the trap also permits a highly efficient and low background variant of collinear laser spectroscopy to be performed [3]. In this variant the signal from the photomultiplier is gated and only opened as an ion bunch traverses the interaction region. The bunching of the ion beam allows fluorescent photons to be counted at a greatly reduced background (which is otherwise domi-nated by a continuous non-resonant scattering of photons from the laser beam).

For ionic spectroscopy the cooling and bunching of the reaction products limits the optical spectroscopy to transitions from the electronic ground state (with all other atomic metastable state populations relaxed during the beam preparation). The laser excitation frequency, electronic spin, hyperfine structure and transition strength are thus limited and often one or more of these parameters is not optimal, or for some elements, fatally restricted. At JYFL we are therefore developing methods for the preparation of cooled and bunched ionic beams in states other than the ground state. The collaboration has now succeeded in efficiently manipulating state populations in ionic ensembles using optical pumping within an RF cooler buncher.

A pertinent example of the capabilities, and the limits, of non state-selected ionic spectroscopy at the IGISOL is the system of yttrium. This odd-Z isotope chain ($Z = 39$) is rich in nuclear isomers. The wealth of isomeric states provide a range of nuclear systems with non-zero spin, thus providing information on magnetic moments, and many with spins $I \geq 1$ which allows the determination of static quadrupole deformation parameters, $\langle \beta_2 \rangle$. In comparison to the neighbouring even-N chains of Sr and Zr, where many of the accessible and interesting isotones have spin 0 or spin 1/2, explorations in the yttrium chain permits a far more detailed investigation of the low-lying nuclear structure.

The nuclear structure around $Z = 39$ displays some of the most rapidly changing and varied phenomena observed anywhere in the nuclear chart. Both optical [5–7] and non-optical [8, 9] measurements on isotopes with $Z = 36 - 40$ have revealed a sudden onset of deformation at $N \approx 60$. While the existence of the onset of deformation has been established, the nature of this transition remains disputed. The behaviour of the 2^+ state lifetimes in Sr and Zr originally suggested an abrupt transition in the nuclear shape [9]. Other nuclear parameters, such as the magnetic

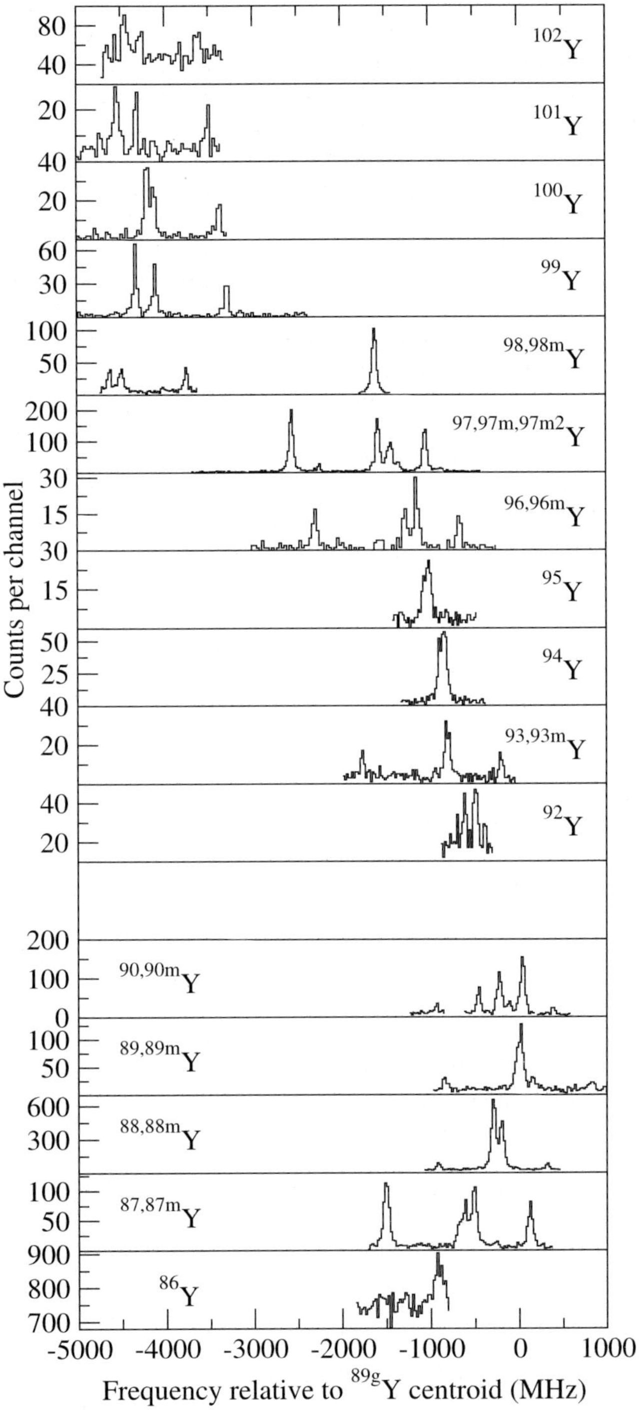

Fig. 1 Sample resonance fluorescence spectra for the studied yttrium isotopes and isomers

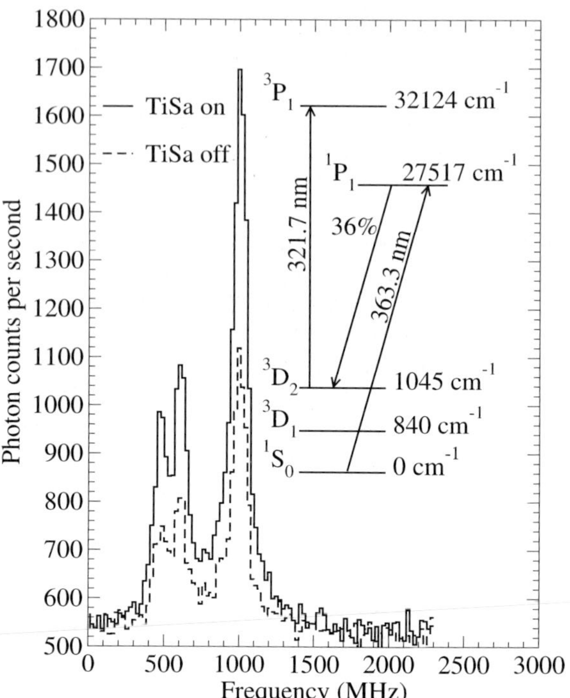

Fig. 2 Fluorescence spectrum of cooled ^{89}Y$^+$. The optical transition is excited from the second metastable state of Y$^+$ and the effect of laser optical pumping is shown

moments and existence of 0^+ first excited levels, also support the idea of a sudden transition from spherical to rigidly deformed shapes. However γ-spectroscopic studies of excited nuclear states have questioned the existence of strongly deformed excited states and have suggested that the onset of deformation is more gradual [10]. Such a deformation trend prior to the $N = 60$ shape change is supported by charge radii data from the optical measurements [5–7] in the region.

At JYFL laser spectroscopy has been used to study $^{86-90,92}$ ^{102}Y ground states and isomeric states of $^{87-90,93,96,97,98}$Y. The yttrium isotopes and isomers with mass numbers from 86 to 90 were produced using 33 MeV ^{89}Y(p,pxn)$^{86-89}$Y and 14 MeV ^{89}Y(d,pxn)^{89,90}Y reactions. The neutron-rich yttrium isotopes and isomers were obtained from 30 MeV proton induced fission of natural uranium. In both cases the recoiling reaction products were thermalised in fast flowing helium gas, electrostatically extracted and mass analysed (the IGISOL technique [4]). Typical ion fluxes of 7,000 s^{-1} for mass $A = 90$ and 3,000 s^{-1} for the $A = 98$ fission fragments were obtained. A stabilised continuous wave, frequency-doubled dye laser provided up to 0.5 mW of 363 nm UV light, which was focused and overlapped collinearly with the ion beam [3]. High resolution spectra of the $5s^2$ ^{1}S$_0$ (ground state) $\rightarrow$ $4d5p$ ^{1}P$_1$ (27,516.691 cm^{-1}) ionic transition were taken for each yttrium isotope by observing the flux of fluorescent photons as the acceleration voltage was ramped. This tuning voltage, applied to the laser-ion interaction region, acts to Doppler-shift the apparent frequency of the laser light producing a laser frequency scan across the ionic hyperfine resonances. In both sets of experiments the absolute background from scattered non-resonant photons was reduced by accumulating and bunching the ions during cycles of 50–200 ms and only counting photons during a 25 µs gate defining the laser-ion bunch interaction time [3]. Sample spectra are shown in Fig. 1.

In order to extract nuclear data from the observed spectra at least one of four nuclear parameters must already be known – either the nuclear spin, magnetic moment, quadrupole moment or $\delta\langle r^2\rangle^{A,A'}$ relative to a known system. This situation arises as the $^1S_0$ ground state term limits the possible upper ionic terms to $J' = 1$ states (reachable by electric dipole transitions) and only three hyperfine components, and thus two splittings, can be observed. Nuclear spins for many of the measured ground states and isomers are already known, or strongly suggested, but for three structurally important systems, the deformed ^{98}Y isomer, ^{100}Y and ^{102}Y, an independent evaluation is required [11].

A general method of sample preparation, capable of alleviating the above restriction in yttrium and providing a range of experimental opportunities, is offered by optically pumping trapped ensembles in the cooler-buncher.

Figure 2 shows two resonance fluorescence spectra observed in a collinear fast beam study of the stable, $A = 89$, yttrium ion (using a continuous 1 pA flux of ions). For both spectra the optical transition was excited from the same metastable state and the effect of optical pumping in the cooler buncher can be observed. During the acquisition of the spectrum labelled 'TiSa on' ~ 200 mW of 363.3 nm light from a pulsed titanium-sapphire (TiSa) laser illuminated the thermalisation axis of the cooler. Population transfer is observed at a level that remains constant while the ion beam is bunched. The fluorescent intensity is likewise constant, unlike that from unpumped ensembles which decreases to zero as the relaxing metastable population is held in the cooler.

In the case of yttrium the optical pumping permits efficient collinear spectroscopy on a $J = 2$ to $J' = 1$ ionic transition. A determination of all four measurable nuclear parameters can be achieved from spectroscopy of such a line. Future work at the JYFL IGISOL facility will now use this technique to independently measure the nuclear spins of the studied yttrium isotopes and isomers.

More generally the ability to manipulate state populations using broad bandwidth pulsed lasers is intended to greatly extend the prospects for collinear laser spectroscopy at JYFL and at other facilities. The pulsed lasers used for the state manipulation can readily access a far greater range of wavelengths, via harmonic generation, than that possible with the high-resolution single mode lasers used in the collinear spectroscopy. It is thus possible to use pulsed lasers to manipulate the ionic state and populate levels from which transitions suitable for single mode CW lasers can be studied. The choice of transition can then be optimised and for many elements previously impossible investigations can then be made.

Acknowledgements This work has been supported by the UK Engineering and Physical Sciences Research Council, the Academy of Finland under the Finnish Centre of Excellence Programme 2000-2005 (Project No. 44875), and by the European Union Fifth Framework Programme 'Improving Human Potential – Access to Research Infrastructure.' Contract No. HPRI-CT-1999-00044.

References

1. Otten, E.W.: Nuclear radii and moments of unstable isotopes. In: Bromley, D.A. (ed.) Treatise on Heavy-ion Science, vol. 8, p. 517. Plenum Press, New York (1989)
2. Billowes, J., Campbell, P.: J. Phys., G **21**, 707 (1995)

3. Nieminen, A., et al.: Phys. Rev. Lett. **88**(9), 094801 (2002)
4. Äystö, J.: Nucl. Phys., A **693**, 477 (2001)
5. Thibault, C., et al.: Phys. Rev., C **23**, 2720 (1981)
6. Buchinger, F., et al.: Phys. Rev., C **41**(6), 2883 (1990)
7. Campbell, P., et al.: Phys. Rev. Lett. **89**, 082501 (2002)
8. Meyer, R.A.: Nucl. Phys., A **439**, 510 (1985)
9. Mach, H., et al.: Nucl. Phys., A **523**, 197 (1991)
10. Urban, W., et al.: Nucl. Phys., A **689**, 605 (2001)
11. Cheal, B., et al.: Phys. Lett., B (2007) (in press)

Hyperfine Interact (2006) 171:149–156
DOI 10.1007/s10751-006-9492-1

LaSpec at FAIR'S low energy beamline: A new perspective for laser spectroscopy of radioactive nuclei

**Wilfried Nörtershäuser · Paul Campbell ·
The LaSpec Collaboration**

Published online: 1 February 2007
© Springer Science + Business Media B.V. 2007

Abstract We describe a laser spectroscopy station that has been proposed to exploit the high yields of radioactive beams at the future FAIR facility at GSI, Darmstadt. Exotic nuclei produced in the Super-Fragment-Recoil-Separator (S-FRS) are stopped in a gas cell, extracted, reaccelerated and transported to the LaSpec setup. Here, collinear spectroscopy on ions and atoms, β-NMR experiments or laser-desorbed resonance ionization can be applied to these rare isotopes.

Key words laser spectroscopy · nuclear charge radius · isotope shift · hyperfine structure

1 Introduction

New generations of radioactive beam facilities have always opened new opportunities and new challenges for laser spectroscopy experiments. A multitude of techniques have been developed over the last few decades since the invention of the laser. While the first experiments were performed off-line having collected radioactive products at the facility, true on-line experiments soon followed. Specialized methods have been developed specifically for these on-line applications, such as collinear fast beam spectroscopy, β-NMR, and resonance ionization in gas cells and these techniques have been applied to a large number of long isotopic chains. An overview

W. Nörtershäuser (✉)
Institut für Kernchemie, Universität Mainz, D-55099 Mainz and Gesellschaft für
Schwerionenforschung, Darmstadt 64291, Germany
e-mail: w.noertershaeuser@gsi.de

P. Campbell
University of Manchester, Schuster Building, Brunswick Street,
Manchester M13 9PL, UK

The LaSpec Collaboration
http://www.gsi.de/LaSpec

over such experiments can be found, e.g., in [1–3]. More recently, laser cooling techniques and trapping of neutral atoms in magneto optical traps have found their way into on-line experiments, e.g., to test the standard model of particle physics with short-lived isotopes [4, 5] or to determine nuclear charge radii [6]. But conventional techniques still provide very interesting and important results if they are linked to beam facilities that produce sufficient yields far from the valley of stability, or in regions that were previously not accessible. Such a case is given at in-flight facilities where gas-cells are used to stop high-energy radioactive ion beams. They provide a rapid extraction of the radioactive ions, combining the intrinsic advantages of in-flight fragmentation, short delay times and universality, with those of the ISOL concept, namely the high-quality, low-energy beams. Such set-ups are currently used or in preparation at MSU [7, 8], Riken [9], and at Argonne [10, 11]. Laser spectroscopy at these facilities is complimentary to similar experiments at ISOL facilities where many elements, like refractory or very short-lived isotopes, cannot be extracted with sufficient yields. The Super Fragment Recoil Separator (S-FRS) at the future Facility for Antiproton and Ion Research (FAIR) [12] at GSI will provide a rich spectrum of isotopes that will not be available at any other facility. A laser spectroscopy station at the low-energy beamline behind this separator is planned and will be realized by the LaSpec collaboration. From the view of optical spectroscopic research the proposed facility will afford unique access to regions of particular nuclear interest that would otherwise remain inaccessible. The accuracy of laser-spectroscopic-determined nuclear properties is very high. Requirements concerning production rates are moderate; collinear spectroscopy has been performed with production rates as few as 100 ions per second [13] and resonance ionization mass spectroscopy (combined with β-delayed neutron detection) has been achieved with rates of only a few atoms per second [14]. At FAIR it will be, for example, possible for our collaboration to greatly extend our knowledge of nuclear sizes, deformation and electromagnetic moments far into the neutron-rich side of the upper part of the nuclear chart. Here we will give a short overview of the proposed laser spectroscopy station.

2 Isotope production and the low-energy beamline at FAIR

The FAIR facility will greatly extend the current GSI facility. The heavy ion synchrotron SIS-18 will be upgraded for higher intensities. The SIS-18 beam will be used directly for in-flight fragmentation in a first phase and later as a driver beam for a larger synchrotron (SIS-100/300) that will produce intense ion beams with energies up to 1.5 GeV/u and up to 10^{12} ions/spill. With these beams, the S-FRS will be the most powerful in-flight separator for exotic nuclei up to relativistic energies. It will be possible to produce and separate rare isotopes of all elements up to uranium and the separated beams will serve three branches of different experimental areas: the ring branch, where they can be stored and cooled in a number of storage rings for different applications, the high-energy branch that is used for reaction studies, and the low-energy branch (LEB) that is primarily dedicated to precision experiments with energy-bunched beams stopped in a gas cell. Figure 1 shows the preliminary layout of the low-energy beam area. The incoming beam from the S-FRS will be used by the HISPEC (High-resolution in-flight spectroscopy) and DESPEC (Decay

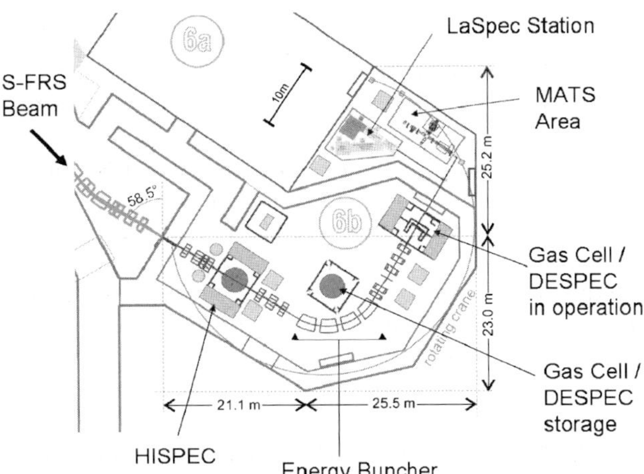

Fig. 1 Proposed layout of the low-energy-beam area at FAIR's super fragment separator (courtesy of M. Winkler, GSI)

spectroscopy) set-ups which are dedicated for γ-ray, charged particle and neutron spectroscopy on slow (3–150 MeV/u) and implanted ions, respectively. The DESPEC set-up is located behind the energy buncher, a dispersive separator stage that has been designed to drastically reduce the energy spread and thus the range straggling of the hot fragments. DESPEC can be replaced by an ion catcher device, with both devices installed on an air cushion system to facilitate an exchange between the two experimental stations.

Two possible ion catcher devices have so far been discussed and are now tested for installation. The first one is a gas-filled stopping cell, based on the standard IGISOL technology but upgraded to handle the high fragment energies of up to 100 MeV/u. A full scale gas catcher chamber with 1.2 m length and 25 cm diameter, designed for projectile energies of up to about 500 MeV/u [11], is presently being tested at GSI. The second possibility is an ion catcher device operated with superfluid helium. It has been recently demonstrated that 100 keV recoiling ^{219}Rn ions, originating from the decay of ^{223}Ra, were stopped and thermalized within 1 µm of this superfluid [15]. The stopped ions formed spontaneously "snowballs" which could be guided with electric fields to the liquid–gas surface, leading to an optimized surface extraction efficiency of 23%. A third type of ion catcher device has been proposed in the literature recently: Injection of the beam into an inverse cyclotron [16, 17]. Whether such a solution might be favorable for the GSI set-up will be carefully investigated and discussed.

3 Beam preparation and transport

The beam transport from the gas cell to the spectroscopy station is approaching its final design. We propose to extract the ions from the gas cell into a transport beamline which is at a negative potential of approximately 10 kV. This provides a beam energy sufficient for efficient beam transfer without the complications of a large gas cell on a high potential. A similar system has already been realized at MSU. The transport beamline will include a sector magnet for mass separation, to isolate those isotopes desired for investigation and also to remove reaction products of the

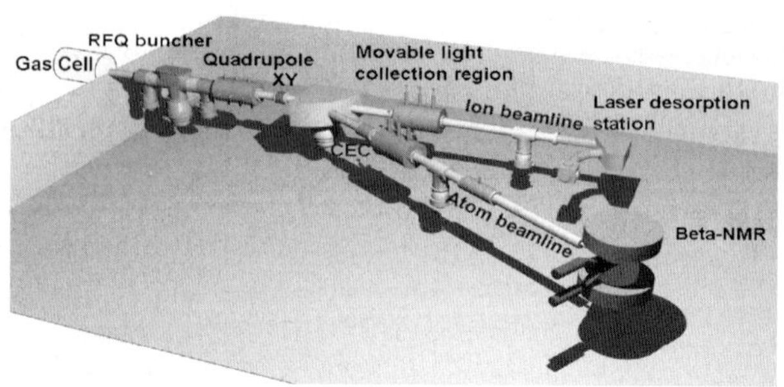

Fig. 2 Schematic layout of the proposed LaSpec spectroscopy station. The separator system, e.g. a magnetic sector field, between gas cell and RFQ cooler and buncher is not shown (CEC = Charge exchange cell)

ions with gas contaminant molecules (such as water, etc.). The beam will then be injected into a radiofrequency quadrupole (RFQ) buncher and cooler unit in which the ions can be stored and cooled by collisions with a neutral buffer gas. After a few ms cooling time the ions are bunched at the point of lowest potential, close to the exit, from which they can be extracted as a short pulse of a few microseconds length. Such cooler and bunchers are already operated routinely, e.g., at the ISOLTRAP facility [18] and at Jyväskylä [19]. The ion bunch is again accelerated into a beamline at lowered potential. A switch box will be used to distribute the beam to the LaSpec station or the MATS (Mass measurements with an advanced trapping system) set-up. As the MATS station needs to stop the ions again in a Penning trap, this is the preferred transport system. Typical collinear spectroscopy requires an ion beam at 40–60 keV. Hence, we will use a fast drift tube, which will be quickly raised in potential to up to +60 kV while the ions are passing through. After leaving the drift tube the bunches are accelerated and the subsequent transport line can be kept at ground potential.

4 The LaSpec station

The LaSpec collaboration intends to construct a number of complementary experimental devices which will provide a complete system with respect to the physics and isotopes that can be studied. A rough sketch of the proposed station is shown in Fig. 2. It includes the following techniques and capabilities, which are all discussed in detail in other contributions to this workshop.

Collinear laser spectroscopy: Collinear laser spectroscopy has been the workhorse for high-precision laser spectroscopy on short-lived isotopes for many years [1–3, 20]. Early work was mainly done on atomic systems, as their absorption lines were often in the visible region and therefore more readily accessible. Fast atomic beams are produced by charge exchange of ion beams in a cell containing a low pressure alkali-metal vapour. In some cases even metastable states are resonantly populated in the exchange process which provided more easily accesible transitions than those from the atomic ground state. Although the absorption lines of ions are typically in the deep blue and ultraviolet regions of the spectrum, suitable laser light can now be produced by frequency doubling or frequency mixing of cw lasers. Thus, a highly robust and universal spectroscopy is provided. The LaSpec set-up will provide one

short beamline dedicated to the spectroscopy of ions and a second, longer one, that is optimized for the spectroscopy of neutral systems. It is longer in order to make efficient optical pumping possible. A common optical detection region will be constructed that can be moved easily between the two beamlines. In combination with the RFQ cooler and buncher for background suppression, optical detection can be applied to ion species provided with yields as low as 100 ions/s as it has been demonstrated recently at Jyväskylä [13]. Another very interesting feature with respect to the RFQ cooler and buncher, is the population of excited, metastable ionic states using pulsed lasers inside the buncher. This will further extend the possibilities of collinear ion spectroscopy to cases where transitions from the ground state are not accessible with cw lasers [21]. In the atomic beamline, other sophisticated detection techniques are combined with collinear spectroscopy and provide single-particle detection and therefore better efficiency, such as collisional or resonance ionization.

Optical pumping and β-NMR: A particularly successful combination that has often been employed is the production of polarized beams of atoms and ions by optical pumping with subsequent β-asymmetry detection and β-NMR. This is a well-established method for the determination of nuclear magnetic dipole and electric quadrupole moments. The technique has been especially favoured for light elements [22] but has also been applied to a range of cases where high precision has been required to explore the nuclear structure (such as measuring small admixtures of intruder components in a wavefunction). The method combines optical pumping with nuclear magnetic resonance and β-decay asymmetry spectroscopy. Recently such spectroscopy has been used to determine the quadrupole moment of ^{11}Li to high accuracy using yields of only a few thousand ions/s [23] and the magnetic dipole and electric quadrupole moments of ^{8,9}Li [24]. The strength of this technique was again recently demonstrated with the determination of the ground-state spin and magnetic moment of ^{31}Mg [25]. The method can be used for all β-decaying (non-zero spin) isotopes with half-lives between approximately 5 ms and 20 s. Generally, β-NMR is able to deliver quadrupole moments with higher accuracy than those obtained from standard laser spectroscopy. However, laser spectroscopy and hyperfine structure investigations are necessarily the first step prior to attempting optical pumping.

Resonance Ionization Laser Ion Source (RILIS): Resonance ionization of atoms combined with the detection of the produced ions is a very efficient method in the study of rare isotopes [26]. During the last decade, it has been used for a broad range of applications, e.g., laser ion sources to produce ion species which were not accessible by other methods [27], in-source laser spectroscopy [28], and ultra-trace detection of cosmogenic and radio-toxic isotopes [29]. Its most prominent features are the efficient ionization process with resonant intermediate and autoionizing states, the elemental, isotopical, and sometimes even isomeric selectivity [30] and the large detection efficiency for the produced, charged particles. The use of laser ion sources at nuclear structural facilities has been demonstrated, advanced and developed to a point where it is the favored production mechanism by the ISOLDE group, CERN (and presently used for over 60% of all experiments). In the case of gas-jet ion sources, a careful systematic development both off-line and on-line has been achieved by the LISOL group of the University of Leuven [31].

 Springer

At the FAIR facility's Low Energy Branch, a resonance ionization laser ion source will add to the selectivity inherent in the production method of the S-FRS and ion catcher device. In the stopping cell, whether it be a gas-filled or a superfluid helium catcher, the formation of singly- or doubly-charged positive ions is expected, as the first ionization potential of the stopping gas – usually helium – lies significantly above the first and often second ionization potential of all other elements. Nevertheless under on-line conditions with plasma present, and any small contaminations of H_2, N_2, O_2 and H_2O in the sub-ppm range, the charge state may be further reduced leading to neutralization by three-body recombination processes. These loss mechanisms to a neutral state can be a significant drawback to the IGISOL technique, however the process can be converted to an advantageous one if subsequently an efficient and selective laser ionization of the neutralized species is achieved. The RILIS will be used to enhance the production of isotopic (or even isomeric) enriched or pure ion beams. A natural development to a laser ion source is the installation of a laser ion source trap (LIST) [32], which may be coupled to the ion catcher. One possible design of a LIST is based on a typical segmented and gas-filled RFQ ion trap. Any neutral species that exit the ion catcher may be selectively ionized with counter propagating high-repetition rate pulsed lasers. The large size of the gas cell, is a challenge to the application of such methods and whether there is a reasonable way in doing so will be carefully investigated.

Laser-Desorption Resonance Ionization (LDRIS): Radioactive ions or atoms can be deposited on an appropriate catcher, laser desorbed, and studied during a secondary (and resonant) laser ionization [33]. When used in conjunction with Time-Of-Flight (TOF) mass-separation [34] and decay-tagged photo-ion detection an extremely sensitive spectroscopy can be achieved. For cases with very low production yields, the technique is superior to fluorescence detection. Such spectroscopy has previously, and notably, been used to study isotopes where the radioactive species was deposited in chemical or cluster form. It was also used for cases where the species of interest was not directly available but precursor nuclei could be deposited [35]. At FAIR the technique will be applied to the study of heavy neutron-rich elements (Pb, Bi, Pt, Au...) and will provide an opportunity to extend the investigation of these elements beyond the neutron-deficient cases previously studied at ISOLDE [36–39]. The laser desorption station will be located at the end of the collinear ion beamline. Here, a commercial but adapted MALDI-TOF (Matrix-assisted laser desorption and ionization - time of flight) apparatus will be used. Two fast kickers will be used to translate the incoming ion beam after deceleration in horizontal or vertical direction in such a way that the extracted beam after implantation, desorption and laser ionization, can be separated and sent into the TOF system.

Spectroscopy in an Electron Beam Ion Trap (EBIT): An additional option for laser spectroscopy is a spectroscopy of highly-charged radioactive ions inside an electron beam ion trap. Such a device is an integral part of the MATS experimental set-up, located very close to the LaSpec spectroscopy station. Precision measurements of the hyperfine splitting of the atomic ground state can be performed to investigate, e.g., details of the nuclear magnetization distribution [40]. This might be particularly interesting in cases of rather short-lived isotopes that cannot be measured at the HITRAP setup, where it takes a few 10 s to prepare the cold highly charged ions [41, 42].

 Springer

5 Summary

The future facility FAIR at GSI, Darmstadt, will provide beams of radioactive isotopes that cannot be produced elsewhere. The LaSpec collaboration (http://www.gsi.de/LaSpec), presently formed by 13 institutes from 7 countries, will exploit the possibilities for on-line laser spectroscopy of these species. Nuclear charge radii, spins, and electromagnetic moments of nuclear ground and long-lived isomeric states will be investigated using collinear laser spectroscopy, resonance ionization, β-NMR, and laser desorption.

Acknowledgements The authors are grateful to all members of the LaSpec collaboration who contributed to the LaSpec Letter of Intent and the Technical Proposal.

References

1. Otten, E.W.: Nuclear radii and moments of unstable isotopes. In: Bromley, D.A. (ed.) Treatise on Heavy-ion Science, pp. 517–638 (1989)
2. Billowes, J., Campbell, P.: J. Phys. G. **21**, 707–739 (1995)
3. Kluge, H.-J., Nörtershäuser, W.: Spectrochim. Acta B **58**, 1031–1045 (2003)
4. Lu, Z.T., Bowers, C. J., Freedman, S.J., Fujikawa, B.K., Mortara, J.L., Shang, S.Q., Coulter, K.P., Young, L.: Phys. Rev. Lett. **72**, 3791 (1994)
5. Behr, J.A., Gorelov, A., Swanson, T., Häusser, O., Jackson, K.P., Trinczek, M., Giesen, U., D'Auria, J.M., Hardy, R., Wilson, T., Choboter, P., Leblond, F., Buchmann, L., Dombsky, M., Levy, C.D.P., Roy, G., Brown, B.A., Dilling, J.: Phys. Rev. Lett. **79**, 375–378 (1997)
6. Wang, L.-B., Mueller, P., Bailey, K., Drake, G.W.F., Greene, J.P., Henderson, D., Holt, R.J., Janssens, R.V.F., Jiang, C.L., Lu, Z.-T., O'Connor, T.P., Pardo, R.C., Rehm, K.E., Schiffer, J.P., Tang, X.D.: Phys. Rev. Lett. **93**, 142501 (2004)
7. Sherrill, B.M.: Nuclear Instruments and Methods in Physics Research Section B **204**, 765 (2003)
8. Bollen, G., Davies, D., Facina, M., Huikari, J., Kwan, E., Lofy, P.A., Morrissey, D.J., Prinke, A., Ringle, R., Savory, J., Schury, P., Schwarz, S., Sumithrarachchi, C., Sun, T., Weissman, L.: Phys. Rev. Lett. **96**, 152501 (2006)
9. Wada, M., Ishida, Y., Nakamura, T., Yamazaki, Y., Kambara, T., Ohyama, H., Kanai, Y., Kojima, T.M., Nakai, Y., Ohshima, N.: Nucl. Instrum. Methods Phys. Res. B **204**, 570-581 (2003)
10. Savard, G., Schwartz, J., Caggiano, J., Greene, J.P., Heinz, A., Maier, M., Seweryniak, D., Zabransky, B.J.: Nucl. Phys. A **701**, 292-295 (2002)
11. Savard, G., Clark, J., Boudreau, C., Buchinger, F., Crawford, J.E., Geissel, H., Greene, J.P., Gulick, S., Heinz, A., Lee, J.K.P., Levand, A., Maier, M., Münzenberg, G., Scheidenberger, C., Seweryniak, D., Sharma, K.S., Sprouse, G., Vaz, J., Wang, J.C., Zabransky, B.J., Zhou, Z., the S258 Collaboration: Nucl. Instrum. Methods Phys. Res. B **204**, 582–586 (2003)
12. Eschke, J.: J. Phys. G **31**, S967–S973 (2005)
13. Campbell, P., Nieminen, A., Billowes, J., Dendooven, P., Flanagan, K.T., Forest, D.H., Gangrsky, Y.P., Griffith, J.A.R., Huikari, J., Jokinen, A., Moore, I.D., Moore, R., Thayer, H.L., Tungate, G., Zemlyanoi, S.G., Äystö, J.: Eur. Phys. J. A **15**, 45–48 (2002)
14. Sebastian, V.: Laserionisation und Laserionenquelle an ISOLDE/CERN. Dissertation, Institute of Physics, Johannes Gutenberg - University, Mainz (1999)
15. Huang, W.X., Dendooven, P., Gloos, K., Takahashi, N., Pekola, J.P., Äystö, J.: Europhys. Lett. **63**, 687–693 (2003)
16. Katayama, I., Wada, M., Kawakami, H., Tanaka, J., Noda, K.: Hyperfine Interact. **115**, 165–170 (1998)
17. Bollen, G., Morrissey, D.J., Schwarz, S.: Nucl. Instrum. Methods Phys. Res. A **550**, 27–38 (2005)
18. Herfurth, F., Dilling, J., Kellerbauer, A., Bollen, G., Henry, S., Kluge, H.-J., Lamour, E., Lunney, D., Moore, R.B., Scheidenberger, C., Schwarz, S., Sikler, G., Szerypo, J.: Nucl. Instrum. Methods Phys. Res. A **469**, 254–275 (2001)
19. Nieminen, A., Campbell, P., Billowes, J., Forest, D.H., Griffith, J.A.R., Huikari, J., Jokinen, A., Moore, I.D., Moore, R., Tungate, G., Äystö, J.: Nucl. Instrum. Methods Phys. Res. B **204**, 563–569 (2003)

20. Neugart, R.: Hyperfine Interact. **24**, 159–180 (1985)
21. Campbell, P.: Optical pumping in an RF cooler buncher (this issue)
22. Neugart, R.: Hyperfine Interact. **127**, 101–110 (2000)
23. Borremans, D.: Precision moments of the ^{11}Li halo nucleus. Dissertation, Instituut voor Kernen Stralingsfysica, Katholieke Universiteit Leuven, Leuven (2004)
24. Borremans, D., Balabanski, D.L., Blaum, K., Geithner, W., Gheysen, S., Himpe, P., Kowalska, M., Lassen, J., Lievens, P., Mallion, S., Neugart, R., Neyens, G., Vermeulen, N., Yordanov, D.: Phys. Rev. C **72**, 044309 (2005)
25. Neyens, G., Kowalska, M., Yordanov, D., Blaum, K., Himpe, P., Lievens, P., Mallion, S., Neugart, R., Vermeulen, N., Utsuno, Y., Otsuka, T.: Phys. Rev. Lett. **94**, 022501 (2005)
26. Payne, M.G., Deng, L., Thonnard, N.: Rev. Sci. Instrum. **65**, 2433–2459 (1994)
27. Köster, U., Fedoseyev, V.N., Mishin, V.I.: Spectrochim. Acta B **58**, 1047–1068 (2003)
28. Weissman, L., Köster, U., Catherall, R., Franchoo, S., Georg, U., Jonsson, O., Fedoseyev, V.N., Mishin, V.I., Seliverstov, M.D., Roosbroeck, J.V., Gheysen, S., Huyse, M., Kruglov, K., Neyens, G., Duppen, P.V.: Phys Rev. C **65**, 024315 (2002)
29. Lu, Z.-T., Wendt, K.: Rev. Sci. Instrum. **74**, 1169–1179 (2003)
30. Fedoseyev, V.N., Huber, G., Köster, U., Lettry, J., Mishin, V.I., Ravn, H., Sebastian, V.: Hyperfine Interact. **127**, 409–416 (2000)
31. Kudryavtsev, Y., Andrzejewski, J., Bijnens, N., Franchoo, S., Gentens, J., Huyse, M., Piechaczek, A., Szerypo, J., Reusen, I., Duppen, P.V., Bergh, P.V.d., Vermeeren, L., Wauters, J., Wöhr, A.: Nucl. Instrum. Methods Phys. Res. B **114**, 350–365 (1996)
32. Blaum, K., Geppert, C., Kluge, H.-J., Mukherjee, M., Schwarz, S., Wendt, K.: Nucl. Instrum. Methods Phys. Res. B **204**, 331–335 (2003)
33. Sauvage, J., Boos, N., Cabaret, L., Crawford, J.E., Duong, H.T., Genevey, J., Girod, M., Huber, G., Ibrahim, F., Krieg, M., Blanc, F.L., Lee, J.K.P., Libert, J., Lunney, D., Obert, J., Oms, J., Péru, S., Pinard, J., Putaux, J.C., Roussière, B., Sebastian, V., Verney, D., Zemlyanoi, S., Arianer, J., Barré, N., Ducourtieux, M., Forkel-Wirth, D., Scornet, G.L., Lettry, J., Richard-Serre, C., Véron, C.: Hyperfine Interact. **129**, 303–317 (2000)
34. Maul, J., Berg, T., Ebrhardt, K., Hoog, I., Huber, G., Karpuk, S., Passler, G., Stachnov, I., Trautmann, N., Wendt, K.: Nucl. Instrum. Methods Phys. Res. B **226**, 644–650 (2004)
35. Le Blanc, F., Obert, J., Oms, J., Putaux, J.C., Roussière, B., Sauvage, J., Pinard, J., Cabaret, L., Duong, H.T., Huber, G., Krieg, M., Sebastian, V., Crawford, J., Lee, J.K.P., Genevey, J., Ibrahim, F.: Phys. Rev. Lett. **79**, 2213–2216 (1997)
36. Le Blanc, F., Lunney, D., Obert, J., Oms, J., Putaux, J.C., Roussière, B., Sauvage, J., Zemlyanoi, S., Pinard, J., Cabaret, L., Duong, H.T., Huber, G., Krieg, M., Sebastian, V., Crawford, J.E., Lee, J.K.P., Girod, M., Péru, S., Genevey, J., Lettry, J., Collaboration, I.: Phys. Rev. C **60**, 054310 (1999)
37. Fedosseev, V.N., Fedorov, D.V., Horn, R., Huber, G., Koster, U., Lassen, J., Mishin, V.I., Seliverstov, M.D., Weissman, L., Wendt, K.: Nucl. Instrum. Methods Phys. Res. B **204**, 353–358 (2003)
38. Andreyev, A.N., Vel, K.V.d., Barzakh, A., Smet, A.D., Witte, II.D., Fedorov, D.V., Fedoseyev, V.N., Franchoo, S., Górska, M., Huyse, M., Janas, Z., Köster, U., Kurcewicz, W., Kurpeta, J., Mishin, V.I., Partes, K., Plochocki, A., Duppen, P.V., Weissman, L.: Eur. Phys. J. A **14**, 63–75 (2002)
39. De Witte, H., Andreyev, A.N., Borzov, I.N., Caurier, E., Cederkall, J., De Smet, A., Eeckhaudt, S., Fedorov, D.V., Fedosseev, V.N., Franchoo, S., Gorska, M., Grawe, H., Huber, G., Huyse, M., Janas, Z., Koster, U., Kurcewicz, W., Kurpeta, J., Plochocki, A., Van de Vel, K., Van Duppen, P., Weissman, L.: Phys. Rev. C **69**, 044305–044306 (2004)
40. Winters, D.F.A., Vogel, M., Segal, D.M., Thompson, R.C., Nörtershäuser, W.: Can. J. Phys. (submitted) (2006)
41. Beier, Th., Dahl, L., Kluge, H.-J., Kozhuharov, C., Quint, W.: Nucl. Instrum. Methods Phys. Res. B **235**, 473–478 (2005)
42. Herfurth, F., Beier, Th., Dahl, L., Eliseev, S., Heinz, S., Kester, O., Kozhuharov, C., Maero, G., Quint, W., The HITRAP Collaboration: Int. J. Mass Spectrom. **251**, 266–272 (2006)

Hyperfine Interact (2006) 171:157–166
DOI 10.1007/s10751-006-9489-9

Nuclei near the closed shells N=20 and N=28

Yu. E. Penionzhkevich

Published online: 30 January 2007
© Springer Science + Business Media B.V. 2007

Abstract The present work reviews the properties of the neutron-rich isotopes near the closed shells N=20 and N=28. The changes in nuclear structure appearing as one goes away from the β-stability line are discussed. The location of the neutron drip line and questions about the stability of nuclides with $Z{\geq}8$ are considered in connection with the weakening or even vanishing of the shell effects at the magic numbers 20 and 28, and the discovery of the new neutron magic numbers at N=16 and N=32. These properties are extremely interesting from the point of view of laser experiments as well as for all other experimental methods giving access to this region.

Key words closed shells · nuclei far from the stability line · nuclear deformation · separation energy · deformed shapes · yrast states · microscopic energy

1 Introduction

Experimental investigations of nuclei in the region of closed neutron shells N=20 and 28 show that their properties (binding energy and deformation) strongly vary as they move away from the stability line [1]. In addition, new effects manifest themselves, because of the change (decrease or increase) in the stability of these nuclei near the boundaries of nucleon stability. The observed effects demand revising the theoretical concepts about the properties of such nuclei. Among the numerous advances made, one may cite the appearance of the so called island of inversion. A striking example of this phenomenon has been observed at the shell closure N=20 for the neutron-rich ^{31}Na [2, 3] and ^{32}Mg [4, 5] which, in spite of our conventional ideas, are strongly deformed. As early as 1975, Thibault et al. [2] found the coexistence of two types of deformation in the magic nucleus ^{31}Na. Within the framework of the shell model, the deformation of the ground state in this nucleus was explained by the strong correlation between the $2p$–$2n$ excitation energies for the sd and pf shells. This fact manifests itself by increasing the binding energy of Na isotopes near the N=20 shell. Subsequent investigations showed that similar effects are also

Y. E. Penionzhkevich (✉)
Joint Institute for Nuclear Research, 141980 Dubna, Russia
e-mail: pyuer@jinr.ru

 Springer

observed for neutron-rich O, F, Ne, Na, and Mg isotopes. For the majority of light nuclei, the so-called coexistence of two shapes (spherical and deformed) manifests itself in the ground state near the shells $N=20$ and $N=28$. This resulted in revising the existing shell models and in prediction of the alteration of the well-known magic numbers near the neutron drip line: $N=14$ and $N=16$ instead $N=20$ for nuclei with $Z=7,8$ and $Z=7-10$ [6, 7], respectively. The problem is important for the future development of our concepts about nuclear-matter properties for large isospin (exotic nuclei). Therefore we try to systematize the currently available information on the properties of neutron-rich nuclei near the boundaries of nucleon stability, close to the shell numbers $N=20$ and $N=28$.

2 Binding energy and shell effects

Historically, three main approaches describing the shell effects have been used. In the macroscopic description, within the framework of the liquid droplet model, the binding-energy of a bound system consisting of A nucleons is given by the well-known semiempirically Bethe-Weizsäcker formula. In this approach, the nuclear binding energy is determined by the incompressibility of nuclear matter and partition of nuclear forces. The formula points out to the strong dependence of the binding energy, respectively, of nuclear stability, on the skin effects and Coulomb repulsion. It predicts that the even-even groupings of protons and neutrons being favoured in stability. The formula provides a good fit to heavier nuclei, and a poor fit to very light nuclei, e.g. ^{4}He. This is because the formula does not consider the internal shell structure of the nucleus. For light nuclei, it is usually better to use a model that takes this structure into account

In the microscopic description based on the model of independent particles, in contrast to the liquid-drop model, each nucleon is located in a certain mean field $V_{(r)}$ formed by other nucleons. In this case, the nuclear potential $V_{(r)}$ is approximated by the harmonic-oscillator potential in the form:

$$V_{(r)} = \frac{1}{2}m\omega^2 r^2 + Dl^2 + Cl \cdot s \,, \tag{1}$$

where l is the orbital angular momentum operator and s is the spin operator. The second term in this expression relates the angular momentum to the harmonic oscillator: the constant D is a negative parameter, determining the change in the $D\hbar^2 l(l+1)$ energy level. The two first factors in (1) describe the binding energy, but reproduce no shell effect. Only the addition of a third factor, which takes into account the "spin-orbit" interaction, made it possible to take into account the irregularity in the potential related to shell manifestation in the microscopic approach. Nuclei with a shell closure number of neutrons or protons are just the magic nuclei with spherical shape and increased stability [8]. In this model, the spin-parity of nuclei defines their stability, and the nuclear energy levels are filled according to the rules for filling of shells and sub-shells.

To combine these two approaches – "macroscopic" and "microscopic", Bohr and Mottelson [9] developed a new so-called "collective-shell" model in which the nuclear energy of an excited state is written as:

$$E_j = \left(\frac{\hbar^2}{2J}\right)j(j+1), \tag{2}$$

 Springer

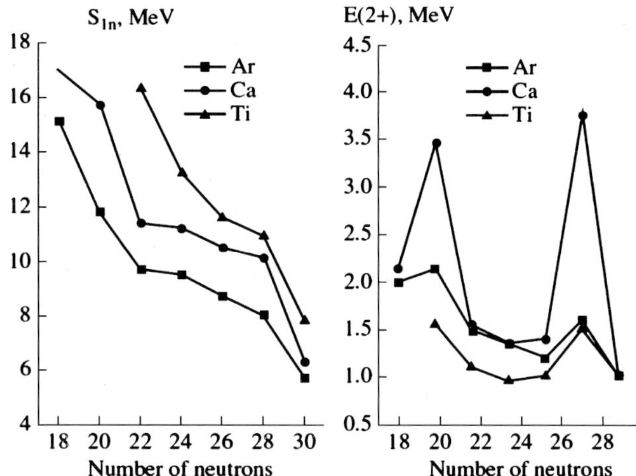

Fig. 1 Neutron separation energy (*on the left*) and the 2^+-level energy (*on the right*) as functions of the neutron number N for argon, calcium and titanium isotopes

where j is the nuclear spin equal to 0, 2, 4, ... for even–even nuclei, and J is the nuclear moment of inertia. This expression describes the individual and collective motion of nucleons in a nucleus that weakly interact with each other.

Magic or doubly magic nuclei are more stable than the neighbouring ions. The effect of shells in nuclei far from the stability line is studied by the regular measurements of the changes in the separation energy of the last neutrons (S_n and S_{2n}) or the energy of the first excited state (2^+), as a function of the number of neutrons. The separation energy S_{2n} of the last two neutrons defines the binding energy and is written as:

$$S_{2n}(A, Z) = [\Delta M(A - 2, Z) - \Delta M(A, Z) + 2\Delta M_n] \cdot c^2 , \tag{3}$$

where $\Delta M(A,Z)$ is the mass excess for nucleus $^A_Z X$, and ΔM_n is the neutron-mass excess equal to 8.071 MeV. The dependence of the binding energy on the number of neutrons is a relatively smooth curve, which changes its trend near the closed shells, because the neutron binding energy of magic nuclei is maximal.

3 Magic nuclei and deformations

In Fig. 1, the one-neutron separation energy and the 2^+-state energy are shown as functions of the number of neutrons in the respective nuclei. These dependencies illustrate the influence of the $N=20$ and $N=28$ shell closures for the isotopes of Ar, Ca and Ti [1]. Let us pointed out that in the case of Ar the sequential addition of neutrons going from the sd- to the $f_{7/2}$-shell gives a nearly smooth change of one-neutron separation energy S_{1n}, which may indicate disappearance of any shell effect. The magicity of the neutron shell closure $N=20$ is questionable as well in other cases. The measured value of the reduced transition probabilities $B(E2,0^+\rightarrow 2^+)$ for the lowest 2^+ states in the neutron rich ^{32}Mg suggests a picture of large deformation and vanishing of the $N=20$ shell gap [5]. Analogous data for the neutron rich "magic" ^{44}S [10, 11] indicates that the 2^+ states have a collective nature.

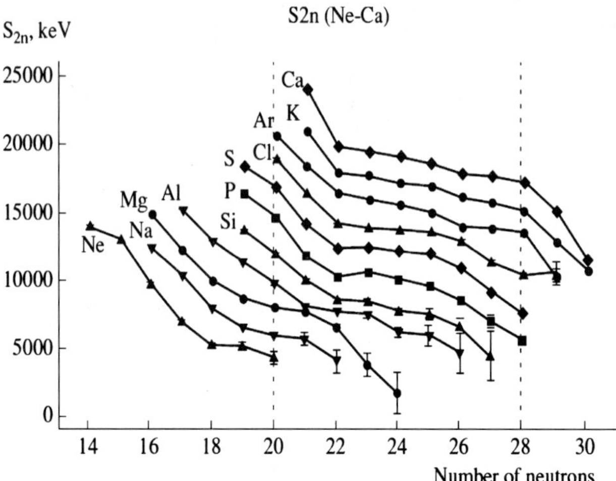

Fig. 2 Separation energy of the last two neutrons for neutron-rich nuclei from neon to calcium in the region of the closed shells $N=20$ and $N=28$ [12]

In addition, the discovery of the isomeric state with transition energy $E=319$ keV and lifetime $T_{1/2}=488\pm48$ ns for the nucleus ^{43}S [11] showed that two shapes (spherical and deformed) can coexist in this nucleus. Such coexistence was first predicted by Lyutostansky [13] for ^{31}Na and later confirmed in the calculations in ref. [14, 15]. In Fig. 2 the experimental trend of the binding energy S_{2n} of the last two neutrons is shown as a function of the number of neutrons. It can be seen that the binding energy of Ca, K and Ar isotopes is affected just by the closed shell $N=28$. However, for Mg, S, P, Si and Cl isotopes, an increase in the binding energy is observed in the region between the neutron numbers $N=22$ and $N=26$. The disappearance of the $N=28$ shell and the appearance of the new shell at $N=26$ is especially evident for the Cl isotopes (see Fig. 3).

The two-neutron binding energy S_{2n} measured experimentally in [11] is much lower for the isotopes ^{41}Si, ^{43}P and ^{44}S than the extrapolated values from the mass tables [16]. This fact also indicates a weakening of the closed shells. Compared to the case of Ca, Na and Ar an increase in the neutron binding energy of the Cl, S and P isotopes near the numbers $N=20$ and $N=26$ is observed. This can be explained by deformation, which forms a nuclear configuration with a stronger binding [8]. Thus, for the neutron-rich nuclei near the neutron numbers $N=22$ and $N=26$, new regions of deformation appear providing the stability of these nuclei.

4 New shell closures

In [19], the effective cross sections for the interaction of the nuclei ^{22}N, ^{23}O and ^{24}F were measured. From the comparison of these cross sections with those for the interaction of other nuclei, a conclusion was made about the existence of a new shell at $N=16$. The existence of new closed shells was assumed on the basis of experiments that measured nuclear masses. In this case, by comparing the experimental mass excess with calculations, using the macroscopic liquid-drop model (FRLDM – Finite Range Liquid Drop Model) [20], it was possible to obtain the shell correction changes depending on the number of neutrons in the nucleus. These dependencies between the shell closures $N=20$ and $N=28$ are shown in Fig. 3 [21] for the Ca, Cl, S, P and Si isotopes. Two minima of the shell

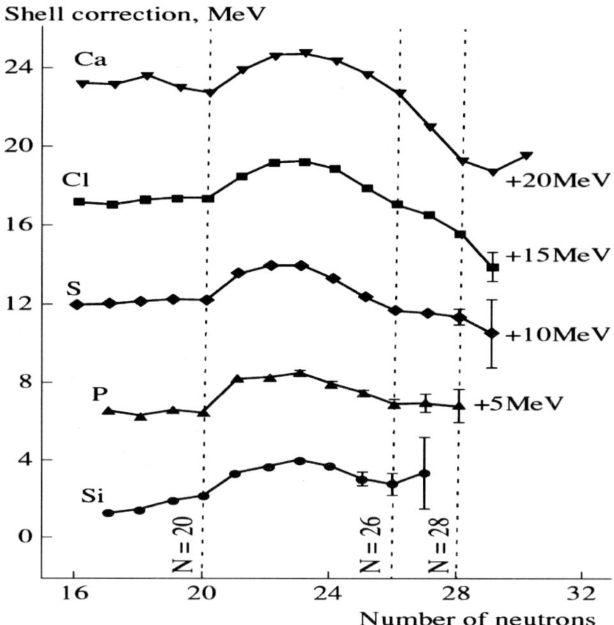

Fig. 3 Variation of the shell correction to the microscopic energy for $Z=20$ and $Z<17$ near the shells $N=20$ and $N=28$. The *points* correspond to experimental values [12] the *lines* to shell model calculation [17, 18]

corrections, as well as of the variation in the microscopic energy for Ca isotopes (see, e.g. [1]), are observed just at $N=20$ and $N=28$. Between these two magic numbers, the variation in the microscopic energy is associated with the filling of the $f_{7/2}$ shell. However, a break of both the shell corrections and microscopic energy dependencies is observed at $N=26$ for the S and P isotopes. This fact led to the conclusion that a new shell closure at $N=26$ appears.

In [22] an attempted theoretical interpretation of the occurrence of new shells for neutron-rich nuclei was made. For this purpose, the Monte Carlo Shell Model (MCSM) [23, 24] was used. This model has a number of advantages. First, it describes nuclear structure by means of large-scale shell model calculations including mixing among normal, intruder and higher intruder configurations. Thus, it is possible to calculate spherical yrast states, deformed rotational bands, as well as undeformed states using the same Hamiltonian. Another advantage consists in the possibility of using a large number of valence particles. This is especially essential for the description of a large number of nuclei aligned along the isospin axis, when the number of nucleons increases. This description is most efficient when passing from the spherical configuration to a deformed state [22]. In such an approach, the change in the shape and the possible coexistence of various shapes in the same nucleus is considered as due to the effect of valence nucleons. For explaining the coexistence of the two shapes in nuclei, effective single-particle energies (ESPEs), which correspond to single-particle orbits, are introduced. It is shown that the energy difference between the $0d_{3/2}$ and $1s_{1/2}$ orbitals for oxygen isotopes with neutron number $N=16$ has a peak and amounts to 6 MeV. This is a rather high value, which is comparable to the distance between the sd and pf shells in the Ca nucleus. By increasing Z, beginning with oxygen, this value decreases and reaches a minimum for Mg and Al. Such behaviour of single-particle levels occurs because of the change of the position of the neutron $0d_{3/2}$ shell when the number of protons varies. The effect can be demonstrated by the example of the two nuclei ^{30}Si and ^{24}O with the same $N=16$. The nucleus ^{30}Si has six valence protons in the sd space above the shell-core with $Z=8$ and is a stable nucleus, while the nucleus ^{24}O

has no valence protons. For the nucleus ^{30}Si, the $0d_{3/2}$ and $1s_{3/2}$ levels are located close enough to each other, while the $0d_{3/2}$ level in ^{24}O is located high and close to the group of pf shells. Owing to this fact, the difference between the $0d_{3/2}$ and $1s_{3/2}$ levels amounts to about 6 MeV. This value for the nucleon stable ^{34}Si is lower, because of the large energy difference (4 MeV) between the $0d_{3/2}$ and pf shells and can be explained by the strong interaction between the proton and neutron orbitals (valence protons are added only to the $0d_{5/2}$ orbital for Z from 8 to 14). Due to the strong interaction between the protons on the $0d_{5/2}$ orbital and neutrons on the $0d_{3/2}$ orbital, a large number of protons added to the $0d_{5/2}$ orbital results in a more bound state for neutrons on the $0d_{3/2}$ orbital. Therefore, the $0d_{3/2}$ level is located lower in ^{34}Si than in ^{24}O. This just explains why the $N=16$ shell is magic.

As suggested in [6, 7, 22], similar tendencies also take place for other nuclei, which results in new magic numbers $N=6$, 16 and 34 for neutron-rich nuclei instead of $N=8$, 20 and 40 for nuclei in the valley of stability. For example, for light p-shell neutron-rich nuclei, $N=6$ appears instead of the magic number $N=8$. As a consequence of this, the nucleus ^{8}He is very well bound, while the isotopes ^{9,10}He are unbound. The same situation takes place for the bound magic nucleus ^{24}O and the unbound nuclei $^{25-28}$O. Neutron-rich nuclei with $Z>8$ are as well of interest from the viewpoint of the manifestation of new shells. As established experimentally, the neutron drip line is reached for fluorine, neon and sodium isotopes at a much higher neutron-to-proton ratio than for oxygen isotopes (the last bound isotope is ^{24}O). The surprising fact is that the addition of one proton to the nucleus ^{24}O makes it possible to hold six additional neutrons in the fluorine nucleus (^{31}F has 22 neutrons in comparison with the ^{24}O nucleus with only 16 neutrons). It is also experimentally established that the doubly magic nucleus ^{28}O ($N=20$) is unbound. All these facts once again indicate a change in the magic numbers 8, 20 and 40 for the nuclei far from the valley of stability.

Anomalous values of mass and half-life were obtained for the first time in ^{31}Na [2]. The nucleus ^{31}Na lying far from the stability valley proved to be more tightly bound than it was expected. Further, the same situation was also found for the magnesium isotopes $^{31-33}$Mg [18, 25–27]. To explain the properties of nuclei in this region, it was assumed that the habitual order of filling the levels by neutrons was violated. It is shown that neutrons do not occupy lower sd orbitals, forming a free hole state but pass to higher pf orbitals forming a filled state. As a result, the nucleus becomes deformed. This is the so-called "intruder" state. The nuclear region where similar effects manifest themselves was named the island of inversion. Such an approach made it possible to describe the experimentally obtained characteristics of the nucleus ^{31}Na. As has been noted, the nucleus ^{31}Na has anomalous values of mass, spin, and magnetic moment in the ground state [2]. Within the conventional shell model with a USD interaction, the spin $5/2^+$ is obtained for the ^{31}Na ground state, if the $0p0h$ configuration is assumed. However, the MCSM predicts the correct spin $3/2^+$ measured experimentally for the ground state. This result is attained by using an essentially larger simulation space in the MCSM model in comparison with the SM USD. It was shown that the experimental values for the ground and first excited ($I = 5/2^+$, $E = 310$ keV) states can be reproduced, if the $2p2h$ configurations are taken into account [28].

An important problem is whether the neutron-rich fluorine isotopes belong to this island of inversion. The main difficulty in describing the properties of fluorine isotopes is that a small number of protons affect the process of filling the neutron shells, which results in anomalous neutron population from the sd to pf shell.

An attempt to describe the properties of the large group of F, Ne, Mg, and Si isotopes was made within the MCSM model by Utsuno et al. [23]. The results are displayed on

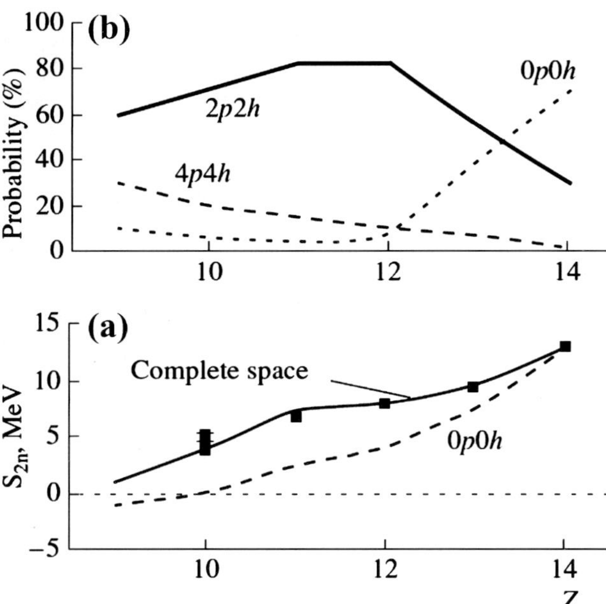

Fig. 4 **a** Two-neutron separation energies for isotopes with $N=20$ [23]. The *full squares* represent the experimental values from [12, 16]. **b** Calculated probabilities of filling the "particle-hole" states: 0p0h, 2p2h and 4p4h [23]

Fig. 4. Figure 4a shows a comparison between the experimental and calculated binding energies S_{2n} as functions of Z for isotopes with $N=20$. As can be seen, the $0p0h$ configuration poorly describes the trend of the experimental dependence. By taking only the $2p2h$ and even $4p4h$ configurations into account, it is possible to achieve a good agreement with the experiment. Our attention is drawn to the fact that the contribution of the two last configurations decreases with increasing Z. This indicates a decreasing deformation in ^{34}Si, whereas the deformation in the nucleus ^{29}F is maximum (the magic neutron number $N=20$).

Meanwhile, it turns out to be difficult to describe the properties of the isotopes ^{29}F and ^{26,28}O only within the MCSM model, because it is impossible to explain the instability of ^{26,28}O and the bound state ^{29}F. An addition of one proton to the oxygen nucleus makes it possible for the neutrons to fill the $0d_{3/2}$ orbital. It results in the fact that neutron-rich fluorine isotopes, $N>16$, become bound, but with a low binding energy. The ^{27}F nucleus has one valence hole, whereas the ^{29}F nucleus does not have a hole state. The nucleus obtains certain strengthening in the binding energy, if a certain vacant valence configuration exists. The authors of [23] assume that the nucleus ^{29}F should be unbound in the $0p0h$ configuration. Thus, for the neutron shell $N=20$, it is improbable to obtain the unbound state ^{26,28}O and the bound state ^{29}F. The latter nucleus starts to become bound only due to the mixing of the $2p2h$ and even $4p4h$ states and, therefore, is rather deformed. The same can be said about other isotopes, ^{30}Ne and ^{32}Mg, with $N=20$, which should also be strongly deformed according to the predictions [23]. As follows from the aforesaid, the calculations of nuclear binding energies can give somewhat increased values due to deformation effects. The mixing of shell configurations entails the mixing of shapes, i.e. it is possible to observe different shapes (deformed and spherical) for the same nucleus (the coexistence of two shapes).

Fig. 5 The potential-energy surface for the nuclei ^{28}Ne, ^{32}Mg, and ^{34}Si [28]

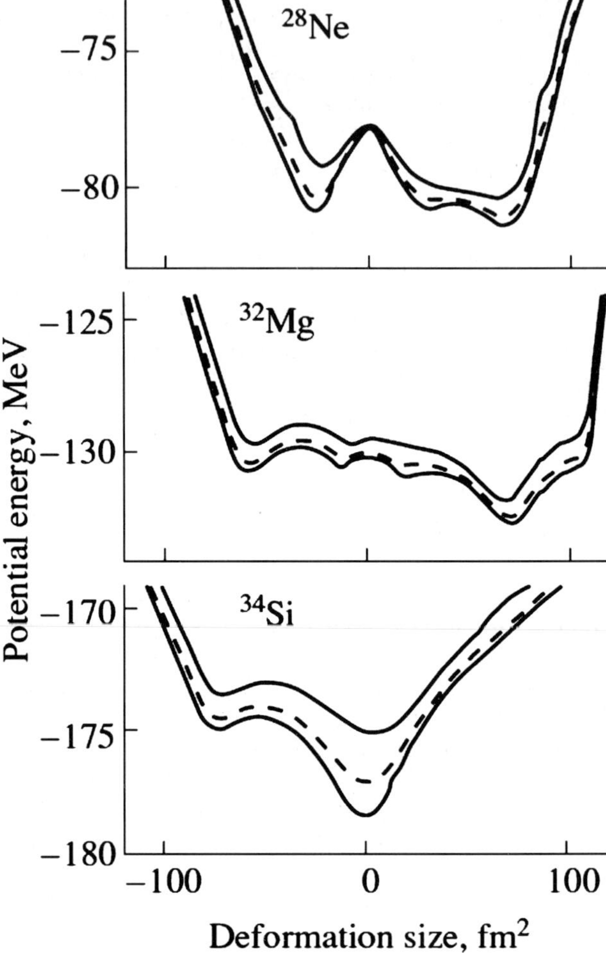

Fig. 6 Average number of neutrons on the *pf* shell as a function of the number of neutrons

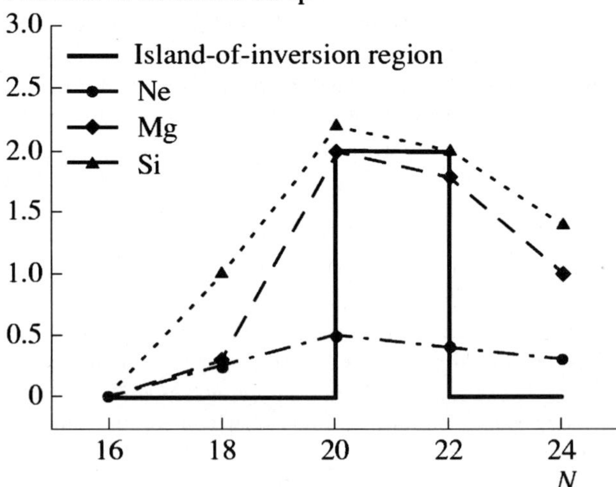

5 Deformation and stability of neutron-rich isotopes of light nuclei

Figure 5 shows the calculated potential-energy curves for the isotopes ^{28}Ne, ^{32}Mg, and ^{34}Si in dependence on the deformation [28]. The presence of local minima of the deformations, which differ from spherical, indicates the so-called coexistence of shapes and, possibly, the existence of shape isomers. Thus, it is possible to re-formulate the definition of the island of inversion in these nuclei as the region where the $2p2h$ configuration makes a greater contribution in the nuclear ground state compared to the $0p0h$ one. The mixing of $0p0h$ and $2p2h$ configurations is illustrated in Fig. 6 [28], where the average number of neutrons on the pf shell is shown as a function of the number of neutrons in the nucleus.

It can be seen that nucleus ^{32}Mg can be considered as a pure $2p2h$ configuration. The largest contribution from the $2p2h$ configuration is observed for $N=20$, and this effect is retained up to $N=24$. For $N=16$, the contribution from the intruder configurations is negligible because, in this case, everything is limited by the $1s_{1/2}$ orbital, which is very far from the pf one.

6 Concluding remarks

As has been shown above, new properties such as the coexistence of shapes, the violation of the shell filling rules, the strong deformation near the shell numbers, and the change in magic numbers manifest themselves in nuclei in the region of $N=20$ and $N=28$. All these facts make this nuclear region extremely interesting for experimental investigations and tests of theoretical models.

The study of light exotic nuclei is at the forefront of nuclear research during the last years. Especially, the very precise laser spectroscopic methods were and still are an important source of information on large number nuclear properties of these nuclei, e.g. deformation, shape coexistence and shell effects. Let us mention for example the absence of any shell effect at $N=20$ observed in the charge radii behavior of Ar, K and Ca isotopic chains [29]; the optical data on ^{31}Na and ^{32}Mg indicating "island of inversion" around shell closure $N=28$ [3, 30, 31]. Thus, the nuclear properties described in this review are extremely interesting from the point of view of laser experiments as well as for all other experimental method giving access to this region. There was and still is a significant impact of the experimental results on the theory and vice versa which lead to new ideas.

In conclusion, let us note that the new information in nuclear region of light exotic would provide a more unambiguous extrapolation of the theoretical and experimental data into the heavier-nuclei region.

References

1. Penionzhkevich, Yu., Lukyanov, S.: Phys. Particles and Nuclei **37**, 5 (2006)
2. Thibault, C., et al.: Phys. Rev. C **12**, 644 (1975)
3. Huber, G., et al.: Phys. Rev. C **18**, 2342 (1978)
4. Guillemaud-Mueller, D., et al.: Nucl. Phys. A **246**, 37 (1984)
5. Motobayashi, T., et al.: Phys. Lett. B **346**, 9 (1995)
6. Otsuka, T., et al.: Phys. Rev. Lett. **87**, 082502 (2001)
7. Samanta, C., et al.: Phys. Rev. C **65**, 037301 (2002)
8. Scheit, H., et al.: Phys. Rev. Lett. **77**, 3967 (1996)
9. Bohr, A., Mottelson, B.: Nucl. Struct., Vol. 1. Benjamin, New York; Mir, Moscow (1971)

10. Glasmacher, T., et al.: Phys. Lett. B **395**, 163 (1997)
11. Sarazin, F., et al.: Preprint GANIL T1999–2003 (1999)
12. Sarazin, F., et al.: Phys. Rev. Lett. **84**, 5062 (2000)
13. Lyutostansky, Yu.S., et al.: Bull. Acad. Sci. S.S.S.R., Ser. Fiz. **53**, 29 (1989)
14. Werner, T.R., et al.: Phys. Lett. B **335**, 259 (1994)
15. Werner, T.R., et al.: Nucl. Phys. A **597**, 327 (1996)
16. Audi, G., et al.: Nucl. Phys. A **624**, 1 (1997)
17. Retamosa, J., et al.: Phys. Rev. C **55**, 1266 (1997)
18. Caurier, E., et al.: Phys. Rev. C **58**, 2033 (1998)
19. Ozawa, A., et al.: Phys. Rev. Lett. **84**, 5493 (2000)
20. Möller, P., et al.: At. Data Nucl. Data Tables **59**, 185 (1995)
21. Otsuka, T., et al.: Eur. Phys. J. A **13**, 69 (2002)
22. Koonin, S.E., et al.: Phys. Rep. **278**, 1 (1997)
23. Utsuno, Y., et al.: Phys. Rev. C **64**, 011301(R) (2001)
24. Utsuno, Y., et al.: Prog. Theor. Phys., Suppl. **138**, 24 (2004)
25. Yoneda, K., et al.: Phys. Lett. B **499**, 233 (2001)
26. Utsuno, Y., et al.: Phys. Rev. C **60**, 054315 (1999)
27. Rodríguez-Guzmán, R., et al.: Phys. Lett. B **474**, 15 (2000)
28. Utsuno, Y., et al.: Nucl. Phys. A **704**, 50 (2002)
29. Blaum, K., et al.: Hyperfine Interact. **162**, 101 (2005)
30. Kowalska, M., et al.: Hyperfine Interact. **162**, 109 (2005)
31. Kowalska, M., et al.: Eur. Phys. J. A **25**, 193 (2005)

Hyperfine Interact (2006) 171:167–172
DOI 10.1007/s10751-007-9512-9

Mg isotopes and the disappearance of magic $N = 20$

Laser and β-NMR studies

Magdalena Kowalska*

Published online: 2 March 2007
© Springer Science + Business Media B.V. 2007

Abstract Collinear laser spectroscopy and β-NMR spectroscopy with optical pumping were applied at ISOLDE/CERN to measure for the first time the magnetic moments of neutron-rich ^{27}Mg, ^{29}Mg, ^{31}Mg and ^{33}Mg, along with the spins of the two latter. The magnetic moment of ^{27}Mg was derived from its hyperfine structure detected in UV fluorescent light, whereas the nuclear magnetic resonance observed in β-decay asymmetry from a polarised ensemble of nuclei gave the magnetic moment of ^{29}Mg. For ^{31}Mg and ^{33}Mg, the hyperfine structure and nuclear magnetic resonance gave the spin and the magnetic moment. The preliminary results for ^{27}Mg and ^{29}Mg are consistent with a large neutron shell gap at $N = 20$, whereas data on ^{31}Mg show that for this nucleus $N = 20$ is not a magic number, which is also the case for ^{33}Mg, based on preliminary analysis. Thus, the two latter isotopes belong to the "island of inversion."

Keywords Properties of nuclei · Electromagnetic moments · Polarised beams · Fine and hyperfine structure

1 Introduction

Laser spectroscopy and similar techniques have played an important role in investigating the ground-state properties of many exotic nuclei. In our recent experiment we have used the collinear laser spectroscopy and the β-NMR method with optical

*Magdalena Kowalska for the IS 427 collaboration at ISOLDE/CERN.

M. Kowalska (✉)
Institut für Physik, Universität Mainz, 55099 Mainz, Germany
e-mail: kowalska@cern.ch

Present Address:
M. Kowalska
CERN, 1211 Geneva 23, Switzerland

pumping to probe an interesting region of the nuclear chart around $Z = 10 - 12$ and $N = 20$ which is characterised by large ground state nuclear deformations inconsistent with a closed $N = 20$ shell [1, 2]. Such behaviour can be explained only by a collapse of the usual filling of the nucleon single-particle levels, thus the region is also known as the "island of inversion." The existence of the "island" carries interesting questions related to its borders, physics mechanism, and connection to other phenomena present far from β stability, such as changing magic numbers [3]. In spite of intensive experimental and theoretical effort the answers to these questions are still not clear.

The nuclei of our interest were neutron-rich Mg isotopes. ^{27}Mg and stable isotopes were studied with classical collinear laser spectroscopy, which uses fluorescence detection. The isotope shifts allowed us to derive changes in their charge radii, whereas the hyperfine structure of ^{27}Mg gave the sign and value of its magnetic moment. The shorter lived odd-neutron ^{29}Mg, ^{31}Mg and ^{33}Mg are well suited for β-NMR investigations with nuclear polarisation achieved by optical pumping. So far, we performed measurements of their hyperfine structure and nuclear magnetic resonance, all observed in the β-decay asymmetry. Such combined studies gave independently the sign and value of the magnetic moment, as well as the spin, which was unknown for ^{31}Mg and ^{33}Mg.

2 Experimental method

The experiments were performed at ISOLDE/CERN where Mg beams were produced by bombarding a thick UC_2 target with 1.4-GeV protons (about 10^{13} protons/s on average). The Mg atoms were selectively laser ionised with the resonance ionisation laser ion source [4] and the ions were accelerated to 60 keV. Next, the mass separated Mg^+ beams were guided to our laser and β-NMR spectroscopy setup, where they were investigated with cw laser beams. A very suitable transition for resonance excitation with fluorescence detection or optical pumping is the 280 nm line of singly charged ions from the atomic ground state $^2S_{1/2}$ to one of the two lowest excited states $^2P_{1/2}$ (D_1 line) and $^2P_{3/2}$ (D_2 line), with about 4 ns lifetime.

Beam energy in the range of 10–100 kV and energy spread around 1–2 eV are the main reasons why laser spectroscopy with a collinear propagation of the ion and laser beams is the most suitable laser-based method in ISOL-type facilities, such as ISOLDE. Its primary advantage is the narrow Doppler width, prerequisite of high resolution and sensitivity, whereas another advantage is the possibility to scan across atomic resonances by changing the acceleration voltage and not the laser frequency. Our experimental setup using this configuration is presented schematically in Fig. 1. To allow the collinear geometry the nuclei of interest are guided into the apparatus via an electrostatic deflector and are overlapped with the laser light which enters straight through a quartz window. After passing beam-shaping elements they reach the section where they are post-accelerated using a tunable voltage of maximum ± 10 kV. This part is followed by the excitation region insulated from both sides by thick plastic flanges, where Mg^+ velocity is tuned into resonance with laser light. The last part of the apparatus hosts the implantation crystal surrounded by the NMR coil with thin scintillation β-detectors and the poles of the NMR magnet placed outside the vacuum chamber.

 Springer

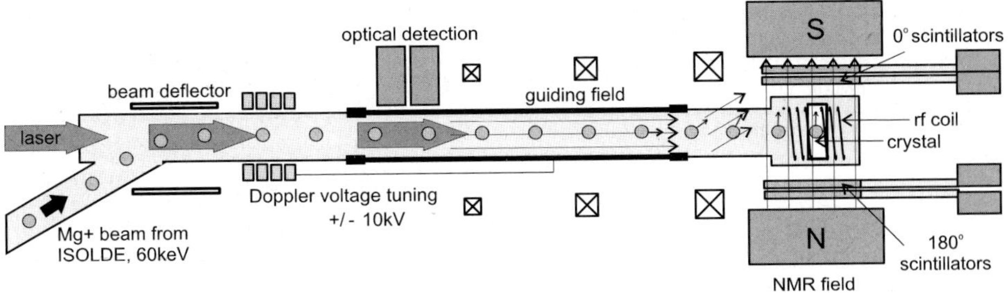

Fig. 1 Collinear laser spectroscopy and β-NMR setup

For fluorescence detection, we used two UV photomultipliers on which the photons were collected by quartz lenses and a cylindrical mirror. With this configuration, we obtained the overall efficiency in the range of 1 detected photon per 10^4 ions, compared to the typical background of 3,000 photons per second per 1 mW laser intensity. The low detection efficiency and high background (mainly due to scattered photons) limits the use of this method to isotopes produced more abundantly than about 10^6 ions/s, corresponding to longer-lived (or stable) Mg isotopes such as ^{27}Mg or ^{28}Mg (see Table 1).

In the case of β-asymmetry detection, the laser light creates nuclear spin polarisation achieved in the optical pumping process [8, 9] in which circularly polarised light polarises the total atomic angular momenta F by inducing transitions between the m_F magnetic sub-levels of the ground and excited state hyperfine multiplets. The theoretical polarisation is lowered due to hyperfine pumping, in which the excited state decays to the other ground state F-level which cannot be excited with the same laser frequency. The optical pumping takes place in the whole isolated section and the quantization axis for σ^+ or σ^- resonance absorption is established by a small longitudinal magnetic field. When the ions reach the fringe field of the NMR magnet, the rotation and subsequent adiabatic decoupling of the nuclear and electron spins takes place. The ions are then implanted into a suitable crystal placed in the centre of the magnet and β-decay electrons are detected in two pairs of thin plastic scintillators, placed at 0 and 180° in front of the magnet poles. When the ions are in resonances with the laser light, the spins are polarised and the intensity of the emitted β particles is asymmetric with respect to the spin direction. For hyperfine structure measurements, the β-decay asymmetry is observed as a function of the acceleration voltage applied to the optical pumping region. On the other hand, for NMR studies the voltage is set at the hyperfine component giving highest β-decay asymmetry and a tunable radio-frequency field (generated by an rf current flowing through a coil placed around the host crystal) is applied perpendicular to the static magnetic field. This detection method is suitable for short-lived Mg isotopes with lifetimes from several milliseconds to several seconds. Due to a very small background and high efficiency for β detection, measurements with down to several hundred ions/s are possible, as shown for ^{29}Mg, ^{31}Mg and ^{33}Mg (see Table 1).

The continuous-wave laser system used for both exciting and optically pumping the ions/atoms of interest consists of an Ar^+ laser (Coherent Innova 400) pumping at 6 W in the multiline visible mode (mainly 488 and 514 nm) a ring dye laser

Table 1 ISOLDE production rates (in ions/s), ground-state spin/parity and electromagnetic moments of neutron-rich Mg isotopes ($Z = 12$) [5–7]

	Yields	$t_{1/2}$	I^{π}	μ_I (μ_N)	Q (mbarn)
^{24}Mg		stable	0^+	0	0
^{25}Mg		stable	$5/2^+$	−0.85546(1)	201(3), 199.4(20)
^{26}Mg		stable	0^+	0	0
^{27}Mg	2×10^8	9.5 min	$1/2^+$	**measured**	0
^{28}Mg	3×10^7	20.9 h	0^+	0	0
^{29}Mg	6.5×10^6	1.3 s	$3/2^+$	**measured**	**measured**
^{30}Mg	1×10^6	335 ms	0^+	0	0
^{31}Mg	1.5×10^5	230 ms	**1/2$^+$**	**−0.88355(15)**	**0**
^{32}Mg	7×10^4	120 ms	0^+	0	0
^{33}Mg	9×10^3	90 ms	**measured**	**measured**	**measured**

In bold: studied by our group.

(Coherent 699-21). The ring output is around 700 mW at 560 nm (with Pyrromethene 556 dye as active medium) and it is additionally frequency-doubled in an external cavity (Spectra-Physics Wavetrain) with angle phase matching and BBO as the non-linear crystal. With 5–10% efficiency, on average 20–50 mW of UV light at 280 nm are available. For optical detection as little as 1 mW is needed, whereas for optical pumping the saturation takes place for about 50 mW power [9] and usual measurements are carried out with 10–20 mW.

3 Results

The hyperfine energy of a level with electron angular momentum J and nuclear spin I is, to the first order, given by

$$E_F = \frac{1}{2}AK + B\,\frac{3/4K(K+1) - I(I+1)J(J+1)}{2I(2I-1)J(2J-1)}\,, \tag{1}$$

where F is the total angular momentum, $K = F(F+1) - I(I+1) - J(J+1)$, and A with B are hyperfine structure parameters proportional respectively to μ_I and Q. These nuclear moments are easily derived from A and B parameters if reference data exists for at least one isotope. In such a case, e.g. for μ_I:

$$\mu_I = \mu_x\,\frac{A\,I}{A_x I_x}\,, \tag{2}$$

where one neglects corrections due to finite charge and magnetisation distribution of the nucleus, which for Mg are in the order of 10^{-3} [10].

The electromagnetic moments of nuclei can be also derived from NMR measurements. For example, in a cubic host crystal with no quadrupole interaction energy, the Zeeman energy of a nucleus with spin I in a strong magnetic field B_0 is given by

$$E(m_I) = -m_I(\mu_I/I)B_0 = -m_I g_I \mu_N B_0 = -m_I h \nu_L\,,$$

where m_I is the magnetic quantum number, g_I is the nuclear g-factor, and ν_L is the Larmor frequency. A radio-frequency field corresponding to ν_L induces transitions

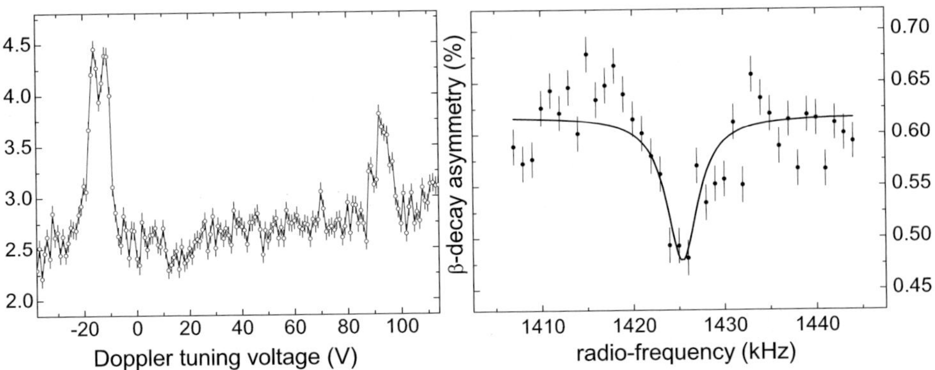

Fig. 2 The hyperfine structure of the D$_2$ transition (*left*) and the NMR resonance (*right*) of ^{29}Mg, observed in the β-decay asymmetry

between Zeeman levels which can be observed as the NMR resonances. By recording the NMR signal on the nucleus of interest and on a reference nucleus with known μ_I, one can determine the unknown g-factor and magnetic moment by using the relation

$$|g_I| = |g_{ref}| \frac{\nu_L}{\nu_{ref}} \, . \tag{3}$$

If the reference is an isotope of a different chemical element, (3) has to include additionally the diamagnetic corrections to account for different shielding of the external magnetic field by the atomic electrons.

In the case when the spin of a given nucleus is not known it can be determined by combining information from the hyperfine structure and NMR. For the splitting between the $J = 1/2$ ground-state sub-levels $F = I + 1/2$ and $F' = I - 1/2$ one obtains $\Delta E = A(I + 1/2)$, where the A-factor can be expressed in g_I and the ratio A/g_I known for the reference isotope

$$I = \frac{\Delta E}{g_I} \frac{g_{ref}}{A_{ref}} - 1/2 \, . \tag{4}$$

By measuring optically the hyperfine structure of ^{27}Mg and ^{25}Mg, and using the known magnetic moment of the latter (see Eq. 2), we were able to determine for the first time the magnetic moment of ^{27}Mg [11], with an uncertainty below 0.5%. In a similar way, but using the β-asymmetry detection, we recorded the hyperfine structure of shorter-lived ^{29}Mg and derived its magnetic moment, whose precision we increased to 0.15% by β-NMR measurements (see Fig. 2) [11]. Since the ground-state spins were not known for ^{31}Mg and ^{33}Mg, the analysis in these two cases combined the hyperfine structure and NMR studies. The results comprise unexpected spins and magnetic moments with uncertainty as low as 0.02% for ^{31}Mg [7] and 0.2% for ^{33}Mg [12].

Comparisons with shell model calculations show that in the ground state of ^{31}Mg the neutron shell is open at $N = 20$ [7], which places this nuclide inside the "island of inversion". The analysis for ^{27,29,33}Mg is still not finalised, therefore no values can be presented at this point. However, preliminary results already show that in the ground state of ^{27}Mg and ^{29}Mg $N = 20$ remains a good shell closure [11], which is not the case for ^{33}Mg [12]. Thus, the latter isotope belongs to the intriguing "island".

Future plans in this region of the nuclear chart include isotope shift measurements up to ^{33}Mg, which will yield changes in charge radii for different Mg isotopes, and will thus show deformation areas. These studies will combine the fluorescence detection with the β-decay asymmetry detection.

The above studies are presented in the Ph.D. theses of M.K. and D. Yordanov.

Acknowledgements   We would like to thank the ISOLDE technical group for their assistance during the experiment, and the following institutions for their financial support: the German Federal Ministry for Education and Research (BMBF) – contract no. 06MZ175 and 06MZ215, the Belgian Fund for Scientific Research – Flanders (FWO), and the European Union in the framework of the HPRI program.

References

1. Thibault, C. et al.: Direct measurement of the masses of ^{11}Li and $^{26-32}$Na with an on-line mass spectrometer. Phys. Rev., C **12**, 644 (1975)
2. Detraz, C. et al.: Beta decay of $^{27-32}$Na and their descendants. Phys. Rev., C **19**, 164 (1979)
3. Grawe, H.: Shell model from a practitioners point of view. Lecture Notes in Physics **651**, 33 (2004)
4. Fedoseyev, V.N. et al.: The ISOLDE laser ion source for exotic nuclei. Hyperfine Interact. **127**, 409 (2000)
5. Raghavan, P.: Table of nuclear moments. At. Data Nucl. Data Tables **42**, 189 (1989)
6. Firestone, R.B., Tauren, J.: LBNL Isotopes Project Nuclear Data Dissemination Home Page, http://ie.lbl.gov/toi.html
7. Neyens, G. et al.: Measurement of the spin and magnetic moment of ^{31}Mg: evidence for a strongly deformed intruder ground state. Phys. Rev. Lett. **94**, 022501 (2005)
8. Kowalska, M. et al.: Laser and β-NMR spectroscopy on neutron-rich magnesium isotopes. Eur. Phys. J., A **25s01**, 193 (2005)
9. Kowalska, M., for the IS427 collaboration: Laser spectroscopy and β-NMR measurements of short-lived Mg isotopes. Hyperfine Interact. **162**, 109 (2005)
10. Kopfermann, H.: Nuclear moments (1969)
11. Kowalska, M. et al.: Publication in preparation
12. Yordanov, D. et al.: Publication in preparation

Hyperfine Interact (2006) 171:173–179
DOI 10.1007/s10751-006-9505-0

Laser spectroscopy measurements of neutron-rich tellurium isotopes by COMPLIS

R. Sifi · F. Le Blanc · N. Barré · L. Cabaret ·
J. Crawford · M. Ducourtieux · S. Essabaa ·
J. Genevey · G. Huber · M. Kowalska · C. Lau ·
J. K. P. Lee · G. Le Scornet · J. Oms · J. Pinard ·
B. Roussière · J. Sauvage · M. Seliverstov · H. Stroke ·
ISOLDE

Published online: 9 February 2007
© Springer Science + Business Media B.V. 2007

Abstract Laser spectroscopy based on resonant ionization of laser-desorbed atoms has been used to study the neutron-rich tellurium isotopes with the COMPLIS facility at ISOLDE-CERN. Isotope shifts and hyperfine structures of several neutron-rich Te isotopes: $^{120-136}$Te and $^{123\,m-133\,m}$Te have been measured. From the hyperfine structure we have extracted magnetic and quadrupole moments. Changes in the mean square charge radii have been deduced and their comparison with the known data for the other elements near $Z=50$ is presented. The experimental $\delta\langle r^2\rangle$ values are compared with those obtained from relativistic mean field calculations.

Key words laser spectroscopy · isotope shift · hyperfine structure · nuclear moments

R. Sifi (✉) · F. Le Blanc · N. Barré · M. Ducourtieux · S. Essabaa · C. Lau ·
J. Oms · B. Roussière · J. Sauvage
Institut de Physique Nucléaire, IN2P3-CNRS, 91406 Orsay cedex, France
e-mail: sifi@ipno.in2p3.fr

L. Cabaret · J. Pinard
Laboratoire Aimé Cotton, 91405 Orsay cedex, France

J. Crawford · J. K. P. Lee
Physics Department, Mc Gill University, H3A2T8 Montréal, Canada

J. Genevey
Institut des Sciences Nucléaires, IN2P3-CNRS, 38026 Grenoble cedex, France

G. Huber · M. Kowalska · M. Seliverstov
Institut für Physik der Universität Mainz, 55099 Mainz, Germany

H. Stroke
Department of physics, New York University, NY103 New York, USA

G. Le Scornet · ISOLDE
CERN, 1211, Geneva 23, Switzerland

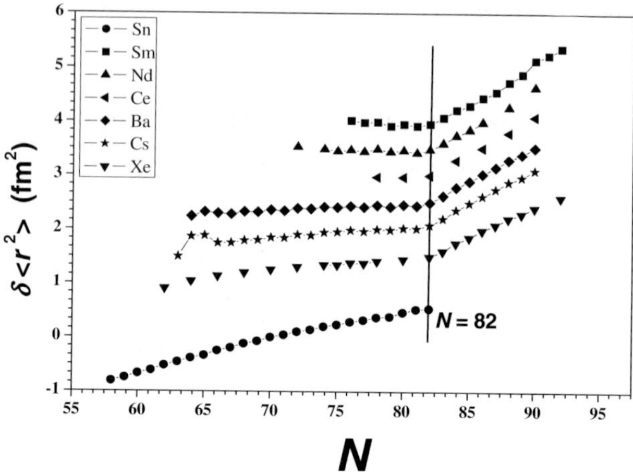

Fig. 1 Changes in the mean square charge radius for elements above $Z=50$

1 Introduction

Laser spectroscopy gives access to the hyperfine structure and the isotope shift. From these we can obtain the properties of nuclei in their ground and isomeric states, namely the nuclear moments and the change in the mean square charge radius.

Tellurium belongs to the region of ^{132}Sn. The variation of the mean square charge radius ($\delta<r^2>$) of several elements in this region shows a change of slope at $N=82$ (see Fig. 1). We also see that, the more Z differs from $Z=50$, the more important is the kink at $N=82$. Moreover, the measurements of the quadrupole and magnetic moments of the odd Te isotopes and isomers inform us about the structure of the states.

In this contribution we present a laser spectroscopy experiment recently performed on neutron-rich Te isotopes. We recorded for the first time the hyperfine spectra of $^{130,\,132,\,134,\,136}$Te and also of $^{123\,m,g,\,125\,m,g,\,127\,m,g,\,129\,m,g,\,131\,m,g,\,133\,m,g,\,135\,g}$Te. The changes in the mean square charge radii; $\delta<r^2>$, of the even–even isotopes has been deduced with preliminary values of the odd isotopes too. These results are compared with the other charge radii changes known in the $Z\sim50$ and $N\sim82$ regions and with those from relativistic mean field calculations. The nuclear moments were deduced also for $^{125\,m,\,127\,m,\,129\,m,\,131\,m,\,133\,mg,\,135\,g}$Te and were compared to the existing values.

2 The set up

The COMPLIS set up (COllaboration for spectroscopy Measurement using a Pulsed Laser Ion Source) at ISOLDE-CERN (Fig. 2) is based on resonant ionization laser spectroscopy. The Te beam delivered by ISOLDE is produced by a uranium carbide target associated to a hot plasma ion source. The beam is accelerated to 60 keV, enters the incident line of COMPLIS, crosses the magnet and is electrostatically decelerated to 1 keV. It is finally collected on the first atomic layers of a graphite substrate. The collected atoms are then laser desorbed and ionized. The ions produced are accelerated, deflected by the magnet and detected by micro channel plates with time-of-flight identification. The COMPLIS setup is described in details in ref. [1].

 Springer

Fig. 2 The COMPLIS set up

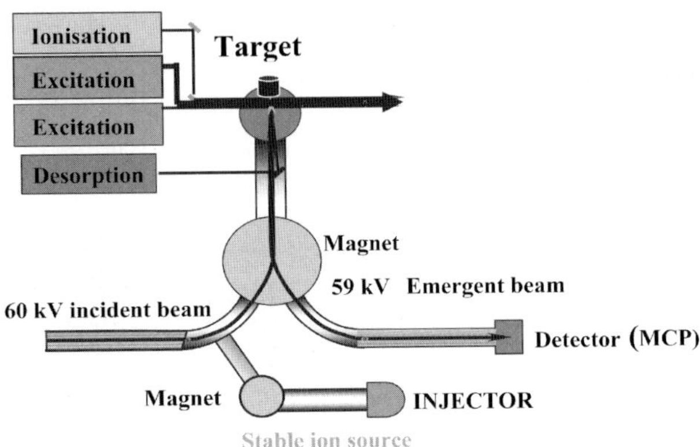

COMPLIS: Collaboration for spectroscopy Measurement Using a Pulsed Laser Ion Source

An injector, a small separator which delivers stable beams, is linked to the incident line of COMPLIS (see Fig. 2). It is used to determine the best experimental conditions of desorption and ionization and to measure the overall efficiency.

With this injector we have recorded the hyperfine spectra of all the stable Te isotopes. The hyperfine structure of the 2 odd stable isotopes was obtained for the first time. The overall efficiency was determined to be around 5.10^{-7} and the resolution was about 250 MHz.

3 Procedure

To ionize Te we have used a 3 step scheme (Fig. 3a): 214 nm for the first excitation, 591 nm for the second excitation and 1064 nm to ionize. For this purpose we used 4 lasers at COMPLIS: 2 Nd:YAG, one for desorption and the other is used to pump the 2 dye lasers.

In this experiment we need to separate the different hyperfine transitions, so the first excitation laser must be tunable at high resolution. For this reason the Aimé Cotton laboratory at Orsay has developed a laser system that gathers these 2 characteristics. It is based on an injection process [2]. This comprises of a tunable continue wave monomode dye laser lock injected by a 10 Hz 532 nm pulsed beam. It delivers a 50 MHz line width pulsed beam at 642 nm, that is frequency tripled to obtain 214 nm.

The second excitation step is given by a commercial lambda-physik FL3002 dye laser pumped by 532 nm from the Nd:YAG. To obtain the 591 nm we used only the oscillator of this laser and the dye used was Rhodamine 6 G dissolved in methanol. The amplifier wasn't used in this case because the intensity of the laser was sufficient to saturate the transition. The IR for ionization was directly obtained from 1,064 nm Nd:YAG beam (Fig. 3a).

To measure the hyperfine structure we perform a frequency scan of the first excitation frequency. When the laser frequency corresponds to a resonant transition, the atoms are excited, ionized and detected. We finally get a frequency spectrum where every peak corresponds to an ion number proportional to the intensity of the hyperfine transition (Fig. 3b, c).

From the hyperfine spectra obtained, we have determined the center of gravity for each mass. The isotope shift is the difference in frequency of these centers of gravity between two isotopes. From the isotope shift measured we can deduce the mean square charge radius. As an example we present in Fig. 4 the spectra obtained for $^{130,\ 132,\ 134,\ 136}$Te and $^{127\ m}$Te. The displacement of the center of gravity of the resonant peak is here clearly observed.

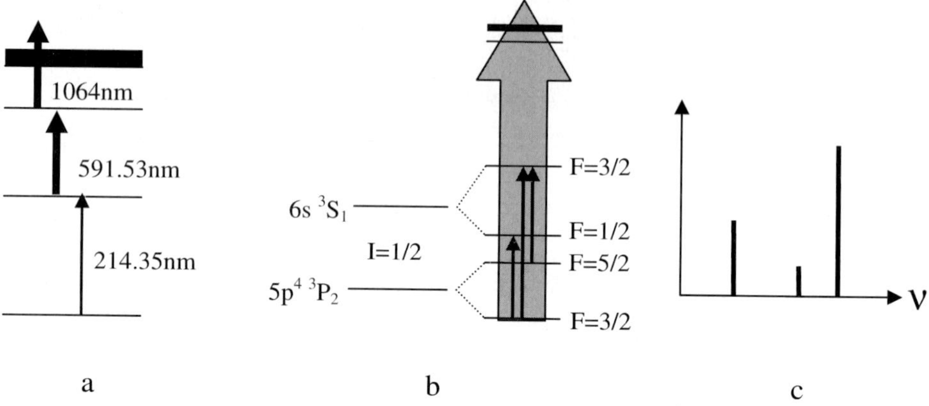

Fig. 3 **a** The laser configuration used to ionize Te. **b** & **c** The frequency scan of the first excitation step and a typical frequency spectrum obtained after the scan

4 The isotope shift and charge radii

The isotope shift includes the contribution of the mass effect and the volume effect [3], and is given by this equation:

$$\Delta v^{AA'} = F\delta < r^2 >^{AA''} +M \cdot \frac{A' - A}{A'A},$$

where F is an atomic factor depending on the transition and M depends on the correlation between the electrons. We can thus determine the mean square charge radius change if F and M are known. F and M can be calculated, or deduced from experimental data using a King-plot. A King-plot consists of plotting the isotope shift versus the change in the mean square charge radius for all the possible pairs of isotopes [4]. They are related by a linear variation where the slope gives the F factor and the intercept the mass effect M.

To perform the King-plot, we used the charge radius values known for the stable isotopes from muonic atoms experiment [5] and our measured isotope shifts. The slope gives F=4.51±0.88 GHz/fm^2 and the intercept gives M=−675±239 MHz.

With these 2 parameters we have extracted the charge radius variation of the neutron rich isotopes 125, 127, 129, 131, 132, 133 m, 134, 135 and 136 g. We present in Fig. 5a the charge radius changes of all the Te isotopes (our results and ref [5]). As can be seen, the evolution of $\delta<r^2>$ is very regular from A=120–134 and a kink appears at N=82.

We have compared our results with relativistic mean field calculations using the Lagrangian parameterization NL3 of Lalazissis et al. [6] (Fig. 5b). The agreement is good for the mass 134 and 136, but beyond A=134, this calculation under estimates the kink of the charge radius at N=82. Other theoretical calculations are in progress including dynamic models and their comparison with our results will be made later.

5 Nuclear moments

In this experiment we measured the hyperfine spectra for all the odd isotopes and isomers, from which we extracted the hyperfine constants A and B. A and B are related

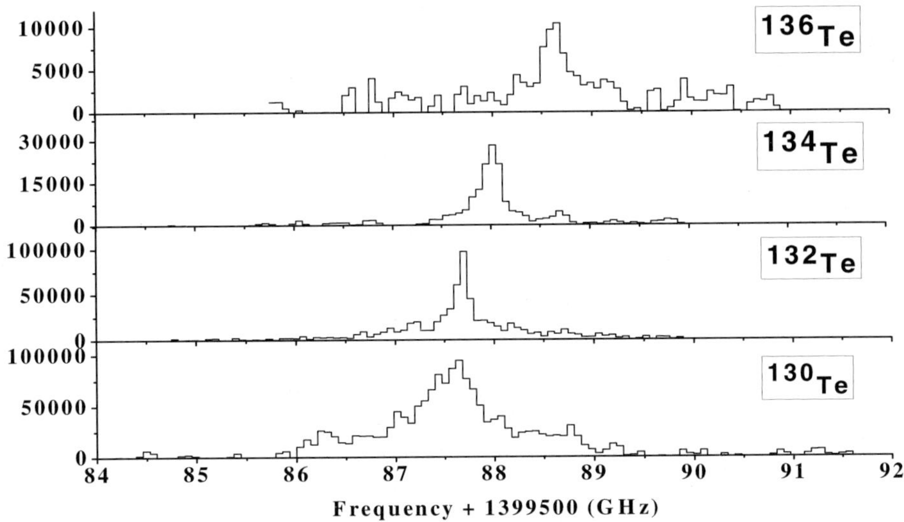

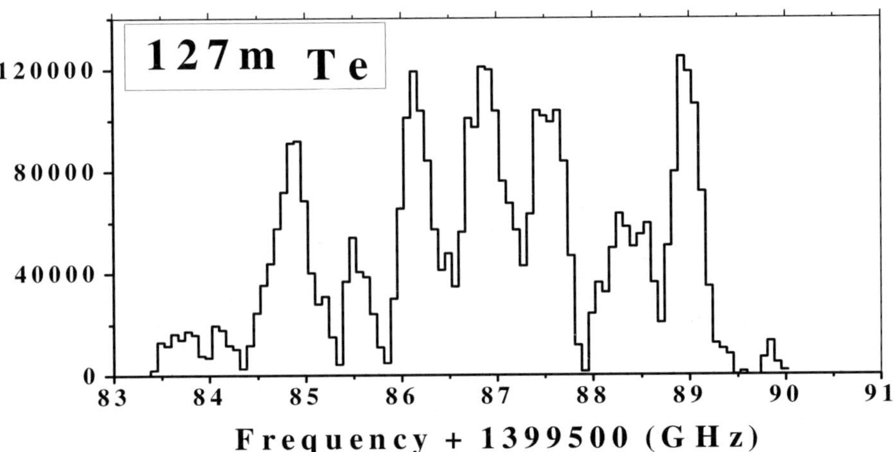

Fig. 4 Spectra obtained for $^{130,\ 132,\ 134,\ 136}$Te (**a**) and $^{127\ m}$Te (**b**)

to the magnetic moment and the quadrupole moment, respectively by the following equations [7]:

$$\mu_I = A\,\frac{IJ}{H(0)} \quad \text{and} \quad qQ_s = \frac{B}{\varphi_{JJ}(0)}$$

Here $H(0)$ and $\varphi_{JJ}(0)$ are, respectively, the magnetic field and the potential created by the electrons in the nucleus. So we can deduce the nuclear moments from the measurements of hyperfine spectra and the extraction of the hyperfine constants A and B, knowing $H(0)$ and $\varphi_{JJ}(0)$.

$H(0)$ was measured from the experimental data that we obtained with stable Te isotopes for which the magnetic moment was known with a good accuracy. For this we used the hyperfine spectrum that has been recorded with COMPLIS for the stable isotope ^{125}Te. The

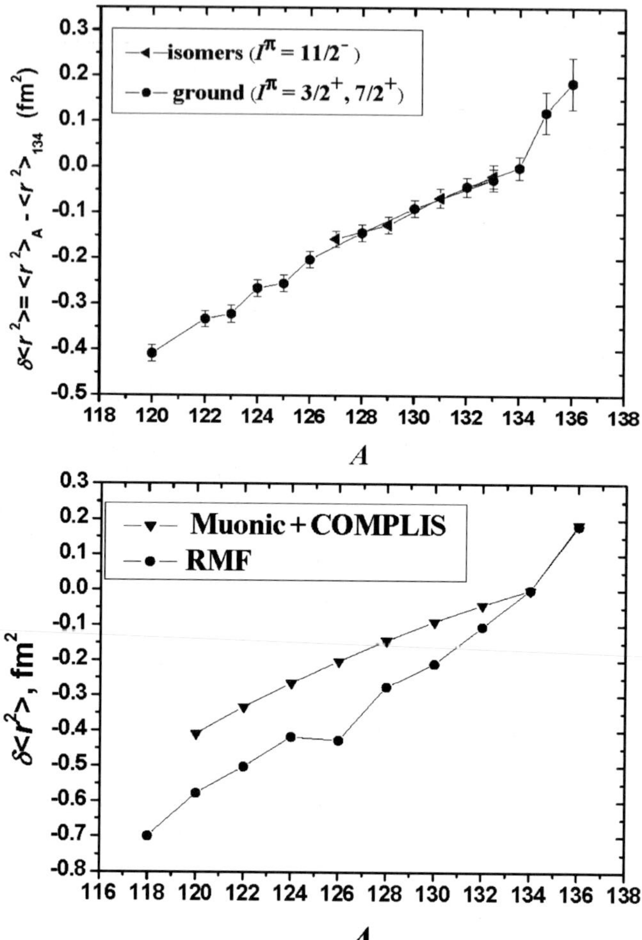

Fig. 5 The mean square charge radius change measured (**a**), compared to those calculated by the RMF theory (**b**)

magnetic moment of this isotope is $\mu(^{125}\text{Te}) = -0.8885051(4)$ nm [8]. The field calculated is $H(0) = 1.12 \pm 0.02$. While $H(0)$ depend only on the electronic cloud, it is constant along the isotopic chain. We used this value of $H(0)$ to calculate the magnetic moments for all the Te isotopes from the hyperfine structure measured. Hyperfine anomaly was neglected with respect to the experimental uncertainties. The errors on the magnetic moments take into account the errors from the peak fitting and from the calibration.

No accurate value for the quadrupole moment was available to calculate $\varphi_{JJ}(0)$ the same way as for $H(0)$. So we calculate it by semi-empiric methods using the fine structure of the ground state [7]. We found $\varphi_{JJ}(0) = 5.83 \cdot 10^{22}$ V/m^2. Then, $Q_s = 0.71B$ [b] with B in [GHz] for all the other isotopes. The model uncertainties are not taken into account. The errors reported on Table 1 are due to the errors on the hyperfine constant B only.

In Table 1 are summarized the experimental values of nuclear moments that we obtained compared to the values in the literature. We see that we are in good agreement with the nuclear moments already obtained. We compared our magnetic moments to the Schmidt values calculated with a $g_l = 0$ and $g_s = -3.826$. The Schmidt magnetic moment is higher than what we obtained.

 Springer

Table 1 Comparison of experimental nuclear moments

Mass	I^{π}	μ_{ref}/μ_N	μ_{Schmidt} [μ_N]	μ_{exp}/μ_N	Q_{sref}, b	Q_{sexp}, b	β
135 g	$7/2^-$		-1.913	$-0.69(5)$		$0.29(9)$	$0.04(1)$
133 g	$3/2^+$		1.148	$0.85(2)$		$0.23(9)$	$0.04(2)$
133 m	$11/2^-$	$(-)1.129$	-1.913	$-1.15(09)$		$0.28(14)$	$0.03(2)$
131 m	$11/2^-$	$-1.04(4)$ [9]	-1.913	$-1.20(14)$		$0.25(14)$	$0.03(16)$
129 m	$11/2^-$	$-1.091(7)$	-1.913	$-1.10(3)$		$0.40(3)$	$0.045(3)$
127 m	$11/2^-$	$-1.041(6)$ [9]	-1.913	$-1.05(12)$		$0.17(12)$	$0.02(1)$
125 m	$11/2^-$	$-0.985(6)$ [9]	-1.913	$-0.98(16)$	$-0.06(2)$ [10]	$0.02(17)$	$-0.007(2)$

Concerning the quadupole moments no comparison with the literature is available because they were not measured before except for $^{125\ m}$Te. For this element the difference between the experimental value that we measured and the value from reference [10] is big and can be included in the error bars. The deformation parameter β deduced from these quadrupole moments is weak which confirm the spherical shape of Te nuclei as predicted.

These results are preliminary until now; they need to be confirmed in the near future with a detailed interpretation.

6 Conclusion

The laser spectroscopy experiment performed on n-rich Te isotopes with the COMPLIS setup enabled us to measure the hyperfine structure and the isotope shift of a large Te isotopes and isomers. The charge radii has been determined and a kink is observed at $N=82$ as for Xe, Cs, Ba, Nd and Sm. The analysis of the odd isotopes and isomers allowed the determination of the nuclear moments which are in good agreement with the existing values.

References

1. Sauvage, J., et al.: Hyperfine Interact. **129**, 303 (2000)
2. Pinard, J., Liberman, S.: Opt. Commun. **20**, 344 (1977)
3. King, W.H.: Isotope Shift in Atomic Spectra. Plenum, New York (1984)
4. King, W.H., et al.: Z. Phys. **265**, 207 (1973)
5. Shera, E.B., et al.: Phys. Rev., C **39**, 195 (1989)
6. Lalazissis, G.A., et al.: Atom. Data and Nucl. Data Tables **71**, 1 (1999)
7. Kopfermann, H.: Nuclear Moments. Academic, New York (1958)
8. Weaver, H.E. Jr: Phys. Rev. **89**, 923 (1953)
9. Lhersonneau, G., et al.: Phys. Rev., C **12**, 609 (1975)
10. Berkes, I., et al.: Hyperfine Interact. **35**, 1023 (1987)

Hyperfine Interact (2006) 171:181–188
DOI 10.1007/s10751-007-9507-6

Nuclear charge radius of ^{11}Li

Rodolfo Sánchez · Wilfried Nörtershäuser ·
Andreas Dax · Guido Ewald · Stefan Götte ·
Reinhard Kirchner · H.-Jürgen Kluge ·
Thomas Kühl · Agnieszka Wojtaszek ·
Bruce A. Bushaw · Gordon W. F. Drake ·
Zong-Chao Yan · Claus Zimmermann ·
Daniel Albers · John Behr · Pierre Bricault ·
Jens Dilling · Marik Dombsky · Jens Lassen ·
C. D. Phil Levy · Matthew R. Pearson ·
Erika J. Prime · Vladimir Ryjkov

Published online: 1 February 2007
© Springer Science + Business Media B.V. 2007

R. Sánchez (✉) · W. Nörtershäuser · A. Dax · G. Ewald · S. Götte · R. Kirchner ·
H.-J. Kluge · T. Kühl · A. Wojtaszek
Gesellschaft für Schwerionenforschung, D-64291 Darmstadt, Germany,
e-mail: R.Sanchez@GSI.DE

B. A. Bushaw
Pacific Northwest National Laboratory, P.O. Box 999, Richland, WA 99352, USA

G. W. F. Drake
Department of Physics, University of Windsor, Windsor, Ontario, Canada N9B 3P4

Z.-C. Yan
Department of Physics, University of New Brunswick, Fredericton, New Brunswick,
Canada E3B 5A3

C. Zimmermann
Eberhard Karls Universität Tübingen, Physikalisches Institut, D-72076 Tübingen, Germany

D. Albers · J. Behr · P. Bricault · J. Dilling · M. Dombsky · J. Lassen · C. D. Phil Levy ·
M. R. Pearson · E. J. Prime · V. Ryjkov
Tri-University Meson Facility, Vancouver, British Columbia, Canada V6T 2A3

Present Address:
A. Dax
CERN, CH-1211 Geneva 23, Switzerland

Present Address:
A. Wojtaszek
Institute of Physics, Swietokrzyska Academy, PL-25-406 Kielce, Poland

Abstract We have determined the nuclear charge radius of ^{11}Li by high-precision laser spectroscopy. The experiment was performed at the TRIUMF-ISAC facility where the ^{7}Li-^{11}Li isotope shift (IS) was measured in the $2s \to 3s$ electronic transition using Doppler-free two-photon spectroscopy with a relative accuracy better than 10^{-5}. The accuracy for the IS of the other lithium isotopes was also improved. IS's are mainly caused by differences in nuclear mass, but changes in proton distribution also give small contributions. Comparing experimentally measured IS with advanced atomic calculation of purely mass-based shifts, including QED and relativistic effects, allows derivation of the nuclear charge radii. The radii are found to decrease monotonically from ^{6}Li to ^{9}Li, and then increase with ^{11}Li about 11% larger than ^{9}Li. These results are a benchmark for the open question as to whether nuclear core excitation by halo neutrons is necessary to explain the large nuclear matter radius of ^{11}Li; thus, the results are compared with a number of nuclear structure models.

Key words laser spectroscopy · nuclear charge radius · isotope shift · halo nucleus · lithium

1 Introduction

Some of the lightest neutron-rich nuclei have been found to have much larger root mean squared (rms) radius than their neighboring isotopes [1]. Very well known is the neutron-drip-line nucleus ^{11}Li. Its large nuclear matter radius has been explained in terms of the small two-neutron separation energy (375 keV [2]), which allows the neutron wavefunction to extend far outside the nuclear core as a result of quantum mechanical tunneling. Thus, ^{11}Li is pictured as a ^{9}Li-like core plus the two weakly-bound neutrons forming a "halo" around it. The size of this halo can be as large as the matter radii found in heaviest naturally occurring elements. In order to understand this fascinating structure, extensive experimental and theoretical studies have been performed on this halo nucleus. However, the distribution of the three protons inside the core is not yet completely understood. Experimentally, this can be probed by observing a change in the nuclear charge radius between ^{9}Li and ^{11}Li. Combining isotope shifts determined by high-resolution laser spectroscopy with high-accuracy atomic theory calculations has lead to a new measurement principle for the determination of nuclear charge radii for short-lived isotopes of very light nuclei [6]. This method has already successfully been applied in two- (^{6}He [3]) and three-electron (^{8}Li, ^{9}Li [4]) systems, where recent advances in atomic theory calculations have provided sufficiently accurate values for the mass effect. In this contribution, we discuss how this technique was used for the determination of the ^{11}Li nuclear charge radius [5], and the results are compared with a variety of nuclear structure models.

2 Isotope shift and nuclear charge radius

The extraction of nuclear charge radii from optical isotope shifts (IS) is based on the fact that the frequency of an atomic transition is shifted between two isotopes A and A' according to the change in their mean-square nuclear charge radii ($\langle r_\mathrm{c}^2 \rangle$) as,

$$\delta \nu_{\mathrm{FS}}^{A,A'} = -\frac{2\pi}{3} Z e^2 \Delta |\Psi(0)|^2 \left(\langle r_\mathrm{c}^2 \rangle_A - \langle r_\mathrm{c}^2 \rangle_{A'} \right), \tag{1}$$

 Springer

Table 1 Values for the experimental isotope shift ($\delta\nu^{A,7}$), calculated mass shift ($\delta\nu_{MS}^{A,7}$)*, changes in the mean-square nuclear charge radii ($\delta\langle r_c^2\rangle$) [5] and root-mean-square nuclear charge radii (r_c) for the lithium isotopes

Quantity	^{6}Li	^{8}Li	^{9}Li	^{11}Li
$\delta\nu^{A,7}$, MHz	-11,453.983(20)	8,635.782(44)	15,333.226(39)	25, 101.226(124)
$\delta\nu_{MS}^{A,7}$, MHz	-11,453.010(56)	8,635.113(42)	15,332.025(75)	25,101.812(123)
$\delta\langle r_c^2\rangle^{A,7}$, fm^2	0.622(38)	-0.427(39)	-0.796(54)	0.374(54)
r_c, fm	2.517(30)	2.299(32)	2.217(35)	2.467(37)

Uncertainties for r_c are dominated by the uncertainty of the reference radius $r_c(^7\text{Li}) = 2.39(3)$ fm.

*After completing this report, more accurate mass shift calculations were published by M. Puchalski et al. [7], which differ from those given in [6], and slightly shift the nuclear charge radii. The largest difference is for ^{11}Li, where it is reduced by 0.045 fm.

where $\Delta|\Psi(0)|^2$ is the change of the expectation value of the electron density at the nucleus of charge Ze. $\delta\nu_{FS}^{A,A'}$ is known as the field shift (FS). For the low-Z elements, like lithium, the purely mass-based portion of the IS, the mass shift (MS), can be calculated with sufficient accuracy such that the residual difference between experimental IS and calculated MS can be attributed to the FS, and thus be related to the changes in nuclear charge radii (1). These calculations solve the three-electron non-relativistic Schrödinger equation with high accuracy and include relativistic and quantum electrodynamic effects by perturbation theory. Results of such calculations [6] are given for all lithium isotopes in Table 1.

3 Experimental

To determine the nuclear charge radius of ^{11}Li, the experimental setup was installed at the TRIUMF-ISAC facility in Vancouver, Canada where one of the world's highest yield ($\sim$ 30,000/s) of low-energy ($\approx$ 40 keV) ^{11}Li ions is produced by bombarding a tantalum target with a 40 μA, 500 MeV continuous proton beam from a cyclotron. In our experiment (see Fig. 1) the ^{11}Li ions are stopped inside a thin carbon foil (thickness $\approx$ 300 nm), and then thermally released as neutral atoms by heating the foil with a CO_2 laser. The atoms cross the focus of two overlapping laser beams: a titanium–sapphire (Ti:Sa) laser at 735 nm induces Doppler-free two-photon transitions from the $2s$ ground levels to the $3s$ excited levels. The atoms then decay to the $2p$ levels, where they are resonantly excited to $3d$ levels by a dye laser at 610 nm, and then photoionized by absorption of another photon from either of the lasers. The resulting ions are mass separated with a quadrupole mass filter and detected with a continuous-dynode electron multiplier (CDEM). To maximize excitation and ionization efficiency, both laser powers of several 100 mW are increased by a factor of 100 in a resonant optical cavity (30 cm length) built around the excitation region. To couple both lasers into the same cavity, the cavity is locked to the Ti:Sa laser while the dye laser is locked to the cavity. The Ti:Sa laser is stabilized by frequency-offset locking to a reference diode laser, which is locked to an I_2 hyperfine line by Doppler-free saturated absorption. The overall detection efficiency is greater than 10^{-4}.

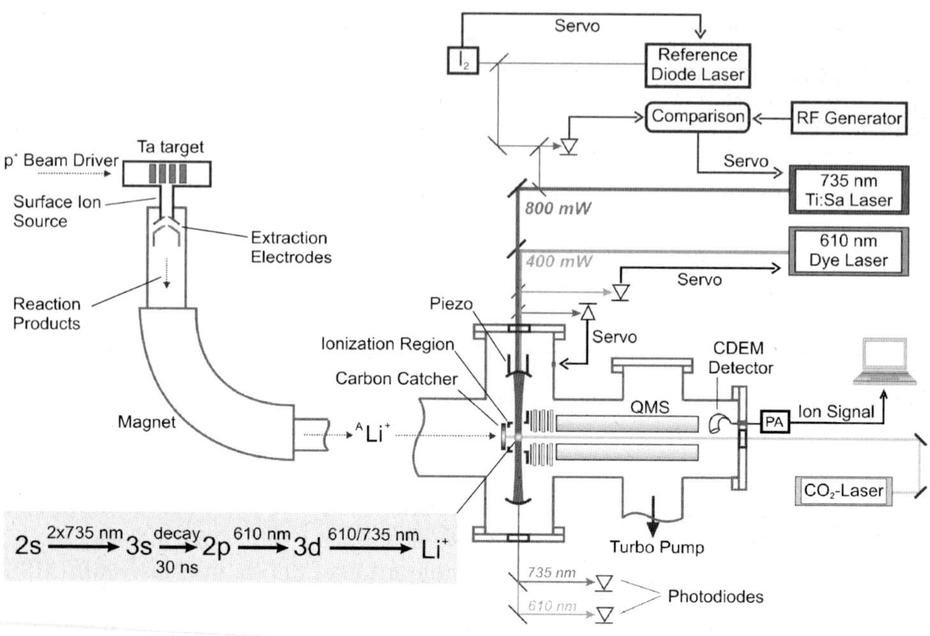

$$2s \xrightarrow{2\times735\ nm} 3s \xrightarrow[30\ ns]{decay} 2p \xrightarrow{610\ nm} 3d \xrightarrow{610/735\ nm} Li^+$$

Fig. 1 Experimental setup and excitation scheme for the resonance ionization of lithium

4 Results

A typical spectrum recorded for ^{11}Li is shown in Fig. 2. The nuclear spin of ^{11}Li ($I = 3/2$) splits the $2s$ and $3s$ atomic energy levels into $F = 1$ and $F = 2$ hyperfine components. Selection rules for $s \rightarrow s$ two-photon transitions allow only $\Delta F = 0$ and thus only two transitions are observed. The isotope shift is calculated as the center of gravity (cg) of the two hyperfine components, relative to the cg of a reference isotope, ^{7}Li. In total, twenty-four spectra like that in Fig. 2 were obtained for ^{11}Li. Spectra were also obtained for the other lithium isotopes and resulting isotope shifts are given in Table 1.

Root-mean-square nuclear charge radii extracted from these measurements in combination with mass shift calculations are also given in Table 1, where the nuclear charge radius of ^{7}Li ($r_c = 2.39(3)$ fm) measured by electron scattering [8] has been used as the reference. These radii are represented by the black points in Fig. 3. Experimental uncertainties in nuclear charge radii for ^{6,8,9}Li are dominated by the uncertainty in the ^{7}Li reference radius, while the spectroscopic measurements contribute an uncertainty of $\sim$0.022 fm for ^{11}Li. The values obtained for $^{6-9}$Li are in excellent agreement with our previous measurements at GSI [4] but have improved precision. In Fig. 3 one can clearly see that the nuclear charge radius decreases continuously from ^{6}Li to ^{9}Li. This decrease can be understood in terms of clustering of the different nuclei. For example, ^{6}Li is known to be strongly clustered into an α-particle and a deuteron. Adding more neutrons, this cluster structure tends to "melt" and a more compact object is formed. Beyond the decrease of nuclear charge radii from ^{6}Li to ^{9}Li, a significant increase is observed for ^{11}Li, indicating a large change in the behavior of the ^{9}Li-like core.

 Springer

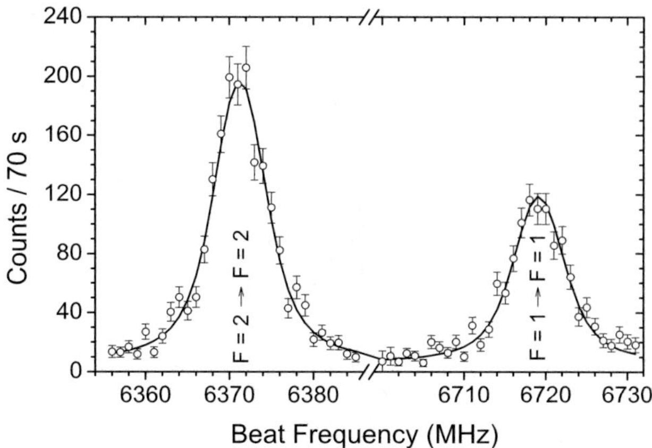

Fig. 2 Typical resonance-ionization spectrum in the $2s \rightarrow 3s$ transition recorded for ^{11}Li. The x-axis represents the beat frequency between the reference diode laser and the titanium-sapphire laser

Nuclear charge radii predicted by different nuclear models are also shown in Fig. 3. These theories construct the equations of motion that describe light nuclei using realistic two- and three- nucleon potentials. These potentials are based on meson-exchange interactions, with parameters that are usually determined by fitting to experimental nucleon-nucleon scattering data. Once these potentials are defined they can be used to numerically calculate wave functions, ground state energies, and density and momentum distributions numerically. Such calculations require rapidly expanding computer resources for every nucleon added and are currently limited to systems of mass number ≤ 12.

The nuclear charge radii predicted by models that treat protons and neutrons as point-like particles were converted from point-proton mean-square radii $\langle r_\mathrm{p}^2 \rangle$ into mean-square nuclear charge radii $\langle r_\mathrm{c}^2 \rangle$ by folding in the proton mean-square charge radius [17], $\langle R_\mathrm{p}^2 \rangle = 0.801(32)$ fm^2, and the neutron mean-square charge radius [18], $\langle R_\mathrm{n}^2 \rangle = -0.117(4)$ fm^2, according to [19]

$$\langle r_\mathrm{c}^2 \rangle = \langle r_\mathrm{p}^2 \rangle + \langle R_\mathrm{p}^2 \rangle + \frac{N}{Z}\langle R_\mathrm{n}^2 \rangle + \frac{3\hbar^2}{4M_\mathrm{p}^2 c^2}. \tag{2}$$

The last term, $3\hbar^2/4M_\mathrm{p}^2 c^2 \sim 0.033$ fm^2, where M_p is the proton mass is the Darwin–Foldy correction which accounts for "Zitterbewegung" of the protons.

As shown in Fig. 3 most of the models give good predictions of nuclear charge radius from ^{6}Li to ^{9}Li; however, the predictions for ^{11}Li spread over a large range.

The No-Core Shell Model [10] ($\square$) and the Large-Basis Shell Model [9] ($\diamond$) are essentially the same models. In both cases calculations have been performed using realistic nucleon-nucleon potentials. While early calculations for $^{7-11}$Li treated the three-body interactions as an effective phenomenological potential [9], the model has been revised for ^{6,7}Li to include microscopic three-body potentials [10]. As shown in Fig. 3, neither the absolute charge radii nor the trend along the isotope chain agree with the experimental results; in particular, the ^{9}Li and ^{11}Li charge radii are predicted to be nearly the same.

Greens-Function Monte Carlo Calculations [11, 12] ($\triangle$) are in good agreement with the experimental results for ^{6}Li to ^{9}Li, but the halo nucleus ^{11}Li has not yet been treated successfully by this technique. Due to the weak two-neutron binding energy

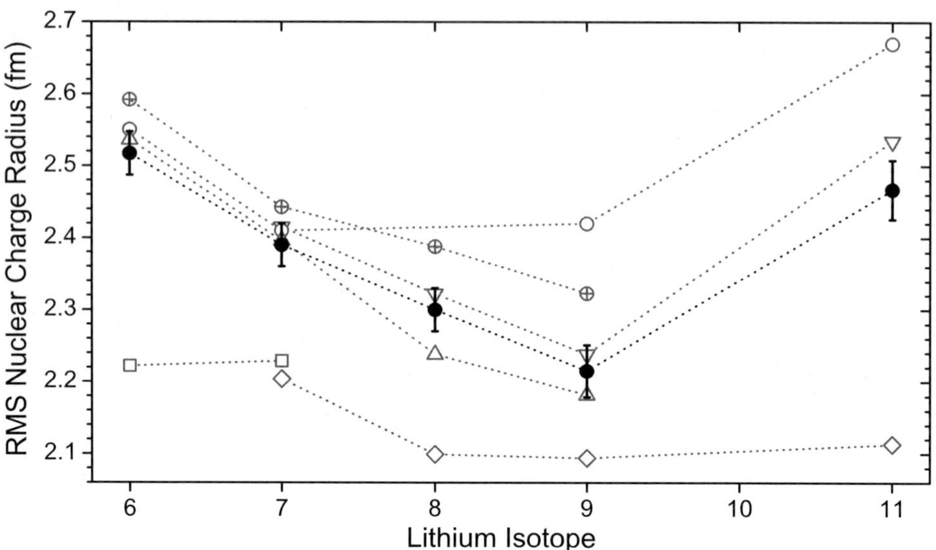

Fig. 3 Root-mean-square nuclear charge radii of the lithium isotopes: $\cdots\bullet\cdots$ this work, $\cdots\square\cdots$ ab-initio no-Core Shell Model [10], $\cdots\diamond\cdots$ Large-Basis Shell Model [9], $\cdots\triangle\cdots$ Greens-Function Monte-Carlo Model [11, 12], $\cdots\triangledown\cdots$ Stochastic Variational Multi-Cluster Model [13, 14], $\cdots\oplus\cdots$ Fermionic Molecular Dynamics Model [15], $\cdots\bigcirc\cdots$ Dynamic Correlation Model [16]

of 375 eV [2], the calculated trial function in this approach usually does not converge into a bound state during the subsequent propagation procedure in the variational Monte Carlo calculations.

Fermionic Molecular Dynamic Model [15] ($\oplus$) values almost agree with the experimental results for $^{6-9}$Li, but, so far, the nuclear charge radius for ^{11}Li could not be obtained since the model does not deliver the right binding energy for this nucleus.

The values for ^{6,7}Li derived from the Dynamic Correlation Model [16] ($\bigcirc$) agree with the measurements but the values for ^{9,11}Li are clearly overestimated. However, it does reproduce the increase from ^{9}Li to ^{11}Li.

The best overall agreement is observed for the Stochastic Variational Multi-Cluster Model [13, 14] ($\triangledown$), and the predicted ^{11}Li charge radius is close to the experimental value. It is interesting to note that a simple three-body model of ^{11}Li, which does not include ^{9}Li core excitation, gives a value of $r_c = 2.40(6)$ fm [20], (Zhukov, 2005, private communication) in good agreement with our result.

5 Summary and outlook

Laser spectroscopy measurements of isotope shifts combined with accurate theoretical mass shift calculations have allowed the determination of the nuclear charge radius of the lightest short-lived isotopes.

Figure 4 schematically shows measured nuclear charge radii in the lower part of the chart of nuclides. Helium and Lithium are now the only elements with $Z < 10$

 Springer

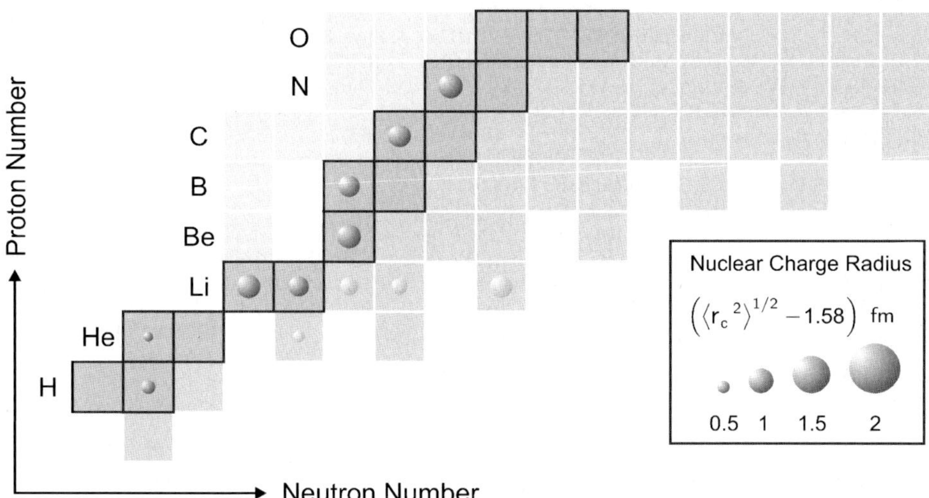

Fig. 4 Root-mean-square nuclear charge radii of the lightest stable and unstable isotopes. For the naturally occurring isotopes nuclear charge radii were determined by electron scattering while for the radioisotopes, like, $^{8-11}$Li [this work] and ^{6}He [3] the isotope shift method has been applied

for which nuclear charge radii have been determined for radioactive isotopes. With ^{11}Li and ^{6}He, two systems of two-neutron halos have been investigated. Further experiments are planned to measure the nuclear charge radius of other short-lived light isotopes. The BeTINa collaboration [21] will investigate the nuclear charge radius of the one-neutron halo ^{11}Be by laser spectroscopy of Be ions in a Paul trap, while the group from Argonne National Laboratory will extend their measurements to the four-neutron halo nucleus ^{8}He.

Acknowledgements This work is supported from BMBF Contract No. 06TU203 and EURONS (European Commission Contract No. 506065). Support from the U.S. DOE Office of Science (B.A.B.), NRC through TRIUMF, and NSERC and SHARCnet.(G.W.F.D. and Z.-C.Y.) is acknowledged. A.W. was supported by a Marie-Curie Fellowship of the European Community Programme IHP under contract number HPMT-CT-2000-00197.

References

1. Tanihata, I., Hamagaki, H., Hashimoto, O., Shida, Y., Yoshikawa, N., Sugimoto, K., Yamakawa, O., Kobayashi, T., Takahashi, N.: Phys. Rev. Lett. **55**, 2676 (1985)
2. Bachelet, C., Audi, G., Gaulard, C., Guénaut, C., Herfurth, F., Lunney, D., De Saint Simon, M., Thibault, C., I Collaboration: Eur. Phys. J. **A 25**(Supp. 1), 31 (2005)
3. Wang, L.-B., Mueller, P., Bailey, K., Drake, G.W.F., Greene, J.P., Henderson, D., Holt, R.J., Janssens, R.V.F., Jiang, C.L., Lu, Z.-T., O'Connor, T.P., Pardo, R.C., Rehm, K.E., Schiffer, J.P., Tang, X.D.: Phys. Rev. Lett. **93**, 142501 (2004)
4. Ewald, G., Nörtershäuser, W., Dax, A., Götte, S., Kirchner, R., Kluge, H.-J., Kühl, T., Sanchez, R., Wojtaszek, A., Bushaw, B.A., Drake, G.W.F., Yan, Z.-C., Zimmermann, C.: Phys. Rev. Lett. **93** 113002 (2004); Phys. Rev. Lett. **94**, 039901 (2005)
5. Sánchez, R., Nörtershäuser, W., Ewald, G., Albers, D., Behr, J., Bricault, P., Bushaw, B.A., Dax, A., Dilling, J., Dombsky, M., Drake, G.W.F., Götte, S., Kirchner, R., Kluge, H.-J., Kühl, T., Lassen, J., Levy, C.D.P., Pearson, M.R., Prime, E.J., Ryjkov, V., Wojtaszek, A., Yan, Z.-C., Zimmermann, C.: Phys. Rev. Lett. **96**, 033002 (2006)

6. Drake, G.W.F., Nörtershäuser, W., Yan, Z.-C.: Can. J. Phys. **83**, 311 (2005)
7. Puchalski, M., Moro, A.M., Pachucki, K.: Phys. Rev. Lett. **97**, 133001 (2006)
8. de Jager, C.W., deVries, H., deVries, C.: At. Data Nucl. Data Tables **14**, 479 (1974)
9. Navrátil, P., Barrett, B.R.: Phys. Rev., C **57**, 3119 (1998)
10. Navrátil, P., Ormand, W.E.: Phys. Rev., C **68**, 034305 (2003)
11. Pieper, S.C., Pandharipande, V.R., Wiringa, R.B., Carlson, J.: Phys. Rev., C **64**, 014001 (2001)
12. Pieper, S.C., Varga, K., Wiringa, R.B.: Phys. Rev., C **66**, 044310 (2002)
13. Varga, K., Suzuki, Y., Tanihata, I.: Phys. Rev., C **52**, 3013 (1995)
14. Varga, K., Suzuki, Y., Lovas, R.G.: Phys. Rev., C **66**, 041302 (2002)
15. Neff, T., Feldmeier, H., Roth, R.: In: Bauer, W., Bellwied, R., Panitkin, S. (eds.) Proceedings of the 21st Winter Workshop on Nuclear Dynamics. EP Systema, Budapest, Hungary (2005)
16. Tomaselli, M., Fritzsche, S., Dax, A., Egelhof, P., Kozhuharov, C., Kühl, T., Marx, D., Mutterer, M., Neumaier, S.R., Nörtershäuser, W., Wang, H., Kluge, H.-J.: Nuc. Phys., A **690**, 298 (2001)
17. Sick, I.: Phys. Lett. **B576**, 62 (2003)
18. Kopecky, S., Harvey, J.A., Hill, N.W., Krenn, M., Pernicka, M., Riehs, P., Steiner, S.: Phys. Rev., C **56**, 2229 (1997)
19. Friar, J.L., Martorell, J., Sprung, D.W.L.: Phys. Rev., A **56**, 4579 (1997)
20. Zhukov, M.V., et al.: Phys. Rep. **231**, 151 (1993)
21. Zakova, M., et al.: (this issue)

Hyperfine Interact (2006) 171:189–195
DOI 10.1007/s10751-006-9490-3

Towards a nuclear charge radius determination for beryllium isotopes

M. Žáková · Ch. Geppert · A. Herlert · H.-J. Kluge ·
R. Sánchez · F. Schmidt-Kaler · D. Tiedemann ·
C. Zimmermann · W. Nörtershäuser

Published online: 1 February 2007
© Springer Science + Business Media B.V. 2007

Abstract We propose determination of isotope shifts for radioactive beryllium isotopes using laser cooled ions in a linear radio frequency (RF) trap. Based on these measurements, combined with precise mass shift calculations, it will be possible to extract model-independent nuclear charge radii of ^{7,9,10}Be and the one-neutron halo ^{11}Be with precision better than 3%. Radioactive beryllium isotopes produced at ISOLDE and ionized with a laser ion source will be cooled and bunched in the radio frequency quadrupole buncher of ISOLTRAP. Ion temperatures will be reduced to the mK range by sympathetic cooling with co-trapped laser cooled ions in a specially designed two-stage linear RF trap. Resonances will be detected via fluorescence and frequencies measured with a femtosecond frequency comb.

Key words laser spectroscopy · isotope shift · nuclear charge radius ·
linear radio frequency trap · halo nuclei

M. Žáková (✉) · R. Sánchez · D. Tiedemann · W. Nörtershäuser
Johannes Gutenberg-Universität Mainz, Mainz, Germany
e-mail: m.zakova@gsi.de

Ch. Geppert · H.-J. Kluge · W. Nörtershäuser
GSI, Darmstadt, Germany

A. Herlert
CERN, CH-1211 Genéve 23, Genéve, Switzerland

H.-J. Kluge
Universität Heidelberg, Heidelberg, Germany

F. Schmidt-Kaler
Universität Ulm, Ulm, Germany

C. Zimmermann
Eberhard-Karls-Universität Tübingen, Tübingen, Germany

1 Introduction

More than 20 years ago Tanihata [1] discovered that some isotopes of light elements close to the neutron drip line have nuclear matter radii much larger than their neighbors. These nuclear matter radii were extracted from breakup cross-section measurements. It was soon proposed that the observed large radii are caused by weakly bound nucleons that form a so-called halo around a compact nuclear core. This discovery inspired a large variety of experiments in which properties of these halo nuclei were studied [2].

One of the early arising questions was to which extent the structure of the nuclear core is modified by the orbiting halo nucleon(s). Laser spectroscopy provides a model-independent way to access nuclear ground state properties, like spins, electromagnetic moments and nuclear charge radii. ^{11}Li was the first halo nucleus that was studied in this way. In a series of measurements the COLLAPS collaboration at ISOLDE determined the nuclear magnetic dipole moment and the electric quadrupole moment using the β-NMR technique [3–5]. Concerning the halo structure of ^{11}Li, the most remarkable result is the small difference in the moments in ^{9}Li and ^{11}Li. The magnetic moments are close to the Schmidt value and are less than 10% different, while the nuclear quadrupole moments agreed within their uncertainties. More recently, the magnetic dipole moment of ^{11}Be has been measured [6] by the same technique and the measurements of ^{9,11}Li were improved [7, 8].

The halo-core interaction can also be probed by measuring the change in the nuclear charge radius and therefore a determination of this quantity is of great interest for halo isotopes. However, it was only recently that such measurements became possible for the radioactive isotopes of the lightest nuclei [3–6]. Based on new and very specialized techniques, two groups have succeeded in measuring, for the first time, optical isotope shifts of very light radioactive nuclei ($Z<10$) with a precision that is sufficient to extract nuclear charge radii. At the Argonne National Laboratory the charge radius of ^{6}He was obtained from laser spectroscopy on helium atoms confined in a magneto-optical trap [9], while an international collaboration, lead by GSI Darmstadt and the University of Tübingen, developed a method for high-resolution resonance ionization spectroscopy and measured the charge radii of ^{8,9}Li at GSI [10, 11] and of ^{11}Li [12] at the ISAC mass separator at TRIUMF. Both ^{6}He and ^{11}Li are two-neutron halo nuclei and a charge radius measurement of the one-neutron halo ^{11}Be will provide important complementary information. In this contribution we describe the proposed *Be*ryllium *T*rap for *I*nvestigation of *N*uclear charge r*a*dii (BeTINa).

2 Experimental method

2.1 Optical isotope shift

The isotope shift (IS) in an electronic transition, i.e. the difference in transition frequencies between two isotopes, has two origins: The difference in nuclear mass affects exact electron energies due to the nuclear motion around the center of mass (mass shift, MS), while the spatial extent of the nuclear charge leads to a difference in the binding energy (volume or field shift, FS) for electrons which have non-zero probability of being inside the nucleus. While the MS between two isotopes with mass numbers A and A' is proportional to

$$\delta\nu_{\mathrm{MS}}^{AA'} \propto \frac{M_A - M_{A'}}{M_A \cdot M_{A'}},$$

 Springer

and thus becomes smaller for heavier elements, the nuclear volume effect increases with Z. For example, in helium, lithium and beryllium transitions, the field shift contributes approximately only 1 MHz, while the mass shift is on the order of several 10 GHz. Separation of the two effects can only be achieved if the mass shift in the electronic transition can be reliably calculated with an accuracy of 10^{-5} or better. To reach this accuracy, relativistic and quantum electro-dynamical corrections have to be taken into account. Such calculations were demonstrated for the three-electron system lithium a few years ago [13–15] and since have been considerably improved [16].

The Be^+ ion is also a three-electron system and can be calculated with the techniques already developed for the case of lithium. The nuclear volume shifts can be determined according to

$$\delta\nu_{FS} = \delta\nu_{IS}^{Exp} - \delta\nu_{MS}^{Theorie}.$$

Using the well-known nuclear charge radius dependence of the field shift

$$\delta\nu_{FS} = \frac{2\pi}{3}\Delta|\psi(0)|^2 \cdot \delta\langle r_c^2\rangle,$$

where $\Delta|\Psi(0)|^2$ is the change of the electron probability at the nucleus in the transition (deducible from atomic theory). From these, one obtains an expression for change in the root-mean-square nuclear charge radius of the isotope A'

$$\langle r_c\rangle^{A'} = \langle r_c\rangle^{A} + \frac{\delta\nu_{IS}^{Exp} - \delta\nu_{MS}^{Theorie}}{\frac{2\pi}{3}\Delta|\Psi(0)|^2},$$

provided that the nuclear charge radius of a reference isotope A is known. The charge radius of stable ^9Be, which has been determined with sufficient precision by both electron scattering [17, 18] and spectroscopy on muonic atoms [19], will be used as the reference.

2.2 Laser excitation scheme

Almost all transitions from the Be^+ ground state are in the vacuum-ultraviolet range; the only one that is relatively accessible with available commercial laser systems is the $2s\rightarrow2p$ transition at 313 nm. Stable $^9Be^+$ ions stored and cooled in RF traps are one of the workhorses of quantum computing experiments, and thus provides a well-established starting point for the currently proposed experiments. All odd beryllium isotopes exhibit hyperfine structure due to non-vanishing nuclear spins and are, thus, not two-level systems. To avoid optical pumping into the dark states, and to allow many absorption-emission cycles, an RF frequency electric field will be applied to couple the $2s_{1/2}$ ground state hyperfine levels to mix their populations. The fine structure splitting allows two possible transitions, $2s_{1/2}\rightarrow2p_{1/2}$ and $2s_{1/2}\rightarrow2p_{3/2}$, but hyperfine structure splitting can be resolved only in the $2s_{1/2}\rightarrow2p_{1/2}$ transition. The 8.1(4) ns lifetime of the $2p$ level [20] corresponds to a natural linewidth ($\Gamma=1/2\pi\tau$) of 20 MHz (FWHM). Thus, line centers must be determined to an accuracy of better than 1% (200 kHz) to extract meaningful values for nuclear charge radii.

2.3 Production and transfer of radioactive berillium ions

All radioactive beryllium isotopes can be produced at the ISOLDE facility at CERN. The laser ion source has demonstrated yields $>10^6$ ions/s for the halo isotope ^{11}Be, and these

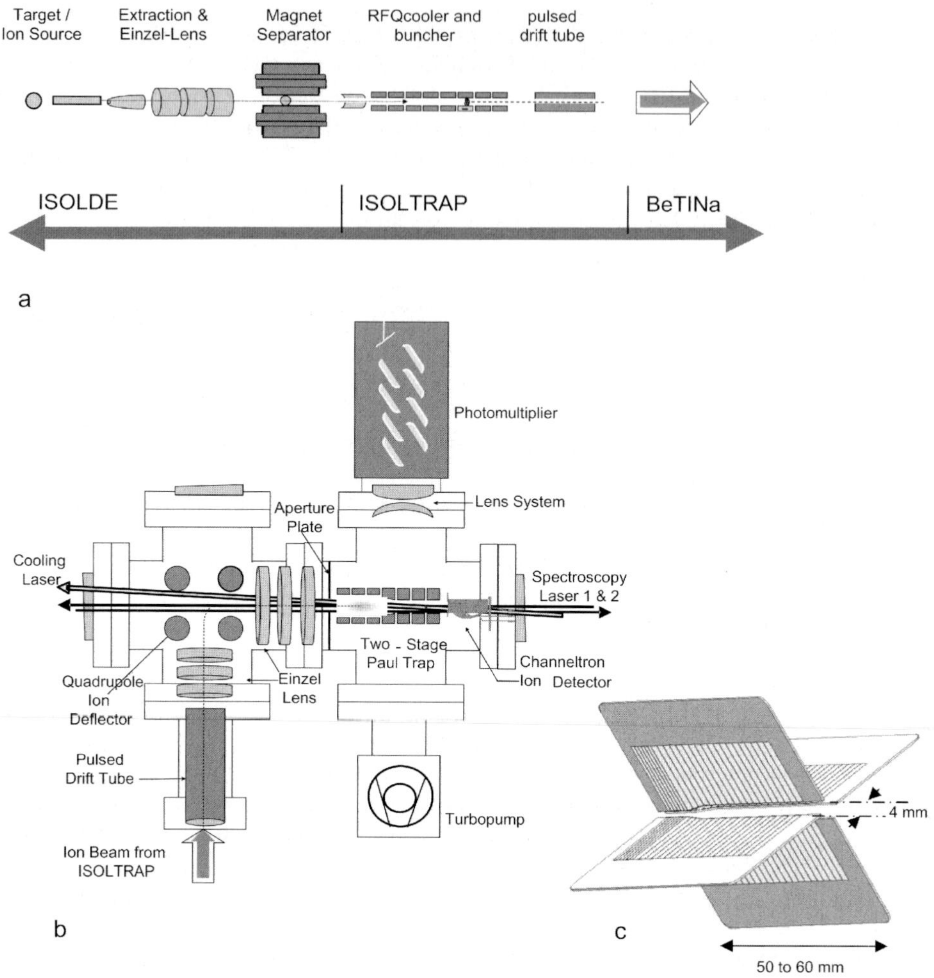

Fig. 1 ISOLDE ion source and radio frequency quadrupole cooler and buncher of ISOLTRAP (**a**); Schematic experimental set-up (**b**) and especially designed RF trap for BeTINa (**c**). Details are described in the text

have been delivered to other laser spectroscopy experiments [6]. Even higher yields have been produced for the lighter radioactive isotopes ^{7,10}Be. The ISOLTRAP collaboration operates a gas-filled radio frequency quadrupole (RFQ) cooler and buncher as illustrated in Fig. 1a [21]. Ions produced by ISOLDE can be accumulated and converted into low-energy, low-emittance ion pulses. Ions captured inside the RFQ lose transverse and longitudinal energy due to collisions with buffer gas. They then accumulate in the potential minimum in front of the trap exit, and are released as a pre-cooled ion bunch with a well defined time structure. A pulsed drift tube is finally used to reduce the bunch energy to approximately 3 keV, a value optimal for transport through the stray fields of the ISOLTRAP magnet to the BeTINa experiment.

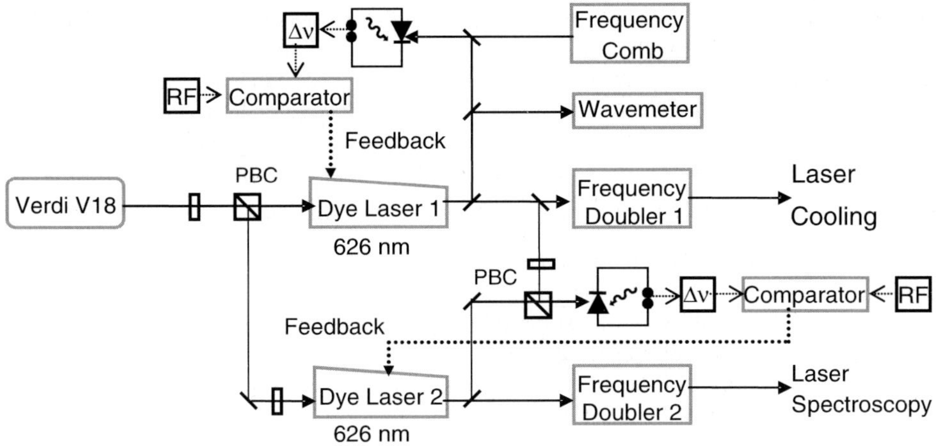

Fig. 2 Scheme of the laser system and the stabilization chain, *RF*: radio frequency generator, $\Delta\nu$: beat frequency signal, *PB*: polarizing beam splitter cube

2.4 Experimental procedure

2.4.1 Linear RF-trap

A sketch of the BeTINa experimental set-up is shown in Fig. 1b. An ion bunch from the ISOLTRAP RFQ system will be delivered through a second pulsed drift tube, where the beam energy is further reduced from 3 keV down to a few 10 eV to facilitate ion capture in the subsequent trap. Two Einzel lenses and a quadrupole ion deflector are used to inject these low-energy ions into a specially designed two-stage linear RF trap, shown in Fig. 1c. A combination of two linear RF traps with different free field radii is driven by the same RF frequency. The first trap (cooler trap) with the larger radius will be used to catch the radioactive Be^+ ions and to cool them sympathetically with simultaneously stored and laser-cooled Ca^+ or Mg^+ ions. Sympathetic cooling is similar to buffer gas cooling, but it reaches lower temperatures by dissipating Be^+ ion energy into an ensemble of laser cooled Ca^+ ions. The light pressure of the laser beam will be used to partially separate the Be^+ ions from the cooling species, and a voltage sequence applied to the multiply segmented DC electrode structure will transfer the Be^+ ions into the narrower spectroscopy trap for additional laser cooling and fluorescence spectroscopy.

2.4.2 Laser set-up

The scheme of the laser system for spectroscopy and laser cooling is shown in Fig. 2. It uses two cw tunable single-mode ring dye lasers operated at 626 nm, which will be frequency doubled to reach the $2s_{1/2} \rightarrow 2p_{1/2}$ transition wavelength at 313 nm. As shown in the figure, the first dye laser, for laser cooling, will be stabilized via frequency offset locking to a femtosecond frequency comb generated with a mode-locked Er-doped fiber laser. The second dye laser, for spectroscopy, will be locked via beat-frequency comparison with the comb-stabilized dye laser. A similar stabilization technique was used in an experimental test of special relativity at GSI Darmstadt [17], where a cw tunable single-

mode titanium:sapphire ring laser was stabilized to the frequency comb. A laser linewidth of 100 kHz was reached with this combination. Acousto-optic modulators (not shown) will be used for fast switching between the spectroscopy and cooling lasers. Spectroscopy will be done with the cooling laser off, and exposure time will be short enough to prevent significant ion heating. Once the spectroscopy laser is turned off, the cooling laser is turned on again to remove any excess energy gained by the ions. Resonance frequencies from the spectroscopy will be determined via fluorescence detection with a photomultiplier and/or a CCD camera. With this set-up we should be able to observe a transition linewidth close to the natural linewidth. To determine the center of a 20 MHz wide line to an accuracy of approximately 1% of the linewidth should be reachable considering the rather high yields and relatively long half-lives of the beryllium isotopes up to the one neutron halo ^{11}Be. However, systematic studies of the exact lineshape and carefully avoiding or correcting all frequency shift effects will be necessary to reach the required precision.

3 Conclusions and outlook

Measurements of isotope shifts by precision laser spectroscopy of trapped and cooled ions will provide valuable information for the determination of the nuclear charge radii of the Be isotopes. We plan to investigate the isotopes ^{7,9,10}Be and the one-neutron halo system ^{11}Be. Assembling of the experimental set-up and off-line test experiments are taking place at the Nuclear Chemistry Department at the University of Mainz, Germany. We plan to move the whole equipment and to install it for on-line measurements at the ISOLDE facility at CERN, Switzerland in 2008.

Acknowledgement This work is supported by the Helmholtz Association under contract VH-NG-148, by BMBF under contract numbers 06TU263I, 06UL264I, and 06MZ215/TP6, and by the European Union under contract number FP-6 EU RII3-CT-2004-506065. We would like to thank Klaus Blaum and Frank Herfurth for informations about the properties of the ISOLTRAP RFQ cooler and buncher and B.A. Bushaw for valuable discussions.

References

1. Tanihata, I., Hamagaki, H., Hashimoto, O., Shida, Y., Yoshikawa, N., Sugimoto, K., Yamakawa, O., Kobayashi, T., Takahashi, N.: Phys. Rev. Lett. **55**, 2676–2679 (1985)
2. Tanihata, I.: J. Phys., G **22**, 157–198 (1996)
3. Arnold, E., Bonn, J., Gegenwart, R., Neu, W., Neugart, R., Otten, E.W., Ulm, G., Wendt, K., ISOLDE Collaboration: Phys. Lett., B **197**, 311–314 (1987)
4. Arnold, E., Bonn, J., Klein, A., Neugart, R., Neuroth, M., Otten, E.W., Lievens, P., Reich, H., Widdra, W., ISOLDE Collaboration: Phys. Lett., B **281**, 16–19 (1992)
5. Arnold, E., Bonn, J., Klein, A., Lievens, P., Neugart, R., Neuroth, M., Otten, E.W., Reich, H., Widdra, W.: Z. Physik, A **349**, 337–338 (1994)
6. Geithner, W., Kappertz, S., Keim, M., Lievens, P., Neugart, R., Vermeeren, L., Wilbert, S., Fedoseyev, V.N., Köster, U., Mishin, V.I., Sebastian, V.: Phys. Rev. Lett. **83**, 3792–3795 (1999)
7. Borremans, D., Balabanski, D.L., Blaum, K., Geithner, W., Gheysen, S., Himpe, P., Kowalska, M., Lassen, J., Lievens, P., Mallion, S., Neugart, R., Neyens, G., Vermeulen, N., Yordanov, D.: Phys. Rev., C **72**(4), 044309 (2005)
8. Neugart, R.: Phys. Rev. Lett. (in preparation) (2006)
9. Wang, L.-B., Mueller, P., Bailey, K., Drake, G.W.F., Greene, J.P., Henderson, D., Holt, R.J., Janssens, R.V.F., Jiang, C.L., Lu, Z.-T., O'Connor, T.P., Pardo, R.C., Rehm, K.E., Schiffer, J.P., Tang, X.D.: Phys. Rev. Lett. **93**, 142501 (2004)

10. Ewald, G., Nörtershäuser, W., Dax, A., Götte, S., Kirchner, R., Kluge, H.-J., Kühl, T., Sanchez, R., Wojtaszek, A., Bushaw, B.A., Drake, G.W.F., Yan, Z.-C., Zimmermann, C.: Phys. Rev. Lett. **93**, 113002 (2004)
11. Ewald, G., Nörtershäuser, W., Dax, A., Götte, S., Kirchner, R., Kluge, H.-J., Kühl, T., Sanchez, R., Wojtaszek, A., Bushaw, B.A., Drake, G.W.F., Yan, Z.-C., Zimmermann, C.: Phys. Rev. Lett. **94**, 039901 (2005)
12. Sánchez, R., Nörtershäuser, W., Ewald, G., Albers, D., Behr, J., Bricault, P., Bushaw, B.A., Dax, A., Dilling, J., Dombsky, M., Drake, G.W.F., Götte, S., Kirchner, R., Kluge, H.-J., Kühl, T., Lassen, J., Levy, C.D.P., Pearson, M.R., Prime, E.J., Ryjkov, V., Wojtaszek, A., Yan, Z.-C., Zimmermann, C.: Phys. Rev. Lett. **96**, 033002 (2006)
13. Yan, Z.-C., Drake, G.W.F.: Phys. Rev., A **61**, 022504 (2000)
14. Yan, Z.-C., Drake, G.W.F.: Phys. Rev., A **66**, 042504 (2002)
15. Yan, Z.-C., Drake, G.W.F.: Phys. Rev. Lett. **91**, 113004 (2003)
16. Puchalski, M., Komasa, J., Moro, A.M., Pachucki, K.: Phys. Rev. Lett. in preparation (2006)
17. de Jager, C.W., de Vries, H., de Vries, C.: Atom. Data Nucl. Data Tables **14**, 479–508 (1974)
18. Jansen, J.A., Peerdeman, R.T., De Vries, C.: Nucl. Phys., A **188**(2), 337–352 (1972)
19. Schaller, L.A., Schellenberg, L., Ruetschi, A., Schneuwly, H.: Nucl. Phys., A **343**, 333–346 (1980)
20. Andersen, T., Jessen, K.A., Sørensen, G.: Phys. Rev. **188**, 76–81 (1969)
21. Herfurth, F., Dilling, J., Kellerbauer, A., Bollen, G., Henry, S., Kluge, H.-J., Lamour, E., Lunney, D., Moore, R.B., Scheidenberger, C., Schwarz, S., Sikler, G., Szerypo, J.: Nucl. Instr. Meth. Phys. Res., A. **469**, 254–275 (2001)

Hyperfine Interact (2006) 171:197–201
DOI 10.1007/s10751-006-9487-y

Investigation of the low-lying isomer in ^{229}Th by collinear laser spectroscopy

B. Tordoff · J. Billowes · P. Campbell · B. Cheal ·
D. H. Forest · T. Kessler · J. Lee · I. D. Moore ·
A. Popov · G. Tungate · J. Äystö

Published online: 15 February 2007
© Springer Science + Business Media B.V. 2007

Abstract A new ion beam of ^{229}Th is available at the Jyväsklyä IGISOL facility, produced from the α decay of ^{233}U. A small branching ratio ($\approx 2\%$) is believed to populate the inferred low-lying (5.5 eV) isomeric state in ^{229}Th. A laser ionization scheme is currently being developed to improve the yield of ^{229}Th from the source. The ion source uses a novel electric field configuration for fast and efficient extraction of α-recoils and is able to provide beams of short lived ($\tau \geq 30$ ms) radioactive nuclei. Identification of the isomeric state by collinear laser spectroscopy will reduce the lower lifetime limit of the state and provide the first direct evidence for its existence.

Key words ^{229}Th · laser ionization · collinear laser spectroscopy

1 Introduction

The existence of a low-lying (5.5 eV) [1] isomeric state in ^{229}Th has been known for some time since it was inferred from gamma ray spectroscopy studies [2] and nucleon

B. Tordoff (✉) · J. Billowes · P. Campbell · B. Cheal
Department of Physics and Astronomy, Manchester University, Manchester, UK
e-mail: bwt@phys.jyu.fi

D. H. Forest · G. Tungate
School of Physics and Astronomy, University of Birmingham, Birmingham, UK

T. Kessler · I. D. Moore · J. Äystö
Department of Physics, University of Jyväskylä, Jyväskylä, Finland

J. Lee
Department of Physics, McGill University, Montréal, Canada

A. Popov
Petersburg Nuclear Physics Institute, Gatchina, Russia

transfer reactions [3] among other techniques. No direct evidence for the isomer's existence has been obtained, however several attempts have been made to observe the photonic emission arising from the partial γ-ray branch to the ground state level [4–7]. Such experiments have been able to set limits on the partial half-life of the isomeric state due to the systematics of the experiment [8]. The isomeric state is also believed to decay via direct alpha emission to excited states in ^{225}Ra. As no evidence for this direct decay has been observed thus far [9], this subject remains an active research area in experimental nuclear physics.

These proceedings describe the method used to produce a beam of ^{229}Th from an α-recoil source of ^{233}U and efforts to selectively ionize the neutral fraction of thorium recoils using high repetition rate pulsed lasers. Because of the low vapour pressure and chemical reactivity of thorium, such beams are not available from traditional ISOL facilities. The Jyväsklyä IGISOL is currently the only laboratory able to deliver ^{229}Th beams to experimental stations. The first experiment with the new beam will attempt to directly measure the spin, isomer shift and nuclear moments of the ground state and low-lying isomer by means of the hyperfine interaction.

2 Production of a ^{229}Th ion beam

A new ion guide has recently been developed and successfully tested for the production of a low energy (≤ 40 keV) beam of ^{229}Th [10]. The nuclei were produced as the daughter products of the alpha decay of a thin ^{233}U source. The daughter recoil energy in the decay (82 keV) was sufficient to eject the ^{229}Th (referred to here as "α-recoils") from the surface of the source of ^{233}U mounted on the inside walls of a large volume gas cell. A small branching ratio ($\approx 2\%$) from the decay of ^{233}U reportedly populates the low-lying isomeric state in ^{229}Th [11], which is of interest in this work. An electric field inside the gas cell is created by a source of electrons near the guide exit hole. The spatial distribution of electrons inside the guide creates a focussing electric field which quickly (≥ 30 ms) evacuates thermalised α-recoils through a 1.2 mm exit hole. These ions are subsequently injected into a conventional mass separator. The overall extraction efficiency of the ion guide is measured to be 0.06% of the source strength. Since this extraction efficiency is low, it is believed that a large proportion of the recoils are in a neutral charge state and are lost either to diffusion to the walls or through molecular formation with impurities. In order to access the neutral fraction of atomic ^{229}Th recoils, a resonant laser ionization scheme is under development using the new FURIOS laser ion source [12].

2.1 Laser ionization scheme development

2.1.1 Atomic beam preparation

As thorium is a refractory metal, it cannot be produced as an atomic vapour through Joule heating in a conventional oven. A new device has been constructed in Jyväskylä in order to universally produce atomic vapours of all elements that may be required as FURIOS beams in the future (Fig. 1). The main purpose of the device is for testing new ionization schemes as well as systematic studies of the effect of different pump lasers, repetition rates and laser gain media on the laser ion yield. An atomic plume

 Springer

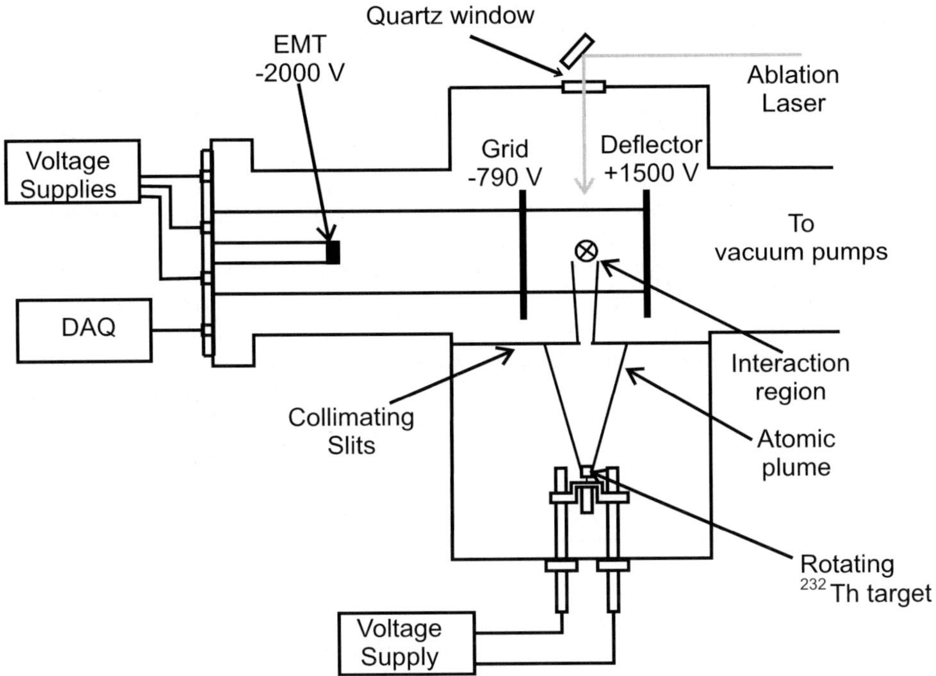

Fig. 1 Schematic view of the atomic beam unit used for laser ionization scheme development

of thorium is produced by laser ablation of a rotating target by a 20 Hz Nd:YAG laser with a typical pulse energy of 25 mJ. The atomic vapour is subsequently collimated by two apertures where it is overlapped in a crossed beam geometry with the FURIOS lasers. Upon resonant ionization following excitation, a laser ion is deflected onto an electron multiplier by means of a plate/grid static electric field system. The amplified signal is then measured as a function of laser parameters (wavelength, power, temporal overlap) allowing for scanning of atomic resonances, investigations of transition saturation and searches for new atomic levels to be performed.

2.1.2 An ionization scheme for thorium

An ionization scheme for thorium has been tested using the apparatus described in Section 2.1.1. The wavelengths used in this work coincide with the lasing range of solid-state Ti:Sapphire lasers and the fluorescent ranges of the dyes DCM and LDS 722. Table 1 lists the ionization scheme and laser information concerning the FURIOS system.

The overlap and layout of the lasers used for the development of this scheme is available in [13]. The time delay between the dye and Ti:Sa lasers is set through external triggering of the Copper Vapour Laser (CVL) thyratron. The first step transition is provided through frequency doubling of the output beam of a double sided pumped Ti:Sa laser. The second and third step transitions are provided by a fundamental Ti:Sa and dye laser, respectively. The second and third step laser beams are spatially overlapped using a polarising cube beamsplitter, taking advantage of the

Table 1 Ionization scheme information for thorium

Lower level (cm^{-1})	J^π	Upper level (cm^{-1})	J^π	λ_{vac} (nm)	Laser power (mW)	Laser source
0	2^+	24,381.323	2^-	410.150	300	Ti:Sapphire
24,381.323	2^-	37,688.202	3^+	751.491	1,200	Ti:Sapphire
37,688.202	3^+	53,576.349	$2^-, 3^-$ or 4^-	629.420	500	DCM dye

Pump lasers operate at 12 kHz.

Table 2 Collinear laser spectroscopy transition efficiencies

Lower level (cm^{-1})	J^π	Upper level (cm^{-1})	J^π	λ_{vac} (nm)	Efficiency (no optical pumping)	Efficiency (optical pumping)	Optical pumping transition (nm)
0	$\frac{3}{2}^-$	30,972.162	$\frac{5}{2}^+$	322.871	1/5,500	-	-
1,521.896	$\frac{5}{2}^+$	35,156.911	$\frac{5}{2}^+$	297.309	1/64,000	1/13,000	374.225

Although optical pumping increases the yield of fluorescent photons arising from a metastable level transition, the yield from the ground state transition is greater.

orthogonal polarisations of the two laser systems. These beams are then coupled to the blue laser beam via a dichroic mirror situated after the polarising beamsplitter. The scheme shown in Table 1 has been successfully tested and will be extended in the future to include Rydberg levels close to the ionization potential of thorium and autoionising states past the ionization potential, using the dye LDS 722.

3 Collinear laser spectroscopy of ^{229}Th

Identification of the low-lying isomeric state in ^{229}Th will be carried out by identification of its hyperfine structure. Information concerning the experimental techniques used in the collinear beams geometry of laser spectroscopy in conjunction with beam bunching and cooling can be found elsewhere [14]. The isotope shifts and hyperfine structure of $^{227-230,232}$Th have been measured previously by means of two step excitation with trapped ions [15]. Changes in the mean-square charge radius between isotopes were deduced from this data and were consistent with the Droplet Model of Myers and Schmidt [16] with the inclusion of deformation.

In this work the hyperfine structure of the isomeric state in ^{229}Th is of primary importance. Testing of different ionic resonances was performed by off-line comparison of transition efficiencies using a discharge source of ^{232}Th. The results are summarised in Table 2.

For ionic spectroscopy, cooling and bunching the reaction products in a rf-cooler device limits the optical spectroscopy to transitions from the electronic ground state or very low-lying metastable states. The laser excitation frequency, electronic spin, hyperfine structure and transition strength are thus fixed by the technique. The JYFL-UK collaboration have succeeded in electronically manipulating state populations in ionic ensembles by optical pumping in the rf-cooler allowing these parameters to vary [17]. However, in this case the ground state transition tested is far more efficient than that of the optically pumped metastable level transition and will be used for the future ^{229m}Th studies.

4 Conclusion and outlook

A new ^{229}Th ion beam is available at the IGISOL facility, University of Jyväskylä. This beam will be used in a collinear laser spectroscopy measurement to attempt to identify hyperfine components of the predicted low-lying isomer. Identification of the isomer by this method will be able to reduce the lower lifetime limit of the state and will begin a series of investigations into the physical properties of this unique system. Efficiencies of ionic resonances by laser spectroscopy have been measured using a discharge source of ^{232}Th. A ground state transition has been identified as a candidate for the measurement and has been shown to be more efficient than using the new technique of optical pumping in a rf-cooler to increase the population of a metastable level. Future experiments will investigate the possible implementation of optical pumping inside the rf-cooler to influence the population of the isomeric level by the process of Nuclear Excitation by non-radiative Electronic Transition (NEET) [18].

References

1. Guimaraes-Filho, Z.O., Helene, O.: Phys. Rev. **C71**, 044303 (2005)
2. Kroger, L.A., Reich, C.W.: Nucl. Phys. **A259**, 29 (1976)
3. Burke, D.G., et al.: Phys. Rev. **C42**, R499 (1990)
4. Irwin, G.M., Kim, K.H.: Phys. Rev. Lett. **79**, 990 (1997)
5. Utter, S.B., et al.: Phys. Rev. Lett. **82**, 505 (1999)
6. Shaw, R.W., et al.: Phys. Rev. Lett. **82**, 1109 (1999)
7. Moore, I.D., et al.: ANL Phys. Dev. Rep. PHY-10990-ME-2004 (2004)
8. Kasamatsu, Y., et al.: Radiochim. Acta **93**, 511 (2005)
9. Dykhne, A.M., Tkalya, E.V.: JETP Lett. **67**, 251 (1998)
10. Tordoff, B., et al.: Nucl. Instrum. Methods **B252**, 347 (2006)
11. Dykhne, A.M., et al.: Pis'ma Zh. Eksp. Toer. Fiz. **64**, 319 (1996)
12. Moore, I.D., et al.: J. Phys. G: Nucl. Part. Phys. **31**, S1499 (2005)
13. Kessler, T., et al.: Nucl. Instrum. Meth. **B** (in press)
14. Billowes, J.: Nucl. Phys. **A682**, 206c (2001)
15. Kälber, W., et al.: Z. Phys. **A334**, 103 (1989)
16. Myers, W.D., Schmidt, K.H: Nucl. Phys. **A410**, 61 (1983)
17. Campbell, P.: Optical pumping in an RF cooler buncher (this issue)
18. Karpeshin, F.F., et al.: Phys. Lett. **B372**, 1 (1996)

Hyperfine Interact (2006) 171:203–208
DOI 10.1007/s10751-006-9502-3

Laser spectroscopy of high spin isomers – a review

Yu. Gangrsky

Published online: 25 January 2007

Abstract This review summarizes the experimental data on charge radii differences among ground state and high spin isomeric states determined by high-resolution laser spectroscopic methods. A comparison is presented between radii changes obtained from the isomeric shifts in the atomic spectra and from the quadrupole moments of both ground and isomeric states under the assumption that the radii changes are determined by the difference of the quadrupole deformations. Special attention is paid to isomers arising from the break-up of nucleon pairs and isomers of odd–odd nuclei. The characteristic features of the radii changes for isomeric states of different origin are discussed.

Key words quasiparticle isomers · two-particle isomers in odd–odd nuclei · isomeric charge radii changes · quadrupole deformation · quadrupole moments

1 Introduction

At present, the laser spectroscopy methods with their high-sensitivity and high-resolution are successfully applied for study of a number of nuclear parameters. All laser spectroscopy techniques, based on hyperfine structure splitting or isotope shifts measurements, yield model-independent information on spins I, magnetic dipole moments μ, electric quadrupole moments Q_s and changes in the mean square (ms) charge radii of nuclei $\Delta \langle r^2 \rangle$ for long chains of stable and radioactive isotopes in ground and isomeric states. These parameters characterise nuclear shape and size as well as the configurations of the valence nucleons.

Long isotopic chains, extending far off stability, are accessible at radioactive beam facilities, on-line isotope separators or in-flight facilities. They enable measurements of the above-mentioned parameters over long sequences of isotopes and large range of mass numbers providing a stringent test of the theory. Information on and systematics of the available data for many nuclides from He to Cm can be found in the review papers [1–4].

Y. Gangrsky (✉)
FLNR Joint Institute for Nuclear Research, 141980 Dubna, Moscow Region, Russia
e-mail: gangr@jinr.ru

The general trend of N- and Z-dependencies of the parameters mentioned is in a good qualitative agreement with the predictions of the droplet model [5]. However, it fails to explain some of the characteristic features observed in the nuclear radii systematics. In particular, deviations of the smooth general increase of the charge radii with the neutron number have been prescribed to the influence of the quadrupole deformation. Characteristic kinks of $\Delta\langle r^2\rangle$ curves are observed at the shell closures or in the regions of nuclear shape transition. In both cases a sharp change of the deformation parameter β_2 occurs. The change of the ms nuclear radius compared to the radius of the homogeneous charged sphere is

$$\Delta\langle r^2\rangle_\beta = \frac{5}{4\pi}\langle r^2\rangle_0 \Delta\langle \beta_2^2\rangle, \tag{1}$$

where $\langle r^2\rangle_0$ is the ms radius of the spherical nucleus with the same volume and $\Delta\langle\beta_2^2\rangle$ – the change of the ms quadrupole deformation parameter. The parameter β_2 can be determined by a useful approximation relating it with the intrinsic quadrupole moment:

$$Q_0 = \frac{3}{\sqrt{5\pi}} ZeR^2\beta_2(1 + 0.156\beta_2 + \ldots) \tag{2}$$

where $R = 1.2A^{1/3}$ is the simple liquid-droplet value of the charge radius. Diverse methods are used for determining the intrinsic quadrupole moment Q_0. One of them is based on measurements of the reduced transition probabilities

$$B(E, 2I_i \rightarrow I_f) = <I_iK_i \rightarrow I_fK_f> Q_0. \tag{3}$$

Here $<I_iK_i{\rightarrow}I_fK_f>$ is the Klebsch–Gordon coefficient of the spins and their projection along the symmetry axis of the initial (I_iK_i) and final (I_fK_f) nuclear states.

The hyperfine splitting of the optical levels measured by laser spectroscopic method gives information on the spectroscopic quadrupole moments Q_s. If the nucleus is well deformed then the conception of an intrinsic quadrupole moment Q_0 is valid. In this case Q_s is the projection of Q_0 along the quantization axis and the two moments are simply related by

$$Q_s = \frac{3K^2 - I(I+1)}{(I+1)(2I+3)} Q_0. \tag{4}$$

In fact, the quadrupole deformation parameter contributes to (2) always according to its ms values and $\beta_2 = \langle\beta_2^2\rangle^{1/2}$, where

$$\beta_2^2 = \langle\beta_2\rangle^2 + \left(\langle\beta_2^2\rangle - \langle\beta_2\rangle^2\right) = \beta_{2,st}^2 + \Delta\beta_{2,dyn}^2 \tag{5}$$

Therefore the radii changes measure not only the static deformation as Q_s does, but as well the dynamic deformation arising from the zero-point fluctuations of the nuclear surface. This plays a fundamental role in the discussion of the experimental results. In Eq. (1), corrections to $\Delta\langle r^2\rangle$ accounting for the dynamic deformation lead to better understanding the experimentally observed isotopic charge radii trend.

The charge radius depends not only on the neutron number but as well as on other nuclear characteristics, for example the excitation energy and the angular momentum. However, the available information on such dependencies is rather poor. This is due to the fact that the excited states of a nucleus usually have very small lifetimes and thus they are inaccessible for the laser methods used at the present. Charge radii changes and nuclear moments can be measured only for isomeric states with sufficiently long life-times. The most short-lived isomeric state for which these parameters have been measured is ^{85m}Rb

with half-life 10^{-6} s [6]. In this case the method of resonance fluorescence in a gas cell was applied. Development of new laser spectroscopic methods is necessary for study of isomers with shorter life-time.

Detailed study of the dependence of the nuclear parameters $<r^2>$, I, μ, Q on the excitation energy and angular momenta is expected to provide information on the physical nature of these states and to stimulate their quantitative description. Laser spectroscopic experiments in this area are in the beginning. The experiments with the shape-isomers of ^{185}Hg [7, 8], ^{240}Am and ^{242}Am [9] giving an evidence of their large quadrupole deformation illustrate the promise held by the development of laser spectroscopic methods.

This work is devoted to a systematics of the known data of the charge radii and electrical quadrupole moments of the isomeric states. The isomeric charge radii changes obtained with two different approaches are discussed and compared: (1) directly measured from the optical isomeric shifts and (2) calculated using the known values of the quadrupole deformations. Thus more definite conclusions can be drawn about the influence of the excitation energy and angular momentum on the charge radii. Two types of isomeric states are chosen: two particles in odd–odd nuclei and quasi-particle spin aligned isomers.

2 Quasi-particle isomer

The occurrence of these isomers is explained by the break-up of neutron or proton pairs and successive spin alignment. The spins of these isomers depend on the number of the break-up pairs and can be sufficiently large. For example, ^{178}Hf has four-quasiparticle isomeric state with energy 2.447 MeV and spin $I^\pi = 16^+$, which arises from the break-up of one neutron and one proton pairs [10].

Table 1 presents the isomeric charge radii changes: the experimental, $\Delta<r^2>_{\text{exp}}$, deduced from the isomeric shifts and, $\Delta < r^2 >_\beta$, calculated according to (1) using the quadrupole deformation parameters. The adopted sign convention is $\Delta<r^2>=<r^2>_{\text{is}} - <r2>_{\text{gr}}$. The spins, parity and excitation energies of the isomeric states are displayed, too. The data refer to four even-even nuclei and three odd–even nuclei. In the case of ^{178}Hf two isomeric states are available: two-quasiparticle 8^- and four-quasiparticle 16^+ states. For ^{87}Y a comparison is made for the single-particle $9/2^+$ and three- quasiparticle $27/2^-$ states.

The β_2 values of ground and isomeric states for odd-Z or odd-N nuclei are obtained from the spectroscopic quadrupole moments Q_s and can be found elsewhere (see e.g. the review paper [11]). The most Q_s have been deduced using laser spectroscopy methods as a rule in experiments giving simultaneously information on the isomer shifts. For this reason the $\Delta < r^2 >_\beta$ values describe radii changes due to the static deformation only. On the contrary, the β_2 values of the ground states of even-even nuclei ($I^\pi = 0$) include as well the dynamic deformation since they are extracted from the intrinsic quadrupole moments. The later have been deduced using the experimental measured probability of the transition from the first excited state with $I^\pi = 2^+$ to the ground state with $I^\pi = 0$ [12]:

$$Q_0 = 7.10\sqrt{B(E2, 2 \rightarrow 0)} \tag{6}$$

The following conclusion can be drawn analyzing the data in Table 1:

1. The radii changes between isomeric and ground states are from the same order of magnitude as those between nuclei with N and $N+2$ neutrons (see for example [2, 3]).
2. The isomeric shifts for all investigated nuclei are negative, $\Delta<r^2>_{\text{exp}}<0$, i.e. the charge radii of the isomeric states are smaller than of the ground states.

Table 1 Energies, spins, intrinsic quadrupole moments of the quasiparticle isomers and their mean square charge radii changes compared to the ground state

Nucleus	E, keV	I^π	Q_0, b	$\Delta < r^2 >_\beta$, fm^2	$\Delta <r^2>_{\exp}$, fm^2	Ref.
^{130}Ba	0	0^+	+3.42(3)			
	2476	8^-	+3.95(43)	+0.06(3)	−0.047(3)	[13]
^{176}Yb	0	0^+	+7.30(13)			
	1051	8^-	+7.55(11)	+0.06(3)	−0.047(3)	[13]
^{178}Hf	0	0^+	+6.96(4)			
	1148	8^-	+7.11(6)	+0.03(2)	−0.0394(1)	
	2447	16^+	+7.20(8)	+0.04(2)	−0.0946(22)	[14]
^{202}Pb	0	0^+	+1.28(2)			
	2170	9^-	+1.40(30)	+0.02(2)	−0.01(2)	[15]
^{97}Y	667	$9/2^+$	−1.40(15)			
	3523	$27/2^-$	−1.50(17)	+0.01(2)	−0.101(1)	[13]
^{135}Cs	0	$7/2^+$	+0.050(2)			
	1627	$19/2^-$	+0.45(5)	+0.06(1)	−0.008(2)	[16]
^{177}Lu	0	$7/2^+$	+7.26(6)			
	969	$23/2^-$	+7.33(6)	+0.02(2)	−0.035(1)	[17]

References refer to the data of $\Delta <r^2>_{\exp}$.

3. In opposite, the intrinsic quadrupole moments of the isomers are larger than of the ground state. This results in positive signs of the calculated radii change, $\Delta < r^2 >_\beta > 0$.

4. The charge radius decrease of the four-quasiparticle isomer in ^{178}Hf is nearly twice larger than of the two-quasiparticle isomer.

Let us note that the calculated $\Delta < r^2 >_\beta$-values for even-even nuclei are underestimated because for the ground state the sum of static and dynamic deformation is taken into account. If the latter is excluded, the inconsistencies between $\Delta < r^2 >_\beta$ and $\Delta <r^2>_{\exp}$ will be larger. The differences between both data sets $\left(\text{on } \Delta < r^2 >_{\exp} \text{ and on } \Delta < r^2 >_\beta\right)$ indicate that other factors exist affecting the nuclear charge radius. Such factors are for example the higher order deformations and/or the surface diffuseness, which decreases at the transition from ground to isomeric state. However, the most reliable explanation seems to be the one used for the odd–even staggering of the charge radii. An odd neutron blocks the zero-point fluctuations of the nuclear surface thus decreasing the dynamic deformation. As a result the increase of the charge radius by adding of a single neutron is smaller than one half of the increase obtained when neutron pair is added. In an analogous way the enhanced number of the unpaired neutrons or protons in quasiparticle isomers influences the amplitude of the zero-point vibrations.

3 Two-particle isomers in the odd–odd nuclei

For comparison with the above discussed case analogous analysis is made also for the case of two-particle isomers in the odd–odd nuclei. The parameters used, the charge radii changes and quadrupole deformations, are obtained from the experimental data on isomeric shifts and hyperfine splitting of the atomic (ionic) levels.

The total angular momentum of these isomers is generated by the spins of two unpaired particles – a neutron and a proton. Their parallel and anti-parallel orientations form an isomeric pair with essentially different total angular momenta.

Table 2 Charge radii changes of two-particle isomers in odd-odd nuclei

Nucleus	I^π	E, keV	Q_s, b	β_2	$\Delta <r^2>_\beta$, fm^2	$\Delta<r^2>_{exp}$, fm^2	Ref
^{158}Ho	5^+	0	+4.10(40)	0.32(2)			
	2^-	67.2	+1.62(17)	0.26(2)	−0.35(20)	−0.158(7)	[18]
^{160}Ho	5^+	0	+3.95(23)	0.31(2)			
	2^-	60.0	+1.78(17)	0.28(2)	−0.18(20)	−0.083(7)	[18]
^{172}Lu	4^-	0	+3.80(4)	0.318(3)			
	1^-	41.9	+0.76(3)	0.32(1)	+0.06(10)	+0.009(1)	[17]
^{174}Lu	1^-	0	+0.773(7)	0.323(3)			
	6^-	171	+4.80(5)	0.321(3)	−0.06(8)	−0.001(1)	[17]
^{176}Lu	7^-	0	+4.92(5)	0.315(5)			
	1^-	127	−1.47(1)	0.310(8)	−0.04(10)	+0.011(1)	[17]
^{178}Lu	1^-	0	+0.708(10)	0.295(4)			
	9^-	~30	+5.39(5)	0.315(3)	+0.065(15)	+0.076(1)	[17]
^{184}Au	5	0	+4.65(26)	0.264(14)			
	2		+1.90(16)	0.221(17)	−0.25(10)	−0.036(3)	[19]

Table 2 displays the charge radii differences between isomeric and ground states – experimentally measured [2] and extracted using the above described procedure from the spectroscopic electric quadrupole moments [11]. The procedure of calculations is the same as for the case of the quasi-particle isomers. For all nuclei $\Delta <r^2>_\beta$ values accounts only for the static quadrupole deformation. Only the well-deformed nuclei Ho, Lu, Au are discussed because for these nuclei the quantities β_2 and $\Delta <r^2>_\beta$ can be calculated with a good approximation using the (1), (2) and (4). In the most cases presented in Table 2, the spin of the isomeric state is smaller than the one of the ground state. The following peculiarities are obvious:

1. Both the measured, $\Delta<r^2>_{exp}$, and the calculated, $\Delta <r^2>_\beta$, absolute values are essentially lower than: (1) the corresponding values for the quasiparticle isomers and (2) the radii change when a neutron is added to the nucleus [2, 3].
2. As a rule, a smaller quadrupole deformation corresponds to states with lower spin.
3. A correlation between $\Delta<r^2>_{exp}$ and $\Delta <r^2>_\beta$ is observed: the charge radii increase at the transition to a state with higher spin (ground or isomeric) is accomplished with a raise of the quadrupole deformation. However, it is not possible to find a quantitative relation between both data sets due to the large uncertainties of the $\Delta <r^2>_\beta$ values.
4. As a rule, the calculated absolute values of $\Delta <r^2>_\beta$ are larger than the experimental $\Delta<r^2>_{exp}$. This indicates a presence of other effects influencing the charge radii.

4 Conclusion

Prominent examples are chosen to demonstrate the general peculiarities of the charge radii in the quasiparticle (Section 2) and two-particle isomers (Section 3). The difference between the natures of these isomers reflects in different behaviour of the nuclear radii.

In accordance with the predictions of the droplet model [20] the general nuclear radii trend can be approximated with a linear dependence on neutron number with a slope 0.2 fm^2/N. In contrast to this, the radii dependence on the angular momentum is more

complex. The data collected in this work indicate remarkable behavior: any alteration of the odd neutron configuration is connected with very small charge radii changes accounting for the nuclear size and of the quadrupole deformation parameter determining the nuclear shape. These are essentially smaller than the changes arising when a neutron is added to the nucleus. The same picture is observed when the configuration of an odd proton undergoes a change. This indicates the negligible influence of the angular momentum of an unpaired nucleon on the nuclear charge radius.

However, when the number of the unpaired nucleons increases, the changes of the charge radii and nuclear deformations are larger and in opposite directions: $<r^2>$ – decreases, while β_2 increases. This might be explained with an increase of the nuclear rigidity relative to the zero point vibration of the nuclear surface.

Let us mention, that the changes of the quadrupole deformation for the collective nuclear states are essential larger than for the ground state. A good example are the rotational bands of ^{192}Hg in the range $I^\pi = 20^+ - 30^+$ with β_2 values up to 0.6 resulting to a large increase of the charge radius [21].

The present work gives an overview on the changes of charge radii and deformation between the ground and isomeric states of nuclei and their dependencies on nuclear structure. The describing effects put new demanding but as well stimulating problem to both the theory and the experiment. From this point of view the importance of the laser spectroscopic studies of nuclear parameters, especially in until now scarce investigated excited states of nuclei, is obvious.

Acknowledgements The author expresses his gratitude to K. Marinova and Yu. Penionzhkevich for the useful discussion and comments. Special thanks are due to J. Billowes for supplying not yet published experimental data on isomeric charge radii. This work was supported by the Russian Foundation for Basic Research grant 04-02-16955.

References

1. Otten, E.W.: Treatise on Heavy Ion Science **8**, 517 (1989)
2. Aufmuth, F., Heulig, K., Steudel A.: Atom. Data Nucl. Data Tables **37**, 455 (1987)
3. Nadjakov, E.G., Marinova, K.P., Gangrsky, Yu. P.: Atom. Data Nucl. Data Tables. **56**, 133 (1994)
4. Marinova, K.P.: Physics Part. Nucl. **35**, 693 (2004)
5. Solovev, V.G.: Theory of atomic nucleus. Moscow, 351 (1981)
6. Shimkaveg, G., et. al.: Phys. Rev. Lett. **53**, 2230 (1984)
7. Alkhazov, G.D., et al.: Z. Phys. A. **316**, 123 (1984)
8. Bonn, J., et al.: Z. Phys. A **276**, 203 (1976)
9. Backe, H., et al.: Phys. Rev. Lett. **80**, 920 (1998)
10. Browne, E., Firenstone, R.B.: Table of Radioactive Isotopes. Wiley, V. Shirley, NY (1986)
11. Stone, N.J.: Atom. Data Nucl. Data Tables. **90**, 75 (2005)
12. Raman, S., Nestor, C.W., Tikkaneen, P.: Atom. Data Nucl. Data Tables. **78**, 1 (2001)
13. Billowes, J. (private communication)
14. Boos, N., et al.: Phys. Rev. Lett. **72**, 2689 (1994)
15. Thompson, R.C., et.al.: J. Phys. G **9**, 443 (1983)
16. Thibault, C., et al.: Nucl. Phys. A. **367**, 1 (1981)
17. Georg, U., et al.: Eur. Phys. J. A **3**, 225 (1998)
18. Alkhazov, G.D., et al.: Nucl. Phys. A **504**, 549 (1989)
19. Pinard, J., et al.: Nucl. Phys. A **512**, 241 (1990)
20. Mayers, W.D., Schmidt, K.H.: Nucl. Phys. A. **419**, 61 (1983)
21. Moore, E.F., et. al.: Phys. Rev. Lett. **64**, 3167 (1990)

Hyperfine Interact (2006) 171:209–215
DOI 10.1007/s10751-006-9488-x

Nuclear charge radii and electromagnetic moments of scandium isotopes and isomers in the $f_{7/2}$ shell

Y. Gangrsky · K. Marinova · S. Zemlyanoi ·
M. Avgoulea · J. Billowes · P. Campbell · B. Cheal ·
B. Tordoff · M. Bissell · D. H. Forest · M. Gardner ·
G. Tungate · J. Huikari · H. Penttilä · J. Äystö

Published online: 1 February 2007
© Springer Science + Business Media B.V. 2007

Abstract Collinear laser spectroscopy experiments on the ScII transition $3d4s\ ^3D_2 \rightarrow 3d4p\ ^3F_3$ at $\lambda \approx 363.1$ nm were performed on the $^{42-46}$Sc isotopic chain using an ion guide isotope separator with a cooler–buncher. Isotope and isomer shifts and hyperfine structures of five ground states and two isomers were measured. Preliminary results on the nuclear moments and charge radii changes deduced from these measurements are reported.

Key words collinear laser spectroscopy · cooled and bunched ion beam · isotope shifts and hyperfine structure of $^{42-46}$Sc · nuclear moments and charge radii in $f_{7/2}$ neutron shell

1 Introduction

The hyperfine structures and isotope shifts of five scandium isotopes ($Z=21$) in the mass region $42 \leq A \leq 46$, with isomeric states in ^{44,45}Sc, have been measured using the collinear-beams laser spectroscopy technique. These isotopes lie in the Ca-region around Z, $N=20$ and $N=28$ shell closures. Several peculiarities of the nuclear radii trends in the Ca-region have been pointed out [1, 2]: (1) the sequential addition of neutrons going from the sd – to the $f_{7/2}$-shell gives a smooth change of the successive mean square charge radii, $<r^2>$, indicating the disappearance of any shell effect (at $N=20$). The effect is observed for Ar, K and Ca; (2) although all curves show a parabolic-like dependence of $<r^2>$ on the

Y. Gangrsky · K. Marinova · S. Zemlyanoi (✉)
FLNR Joint Institute for Nuclear Research, 141980 Dubna, Moscow Region, Russia
e-mail: zemlya@jinr.ru

M. Avgoulea · J. Billowes · P. Campbell · B. Cheal · B. Tordoff
Shuster Building, University of Manchester, Manchester M13 9PL, UK

M. Bissell · D. H. Forest · M. Gardner · G. Tungate
School of Physics and Astronomy, University of Birmingham, Birmingham B15 2TT, UK

J. Huikari · H. Penttilä · J. Äystö
Accelerator Laboratory, University of Jyväskylä, Jyväskylä SF-405 51, Finland

neutron number, the well pronounced symmetric parabolic shape in the calcium case is not reproduced by the neighbouring elements; (3) the shape asymmetry of $<r^2>$ curves for elements above and below calcium is in the opposite direction: for $Z<20$ there is steady increase of the charge radii towards $N=28$; for $Z>20$ the radii increase towards $N=20$; (4) the observed odd–even staggering decreases away from Ca – the largest decrease being at the odd-Z nucleus K. From the point of view of the nuclear radii systematics at neutron numbers $N>28$ [3] such large structural differences are very unusual for the successive isotopic chains. Thus, diverse questions arise concerning the influence of the closed proton shell of calcium and of the pairing effect in the proton number. In this context it is very important to extend the measurements in the calcium region to the not yet investigated isotopes and/or elements. This will give a more detailed picture of the behaviour of nuclear radii in the Ca-region and will greatly contribute to the interpretation of the nuclear structure.

Isotope shifts for the light elements are dominated by the mass shift while the much smaller field shifts provide information about the changes in mean square charge radii. In addition, the production yield of isotopes away from stability is low. For these reasons the experimental situation is very challenging. It requires high sensitivity and selectivity and very high spectroscopic accuracy. The experimental technique, used for the study of the scandium isotopes, has proven to meet these requirements.

2 Experimental details

Isotope shift and hyperfine structure investigations were carried out by collinear laser spectroscopy on cooled and bunched ion beams. Experiments were performed using the online ion guide isotope separator (IGISOL) facility at the Cyclotron Laboratory, University of Jyväskylä. Measurements were performed on the $^{42-46}$Sc isotopes and the ^{44m,45m}Sc isomers. Radioactive isotopes were produced in a fusion ion guide by irradiating a ^{45}Sc target in reactions of the type (d,p), (p,pxn), (p,p') using 15 MeV deuterons and 25–48 MeV protons at 5–10 μA.

The IGISOL facility offers significant advantages over conventional on-line mass separators (for example see [4–8]): (1) the ion guide method can provide isotope beams of any element, regardless of chemistry and with fast extraction times in the range of ~1 ms; (2) the RFQ cooler–buncher installed in the mass-separated line reduces the ion beam energy spread to about 10^{-5} of the total 40 keV energy. At this level the corresponding Doppler broadening of the laser resonances is less than the natural line-width of most allowed optical transitions; (3) the low ion beam emittance after the cooler, estimated to be 3π mm·mrad, results in a better spatial overlap between the ion beam and the counter-propagating laser beam. The overlap is independent of the ion-guide and isotope separator conditions, making the experiment insensitive to their drifts; (4) finally, the bunching capability of the cooler has led to the introduction of a new way of performing laser spectroscopy using ion beams. By accepting the fluorescence signal only while an ion bunch passes through the light collection region a reduction of the background in the laser fluorescence signal of four orders of magnitude has been achieved. Measurements have been made using this method down to ion beam fluxes of 150 ions s^{-1}.

Laser light was provided by a frequency-doubled Spectra-Physics 380D dye laser locked to a chosen molecular iodine absorption line. The isotope shifts and hyperfine structures of the scandium isotopes were measured in the transition $3d4s$ ^{3}D$_2$ (67.7 cm^{-1})$\rightarrow 3d4p$ ^{3}F$_3$ (27602.5 cm^{-1}) of the Sc II spectrum at $\lambda \approx 363.1$ nm. Examples of the recorded spectra are shown in Fig. 1 for all investigated scandium isotopes. As can be seen, most hyperfine

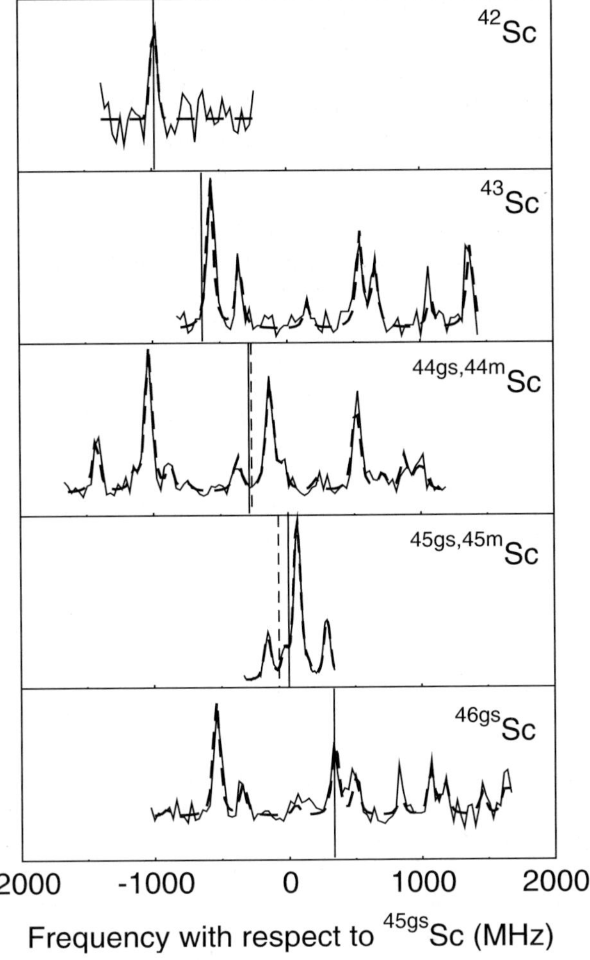

Fig. 1 Laser fluorescence spectra showing the hyperfine structures and isotope shifts of all the investigated scandium isotopes relative to the centroid frequency of the ground state of ^{45}Sc. Only a part of the ^{43}Sc spectrum and that part of ^{45}Sc containing the isomer peaks are presented. The *vertical lines* denote the position of the centroid frequency of the ground states (*full lines*) and isomeric states (*dashed lines*). The *dashed curves* represent fits of the experimental spectra using Voigt profiles

structure components are well resolved. The resonances were fitted by a Voigt profile which was found to describe adequately the ^{45}Sc line shape when measured off-line with high statistical accuracy.

3 Experimental results

For the studied isotopes of the odd-Z element scandium the magnetic dipole and electric quadrupole hyperfine coefficients A and B of both lower, $3d4s\ ^3D_2$, and upper, $3d4p\ ^3F_3$, states are obtained from the hyperfine structures using a χ^2 minimization fitting procedure. The number of observed or resolved lines in some cases was not sufficient for extracting the lower and upper level A and B coefficients independently. In these cases the ratios A_l/A_u and B_l/B_u, which are independent of the nuclear properties, were fixed at $A_l/A_u=+2.469(1)$ and $B_l/B_u=+0.554(22)$ using our off-line high accuracy data for the stable isotope ^{45}Sc: $A_l=+507.67(10)$ MHz, $B_l=-34.7(12)$ MHz, $A_u=+205.61(3)$ MHz, $B_u=-62.6(11)$ MHz. This is

 Springer

Table 1 Spins, magnetic dipole, μ, and electric quadrupole, Q_s, moments of the investigated scandium isotopes

	I^π	μ/μ_N (exp) This work	μ/μ_N (exp) [10]	Q_s (exp), b This work	Q_s (exp), b [10]
^{42}Sc	0^+				
^{43}Sc	$7/2^-$	+4.503(4)	+4.62(4)	−0.208(22)	−0.26(6)
^{44}Sc	2^+	+2.505(3)	+2.56(3)	+0.184(19)	+0.10(5)
^{44m}Sc	6^+	+3.810(6)	+3.88(1)	−0.203(22)	−0.19(2)
^{45}Sc	$7/2^-$	Reference	+4.756487(2)	Reference	−0.220(2)
^{45m}Sc	$3/2^+$	+0.368(5)		+0.318(22)	
^{46}Sc	4^+	+3.040(9)	+3.03(2)	+0.123(17)	+0.119(6)

For comparison the already known values of μ and Q_s as summarised by Stone [10] are shown, too.

Table 2 Life-times $T_{1/2}$ of the investigated scandium isotopes and isotopic shifts as measured in the optical transition $3d4s\ ^3D_2 \rightarrow 3d4p\ ^3F_3$ of ScII. Errors in parenthesis are statistical

A	$T_{1/2}$	$\delta\nu^{45,A}$, MHz
42	681.3 ms	−979(7)
43	3.891 h	−631.2(47)
44	58.6 h	−287.1(30)
44 m	3.927 h	−268.3(38)
45	Stable	0
45 m	318 ms	−72.3(7)
46	83.79 d	338.1(33)

a reasonable procedure because for light elements the hyperfine anomaly is expected to be of the order of or less than 10^{-4} [9], which is below the error limits of the experimental data on nuclear moments.

Calibration of the scandium moments was achieved using the obtained A and B hyperfine coefficients of all investigated isotopes and the known nuclear moments of the stable ^{45}Sc [10]. The results for the magnetic dipole and electric quadrupole moments of ^{43,44,44m,46}Sc isotopes (see Table 1) are in a good agreement with those summarised by Stone [10]. However, the overall accuracy of the magnetic moment is increased. The nuclear moments of the ^{45m}Sc are deduced for the first time.

With known hyperfine coefficients the determination of the isotope and isomer shifts is straightforward. They are evaluated relative to the centroid frequency of the stable ^{45}Sc and are listed in Table 2. The adopted sign convention is $\delta\nu^{45,A} = \nu^A - \nu^{45}$.

In order to extract the charge radii changes from the measured isotope shifts the electronic factor F and the specific mass-shift constant M_s must be determined (see e.g. [11]). The electronic factor F for the investigated optical transition can be calculated following the established semi-empirical approach [12] as was done in our recent paper on titanium [13]. The specific mass shift constant, M_s, which accounts for the correlations of the electronic motion, is much more difficult to calculate reliably (Bauche, private communication). Furthermore, since scandium has only a single stable isotope there are no other experimental charge radii data which would allow a determination of the specific mass shift in a consistent way. Because the charge radii changes are very sensitive to small variations in M_s the preliminary analysis performed to date can only provide estimates of the expected upper and lower limits of the radii changes, which are shown in Fig. 2. A complete analysis will be published in our forthcoming paper [14].

 Springer

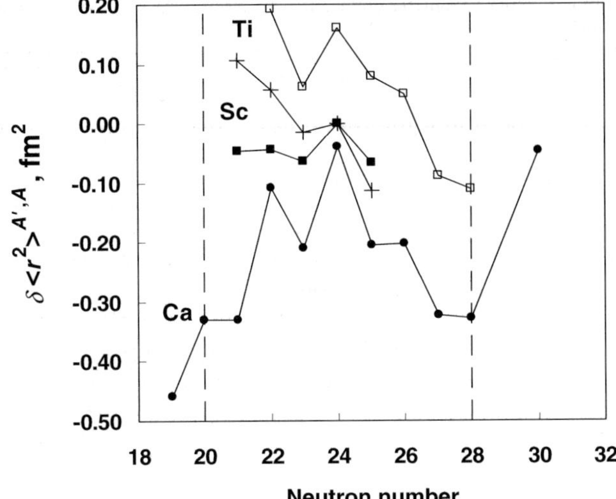

Fig. 2 Predicted limits of possible variation of the mean squared charge radii in the scandium isotopic chains. Comparison between charge radii trends of scandium isotopes and their nearest neighbours titanium and calcium. To avoid the crossing of the curves, the $\delta\langle r^2\rangle$-values for Ti and Ca are shifted respectively with +0.05 fm^2 and −0.33 fm^2

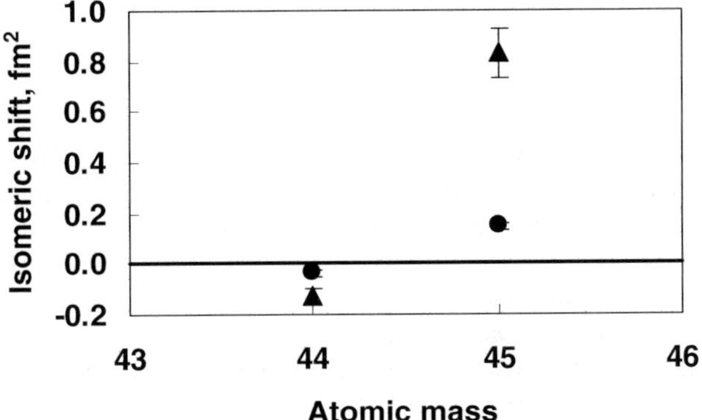

Fig. 3 Changes in mean square charge radii of the isomeric state relative to the ground state as obtained from the isomeric shift (*circles*) and from the deformation parameter β_2 (*triangles*)

Even with the mass shift uncertainty it is evident that no drastic contradiction with the overall radii trend in the $f_{7/2}$ shell is found. While the odd–even staggering is reduced compared with calcium and titanium, it is consistent with that observed in potassium, the only other odd-Z element studied in this region.

There is no mass shift contribution to the isomeric shift between a nuclear ground state and an isomeric state in the same nuclide. Thus, the calculation of the isomeric radii change relative to the ground state is straightforward and does not suffer from the uncertainties due to the mass shift. The isomeric shifts in the case of ^{44}Sc and ^{45}Sc can also be derived from the quadrupole deformation parameter, β_2, using the usual scheme of calculation: spectroscopic quadrupole moment Q_s → intrinsic quadrupole moment Q_0 → quadrupole deformation β_2 → charge radii change $\delta\langle r^2\rangle$ (see for example [11, 15]). It was assumed that Q_s and Q_0 are related by the projection formula in the strong coupling limit. The results are displayed in Fig. 3. It is evident from the figure that ^{44m}Sc isomer mean square charge radius is smaller than its ground state while the ^{45m}Sc isomer is larger than its nuclear

Table 3 Energies and intrinsic quadrupole moments of the isomeric states derived from the experimental $B(E2)$ values for the $I^{\pi} = 3/2^{+}$ bands and averaged over the different rotational transitions – according to [17]

Nucleus	E_{is}, keV	Q_0, b
^{43}Sc	151.9	0.82(5)
^{45}Sc	12.4	0.88(6)
^{47}Sc	766.83	1.01(23)
^{45}Ti	329.5	1.17(11)
^{47}V	259.49	1.34(25)
^{49}V	748.2	1.07(12)

ground state. In both cases the sign of the isomer shift is in accordance with the changes in intrinsic deformation, but the magnitude of the shift, particularly for ^{45}Sc, is substantially smaller than predicted from the quadrupole moments.

4 Discussion

The hyperfine structure and the isotope shift measurements on ground states and some isomeric states in the isotopic sequence $^{42-46}$Sc are the first optical investigation of scandium. The study of these nuclei has become possible due to the ultra-sensitive laser spectroscopy technique on cooled and bunched ion beams developed at the IGISOL facility, University of Jyväskylä.

Among the investigated nuclei the ^{45}Sc isotope deserves closer attention because along with several other odd-A nuclei in the lower $1f_{7/2}$ shell it has a positive parity isomeric state with $I^{\pi} = 3/2^{+}$. Such excited states have been explained [16] in the framework of the Nilsson model: the Nilsson $3/2^{+}$ orbital of the sd shell and the lowest orbital of the $1f_{7/2}$ shell approach each other for increasing deformation, thereby producing a low-lying core excited state. According to the systematics of [16] all these $3/2^{+}$ isomeric states are well-deformed. Their intrinsic quadrupole moments derived from the experimentally measured B(E2)-values are presented in Table 3.

The value of Q_0 obtained from the laser spectroscopic data in this work is nearly two times larger: $Q_0=1.59(11)$ b. The unusually large quadrupole moment of the isomeric state of ^{45}Sc is the most striking feature of the present data. As can be seen in Fig. 3, it leads to a larger charge radius of the isomeric state than follows from the isotope shift. This surprising fact remains so far unexplained.

The investigated isotopes cover the largest part of the $1f_{7/2}$ shell around the mid-shell and toward $N=20$ shell closure. Although qualitative, the new information on Sc charge radii changes constitutes a valuable contribution to the systematics of nuclear charge radii in the calcium region. This is shown in a comparison of the present data on scandium isotopes with the known development of radii for Ca [17] and Ti [13] (see Fig. 2) and can be seen as well in comparison with the more extensive systematics of [2] which includes Ar and K isotopes.

Acknowledgements This work has been supported by a Joint Project grant from the Royal Society, the UK Engineering and Physical Sciences Research Council, the Russian Foundation for Basic Research grant 04-02-16955 and the Academy of Finland under the Finnish Centre of Excellence Programme 2000–2005 (project No 44875).

 Springer

References

1. Klein, A., Brown, B.A., Georg, U., Keim, M., Lievens, P., Neugart, R., Neuroth, M., Silverance, R.E., Vermeeren, L., ISOLDE collaboration: Nucl. Phys., A **607**, 1 (1996)
2. Blaum, K., Geithner, W., Lassen, J., Lievens, P., Marinova, K., Neugart, R.: Hyp. Int. **162**, 101 (2005)
3. Nadjakov, E.G., Marinova, K.P., Gangrsky, Y.P.: At. Data Nucl. Data Tables **56**, 133 (1994)
4. Äystö, J.: Nucl. Phys., A **693**, 477 (2001)
5. Nieminen, A., Campbell, P., Billowes, J., Griffith, J.A.R., Huikari, J., Jokinen, A., Moore, I.D., Moore, R., Tungate, G., Äystö, J.: Phys. Rev. Lett. **88**, 094801 (2002)
6. Campbell, P., Thayer, H.L., Billowes, J., Dendooven, P., Flanagan, K.T., Forest, D.H., Griffith, J.A.R., Huikari, J., Jokinen, A., Moore, R., Nieminen, A., Tungate, G., Zemlyanoi, S., Äystö, J.: Phys. Rev. Lett. **89**, 082501 (2002)
7. Forest, D.H., Billowes, J., Campbell, P., Dendooven, P., Flanagan, K.T., Griffith, J.A.R., Huikari, J., Jokinen, A., Moore, R., Nieminen, A., Thayer, H.L., Tungate, G., Zemlyanoi, S., Äystö, J.: J. Phys., G, Nucl. Part. Phys. **28**, L63 (2002)
8. Billowes, J.: Hyp. Int. **162**, 63 (2005)
9. Büttgenbach, S.: Hyperfine Interact. **20**, 1 (1984)
10. Stone, N.J.: Atomic Data Nucl. Data Tables **90**, 75 (2005)
11. Otten, E.W.: Treatise on Heavy Ion Science 8, p. 517. Plenum, New York (1989)
12. Kopfermann, H., Schneider, E.: Nuclear Moments. Academic, New York (1958)
13. Gangrsky, Yu.P., Marinova, K.P., Zemlyanoi, S.G., Billowes, J., Campbell, P., Flanagan, K.T., Forest, D. H., Griffith, J.A.R., Huikari, J., Moore, R., Nieminen, A., Thayer, H., Tungate, G., Äystö J.: J. Phys., G, Nucl. Part. Phys. **30**, 1089 (2004)
14. Gangrsky, Yu.P., Marinova, K.P., Zemlyanoi, S., Avgoulea, M., Billowes, J., Campbell, P., Cheal, B., Tordoff, B., Bissell, M., Forest, D.H., Gardner, M., Tungate, G., , Huikari, J., Penttilä, H., Äystö J.: Nuclear charge radii and electromagnetic moments of radioactive scandium isotopes and isomers (in press)
15. Thayer, H.I., Billowes, J., Campbell, P., Dendooven, P., Flanagan, K.T., Forest, D.H., Griffith, J.A.R., Huikari, J., Jokinen, A., Moor, R., Nieminen, A., Tungate, G., Zemlyanoi, S., Äystö, J.: J. Phys., G, Nucl. Part. Phys. **29**, 2247 (2003)
16. Styszen, J., Chevallier, J., Haas, B., Schulz, N., Taras, P., Toulemonde, M.: Nucl. Phys., A **262**, 317 (1976)
17. Palmer, C.W.P., Baird, P.E.G., Blundell, S.A., Brandenberger, J.R., Foot, C.J., Stacey, D.N., Woodgate, G.W.: J. Phys. B: At. Mol. Phys. **17**, 2197 (1984)

Hyperfine Interact (2006) 171:217–223
DOI 10.1007/s10751-006-9485-0

Laser spectroscopy of stable Os isotopes

**M. Avgoulea · M. Mahgoub · J. Billowes · P. Campbell ·
A. Ezwam · D. H. Forest · M. Gardner · J. Huikari ·
A. Jokinen · A. Nieminen · G. Tungate · J. Äystö**

Published online: 16 February 2007
© Springer Science + Business Media B.V. 2007

Abstract Doppler-free isotope shift measurements of the stable even $^{184-192}$Os and ^{187,189}Os odd isotopes have been performed for the first time on the $5d^66s^2$ $^5D_4 \rightarrow 5d^66s6p\,^7F_4$ (305.9 nm) transition in the neutral atom by atomic beam laser spectroscopy and on the ionic $5d^66s\,^5D_{9/2} \rightarrow 5d^66p\,^6D_{7/2}$ (228.2 nm) transition by fast collinear ion-laser spectroscopy. The measurements were carried out in Manchester and at the IGISOL facility in Jyväskylä in Finland, respectively. The results presented are the most precise measurements to-date of the absolute isotope shifts.

Key words Osmium · atomic beam spectroscopy · collinear laser spectroscopy

1 Introduction

Osmium lies in the transitional region between the rare-earth isotopic chains of well deformed nuclei and the Pb (Z=82) chain of near-spherical isotopes shown in Fig. 1.

M. Avgoulea (✉) · M. Mahgoub · J. Billowes · P. Campbell · A. Ezwam
Department of Physics and Astronomy, Manchester University, Manchester, UK
e-mail: tma@mags.ph.man.ac.uk

D. H. Forest · M. Gardner · G. Tungate
School of Physics and Astronomy, University of Birmingham, Birmingham, UK

J. Huikari · A. Jokinen · A. Nieminen · J. Äystö
Department of Physics, University of Jyväskylä, Jyväskylä, Finland

Present Address:
M. Mahgoub
Physik Department, E12, Technical University of Munich,
James-Frank St., 85748 Garching, Germany

Present Address:
A. Nieminen
STMicroelectronics, Helsinki, Finland

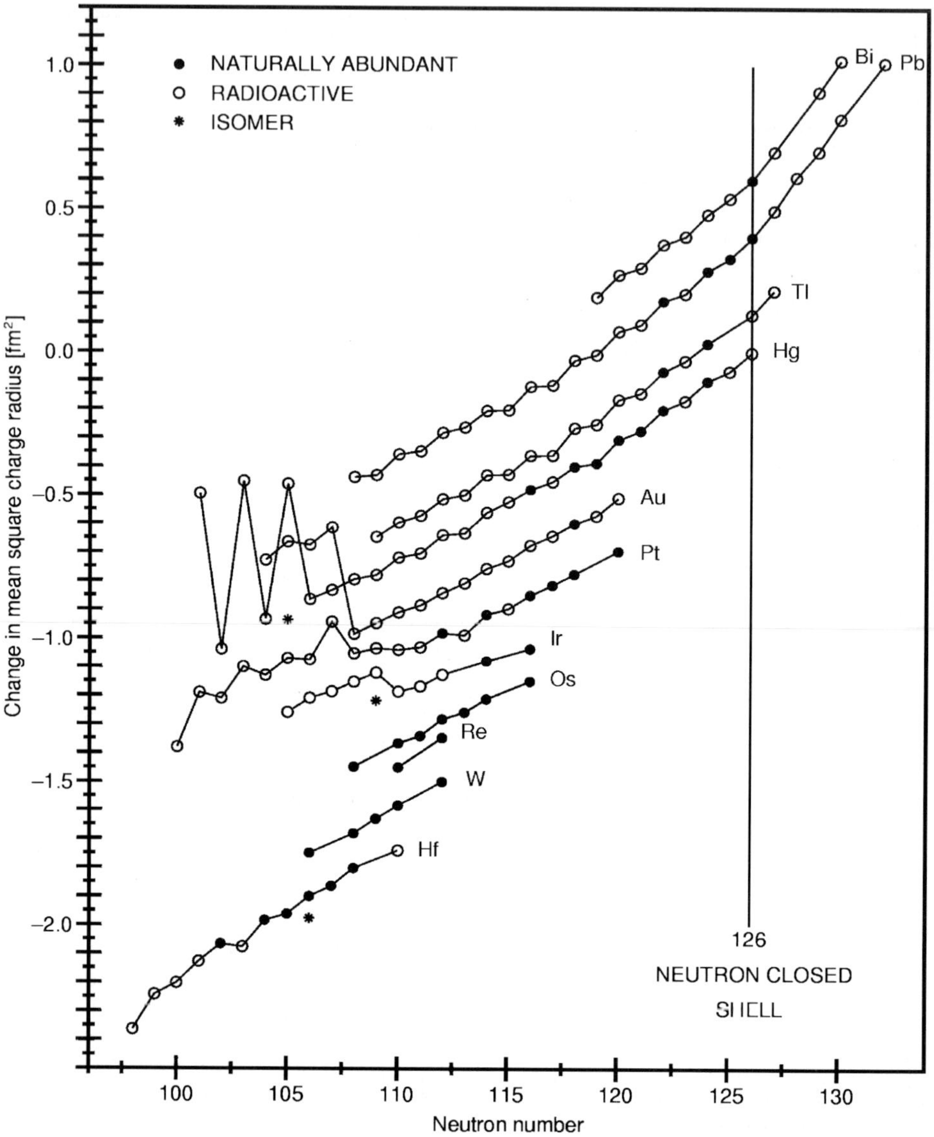

Fig. 1 Changes in the mean square charge radii in the range Z=72–83. The isotopic chains have been shifted from one another for clarity

This region is characterized by very pronounced ground state nuclear shape changes, despite being close to the magic proton number Z=82. In particular, the neutron mid-shell region around N=104 has displayed dramatic changes in the nuclear charge radii of the neutron deficient Hg isotopes. The Hg chain was the first one to be extensively studied by optical methods ([1] and references therein) and it has manifested the largest inverted odd-even staggering (OES) in the nuclear chart as well as shape coexistence, clearly demonstrated in ^{185}Hg, where different deformations exist in states of the same nucleus. The strong deformation effects of the light even and odd

Hg isotopes have inspired many theoretical studies too which attempt to explain the shape staggering effect[2]. A similar shape transition has been observed in the Au chain, although the onset of the deformation takes place two neutrons earlier and the subsequent OES is not large[3]. More complex features are displayed in the Pt chain where a strong inverted OES below $N=110$ is present. The odd Pt isotopes are prolate while the even ones ($A<186$) are rather triaxial[4]. Less dramatic nuclear shape changes are apparent in the Ir chain[3]. Despite the wide interest in this region limited research with optical methods has been carried out in the nearby Os, Re, W and Ta chains as the refractory nature of these elements makes their on-line study at conventional ISOL (isotope separator on-line) facilities difficult due to their low volatility and consequent low production efficiencies. The IGISOL facility in Jyväskylä, where one of the experiments described here was carried out, is of particular benefit as it is the only suitable ion source for on-line production of refractory elements such as Os.

The high resolution Doppler-free investigation of the stable Os isotopes with two laser spectroscopy techniques presented in these proceedings is in preparation for the study of the ground state nuclear properties of the neutron-deficient Os isotopes.

2 Experimental methods

2.1 Crossed beams spectroscopy

The first Doppler-free measurements of the isotope shifts of the stable $^{184,186-190,192}$Os isotopes have been made using the crossed-beams technique in Manchester[5]. Atomic beams of Os were produced by laser ablation of pills of compressed natural Os powder whose diameter and thickness were 5 and 3 mm, respectively. Ablation was achieved with high energy pulses from a pulsed Quantel Brilliant Nd: YAG laser. Energies of up to 100 mJ/pulse with a pulse duration of 6 ns of the 1,064 nm light impinged on a rotating sample in order to spread the surface damage. A plume of atomic Os was released by the Nd:YAG laser. The released atoms travelled 20 cm vertically, passing through two pairs of collimation slits which defined an angular divergence of 3.5°. Atoms were laser-excited on the $5d^66s^2\ ^5D_4\rightarrow 5d^66s6p^7F_4$ 305.9 nm UV line. Resonance fluorescent photons were collected in a direction normal to the plane of the atomic and laser beams by a quartet of silica lenses and focused on a photomultiplier tube. The UV laser light of 0.5 mW power was produced by intracavity frequency doubling in a Spectra Physics 380A dye laser with an anti-reflection-coated $LiIO_3$ crystal of 2 mm thickness. The fluorescence signal was detected in a 400 μs wide gate, delayed by 20–40 μs after the Nd: YAG pulse. The linearity of each laser scan was monitored by a Burleigh CFT-500 confocal etalon with a free spectral range of 149.98(9) MHz[6]. An extended scan was carried out in the region of the least abundant (0.02%) ^{184}Os isotope in order to improve the statistics (Fig. 2).

2.2 Collinear beams spectroscopy

Stable Os ion beams were produced off-line using a discharge source ion guide at the IGISOL (ion guide isotope separator on-line) facility in Jyväskylä[7]. An osmium

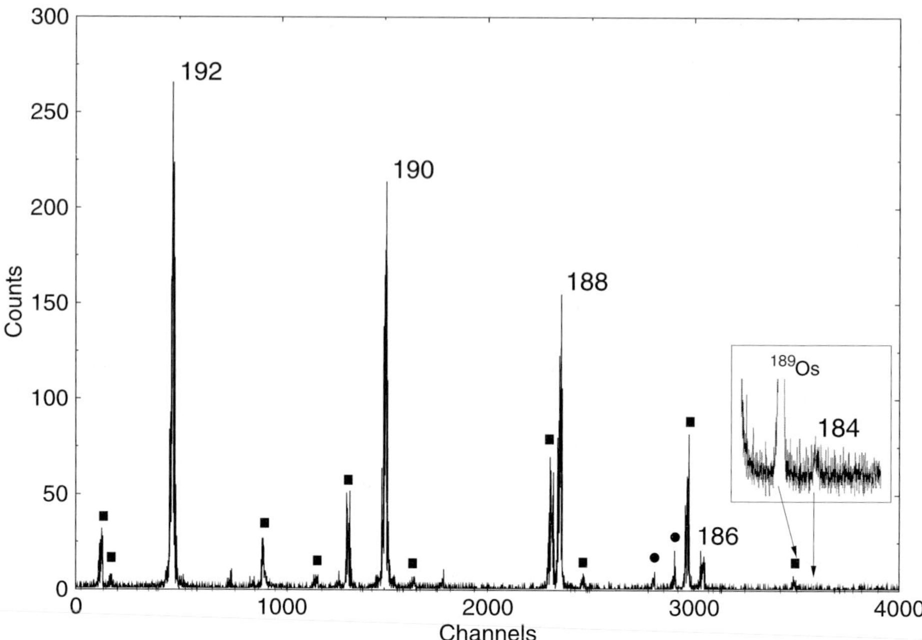

Fig. 2 Atomic resonances of the even $^{184-192}$Os isotopes and hyperfine structures of ^{187}Os (*circles*) and ^{189}Os (*squares*) measured on the 305.86 nm line [5]

Table 1 Upper and lower state measured hyperfine structure parameters for the 305.9 nm (atomic) and the 228.2 nm (ionic) transitions

Hfs parameter	Atomic splitting (MHz) [5] 305.9 nm	Ionic splitting (MHz) 228.2 nm
A_{u}	+730±1	−172±3
B_{u}	+350±6	+587±17
A_{l}	+168±1	+1,235±3
B_{l}	+1,100±6	+1,139±19

sample of ∼0.2 g was placed at the cathode of the source where a voltage of 500 V was applied thereby producing Os$^+$ in the glow discharge. The ions were thermalized in a He gas flow, extracted and electrostatically focused to form an ion beam which was accelerated to around 37 keV before being mass analysed by a 55° dipole magnet. A small energy spread and a reduced ion beam emittance was achieved by cooling in a gas-filled radio frequency ion cooler-buncher [8]. Fundamental laser light was obtained from a Spectra Physics 380D dye laser using stilbene 3 dye pumped by a 5 W Ar$^+$ laser (all UV lines). Intracavity second harmonic generation (SHG) was attempted during preparations of this experiment but it proved unsuccessful with the available optics due to laser instabilities within the cavity and difficulty in extracting the UV beam. In order to avoid all undesirable effects of SHG upon laser stability an optically separated ring resonator was employed. Resonantly enhanced

Table 2 Natural abundance, nuclear spin, atomic and ionic isotope shifts, $\delta v = v^A - v^{192}$ of all stable Os isotopes

A	Abundance(%)	I^π	δv atom (MHz)	δv ion (MHz)
184	0.02	0^+	$+\,9{,}532\pm14$	–
186	1.58	0^+	$+6{,}907\pm7$	$+6{,}953.5\pm4.6$
187	1.6	$1/2^-$	$+6{,}097\pm7$	$+6{,}139.7\pm3.2$
188	13.3	0^+	$+4{,}227\pm2$	$+4{,}261.9\pm3.4$
189	16.1	$3/2^-$	$+3{,}486\pm2$	$+3{,}494.3\pm1.0$
190	26.4	0^+	$+1{,}985\pm5$	$+1{,}993.6\pm3.4$
192	41.0	0^+	0	0

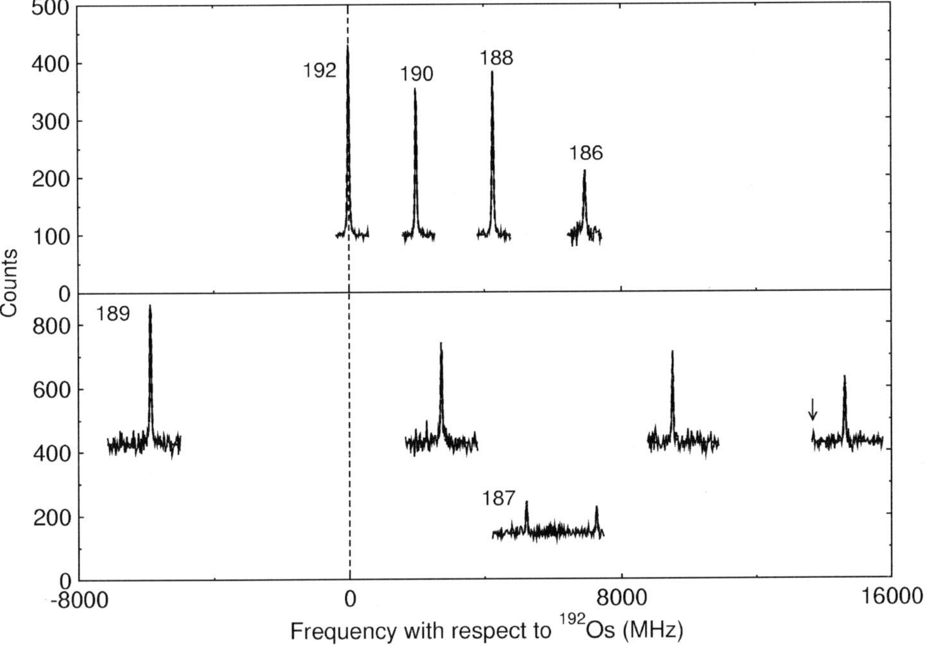

Fig. 3 Ionic resonances of the even $^{186-192}$Os isotopes and hyperfine structures of the ^{187,189}Os isotopes measured in the 228.2 nm line. The *arrow* indicates the position of the fifth peak (see text)

frequency doubling [9] was performed using a Coherent MBD-200 external cavity into which fundamental laser light was coupled. Stabilisation of the fundamental frequency was achieved by active locking of the 380D cavity to the $21{,}886.3048\,\mathrm{cm}^{-1}$ Te_2 absorption line. The estimated UV laser beam power produced by the MBD-200 was 0.1 mW for a power of 40 mW of fundamental light. An Os beam current of 125 pA gave a signal of $\sim$2,700 photons/s with $\sim$750 photons/s background. Isotope shift measurements were made on the $5d^66s^5\mathrm{D}_{9/2}{\rightarrow}5d^66p^6\mathrm{D}_{7/2}$ ionic transition by observing the fluorescence signal imaged on the photomultiplier tube during laser frequency scans. At both the beginning and the end of the experiment, spectra were collected of the most abundant ^{192}Os reference isotope so that any shifts in centroid

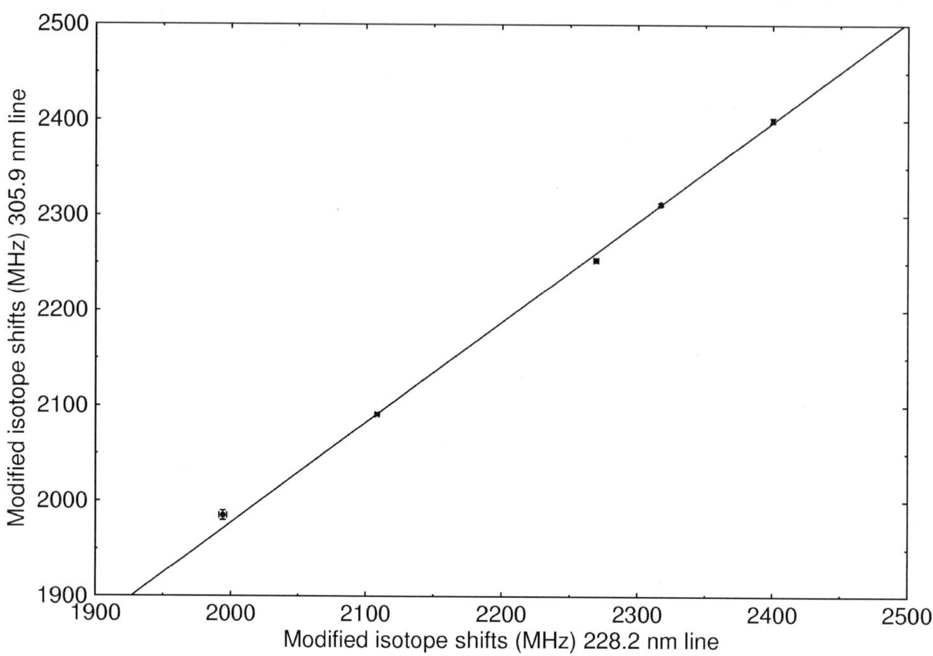

Fig. 4 King plot of the 305.9 nm atomic Os modified isotope shifts against the 228.2 nm ionic Os modified isotope shifts

position could be detected and corrected. A drift of 0.7 V of the total accelerating voltage was observed which was accounted for in the analysis. In this experiment, data were collected for all stable Os isotopes except for ^{184}Os due to its low natural abundance.

3 Results

The experimental spectrum obtained with the crossed-beams technique is illustrated in Fig. 2 where the atomic resonances of the even $^{184-192}$Os isotopes and the hyperfine structures of the ^{187,189}Os isotopes are indicated. Results for the hyperfine parameters of ^{189}Os and the isotope shifts are included in Tables 1 and 2, respectively.

For the collinear laser spectroscopy spectra (Fig. 3) a χ^2 minimisation routine was used to fit Voigt profiles to the ionic resonance peaks. In the case of the odd isotopes a global fit of each hyperfine structure was performed as shown in Fig. 3. Isotope shifts and hyperfine parameters were determined from fits to the data. The hyperfine structure of ^{189}Os consists of nine hyperfine structure components out of which only four were clearly visible in the experimental spectra. Extraction of the unknown hyperfine parameters A_u, B_u, A_l, B_l and of the centre of gravity of the structure requires five peaks to be present in the spectrum otherwise the problem is underdetermined. Four candidate peaks for a fifth peak of the ^{189}Os structure were present in the data. For each candidate a χ^2 value was obtained for the global fit and the one that gave the lowest χ^2 was accepted as the best solution to the problem. In the case ^{187}Os, whose nuclear spin is 1/2, the electric hyperfine parameters are zero.

 Springer

Its structure consists of three components of which only two were observed due to the very low Racah intensity of the third one. The problem again is underdetermined. However, the ^{187}Os interval observed was consistent with the A hyperfine parameters scaled for ^{189}Os assuming zero hyperfine anomaly[10]. The centres of gravity of the hyperfine structures of the ^{187,189}Os isotopes obtained from this analysis are consistent with the King plot discussed later. Table 1 lists the hyperfine parameters of ^{189}Os which are essential for the calibration of the radioactive isotopes. It is necessary to confirm the identity of the fifth peak before proceeding to investigate the radioactive isotopes.

The isotope shifts extracted from both methods are given in Table 2. A consistency check between the two data sets was made with the King plot method[11] (see Fig. 4) which involves modifying each data point by a factor,

$$\mu = \frac{A A'}{A' - A} \frac{A'_{ref} - A_{ref}}{A_{ref} A'_{ref}} = \frac{A A'}{A' - A} \mu_{ref}$$

where A_{ref}, A'_{ref} are the atomic masses of an arbitrarily chosen pair of reference isotopes. Modified isotope shifts of the 305.9 nm transition have been plotted against those of the 228.2 nm transition and a straight line is formed whose gradient is equal to the ratio of their atomic factors $F_{305.9}/F_{228.2}$=1.05±0.02. The similarity of the atomic factors is coincidental although both transitions involve $6s$–$6p$ electrons. Relative isotope shifts, assuming a negligible mass shift have been plotted in Fig. 1 where it can be seen that the stable Os isotopes do not display abrupt changes in their mean square charge radii and that the OES is normal.

4 Conclusion and future plans

Isotope shifts for the stable $^{184-192}$Os and ^{187,189}Os isotopes were measured, relative to ^{192}Os using two Doppler-free techniques. The hyperfine parameters of ^{189}Os, essential for the calibration of the radioactive isotopes have been extracted. Studies of the neutron deficient isotopes down to N=103 are planned using the ^{185}Re(pxn) reaction. With a laser power of 250 µW an efficiency of 1 photon in 12,500 ions can be achieved on cooled Os ion beams which should be sufficient for on-line measurements of the radioactive isotopes.

References

1. Otten, E.W.: In: Bromley, D.A. (ed.) Treatise on Heavy-Ion Science, vol 8, p. 517. Plenum, New York (1989)
2. Talmi, I.: Nucl. Phys., A **423**, 189 (1984)
3. Sauvage, J., et al.: Hyperfine Interact. **129**, 303 (2000)
4. Le Blanc, F., et al.: Phys. Rev., C **60**, 054310-1 (1999)
5. Mahgoub, M.: MSc dissertation. University of Manchester (2002)
6. Cooper, T.G., et al.: J. Phys., G **82**, 99 (1996)
7. Levins, J.M.G., et al.: Phys. Rev. Lett. **82**, 2476 (1999)
8. Nieminen, A., et al.: Nucl. Instrum. Methods A **469**, 244 (2001)
9. Ashkin, A., et al.: IEEE J. Quantum Electron. **2**, 109 (1966)
10. Stone, N.J.: At. Data Nucl. Data Tables **90**, 75 (2005)
11. King, W.H.: Isotope Shifts in Atomic Spectra. Plenum, London (1984)

Hyperfine Interact (2006) 171:225–231
DOI 10.1007/s10751-006-9486-z

Study of the neutron deficient $^{182-190}$Pb isotopes by simultaneous atomic- and nuclear-spectroscopy

**M. Seliverstov · A. Andreyev · N. Barré · H. De Witte ·
D. Fedorov · V. Fedoseyev · S. Franchoo · J. Genevey ·
G. Huber · M. Huyse · U. Köster · P. Kunz ·
S. Lesher · B. Marsh · B. Roussière · J. Sauvage ·
P. Van Duppen · Yu. Volkov**

Published online: 3 February 2007
© Springer Science + Business Media B.V. 2007

Abstract Mean-square charge radii and magnetic moments have been measured for the neutron deficient lead isotopes, $^{182-190}$Pb. The measurement was performed at the ISOLDE online mass separator, using the in-source resonance ionization spectroscopy technique.

M. Seliverstov (✉) · G. Huber · P. Kunz
Institut für Physik, Johannes Gutenberg Universität, 55099 Mainz, Germany
e-mail: selivers@uni-mainz.de

M. Seliverstov · D. Fedorov · Y. Volkov
Petersburg Nuclear Physics Institute, 188350 Gatchina, Russia

A. Andreyev
Department of Physics, Oliver Lodge Laboratory, University of Liverpool,
PO Box 147, Liverpool L69 7ZE, UK

N. Barré · S. Franchoo · S. Lesher · B. Roussière · J. Sauvage
Institut de Physique Nucléaire, IN2P3-CNRS, 91406 Orsay Cedex, France

H. De Witte · M. Huyse · P. Van Duppen
Instituut voor Kern- en Stralingsfysica, IKS-KU Leuven,
Celestijnenlaan 200D, 3001 Leuven, Belgium

V. Fedoseyev · U. Köster · B. Marsh
ISOLDE, CERN, CH-1211 Geneva 23, Switzerland

J. Genevey
Laboratoire de Physique Subatomique et de Cosmologie,
IN2P3-CNRS/Université Joseph Fourier, 38026 Grenoble Cedex, France

U. Köster
Institut Laue-Langevin, 6 rue Jules Horowitz,
38042 Grenoble Cedex 9, France

B. Marsh
The University of Manchester, Manchester M13 9PL, UK

The wavelength of the first excitation step for the resonance ionization laser ion source (RILIS) was scanned over the resonance(s) whilst the α- and γ-ray spectra from the decay of the Pb isotopes were recorded as a function of the wavelength. The isotope shift and, in the case of odd-A isotopes, the hyperfine splitting were deduced. The rms-charge radii of the very neutron deficient Pb isotopes follow the smooth trend of the heavier isotopes. This finding indicates a spherical shape for the lead ground states at the neutron mid-shell ($N=104$), where the excitation energy of the oblate 0^+ state in the even isotopes reaches its minimum.

Key words laser spectroscopy · nuclear structure · resonance ionization · laser ion source · isotope shift · hyperfine structure · nuclear charge radius · nuclear magnetic moment

1 Introduction

In the region of neutron deficient lead isotopes, intruder states and shape coexistence phenomena have been widely investigated around neutron mid-shell at $N=104$. In ^{186}Pb at $N=104$, it has been found that the three lowest lying states have 0^+ as spin and parity values [1]. They have been associated with different shapes in the nucleus: spherical and deformed (oblate and prolate). In a shell model picture these states can be described as 0p–0h, 2p–2h and 4p–4h states, respectively [2]. Significant mixing of the deformed states into the spherical ground state would result in an increased nuclear charge radius. It is therefore of great interest to extend the study of nuclear radii in the lead chain to the very neutron deficient isotopes around neutron mid-shell.

2 Experimental method and set-up

The study of isotopes lying far from β-stability is dominated by the need for an increase in experimental sensitivity. For example, at ISOLDE the Pb ion yield drops by six orders of magnitude from ^{187}Pb to ^{183}Pb [3]. For this reason we used the in-source photoionization spectroscopy technique, which is one of the most sensitive laser spectroscopy methods. The use of the RILIS (Resonant Ionization Laser Ion Source) for atomic spectroscopy studies of exotic nuclei was pioneered at the Leningrad/Petersburg Nuclear Physics Institute (Gatchina, Russia), where it was applied to short lived $^{154-156}$Yb isotopes [4, 5]. At ISOLDE studies of short-lived Be, Cu, Pb isotopes by in-source RILIS laser spectroscopy have been also implemented (see reviews [6, 7] and references therein). These experiments have shown the high sensitivity of the method. Following the upgrade of the laser system a new series of such experiments have been performed at the ISOLDE facility at CERN.

The Pb nuclei were produced in spallation reactions induced by a 1.4 GeV proton beam impinging on a UC$_x$ target (50 g/cm^2 of ^{238}U and $\approx$10 g/cm^2 of carbon) connected to the ISOLDE mass separator. The spallation products diffuse out of the high temperature target ($T\approx$2,050°C) and effuse as neutral atoms into the laser ion source cavity. The cavity is a niobium tube of 30 mm length and 3 mm internal diameter and is heated to approximately 2,100°C. The laser beams were focused inside the cavity and the ionization of Pb isotopes was performed with a three-step ionization scheme (Fig. 1). The same scheme was used for earlier Pb experiments [8]. Laser light for the resonant excitation of the first two atomic

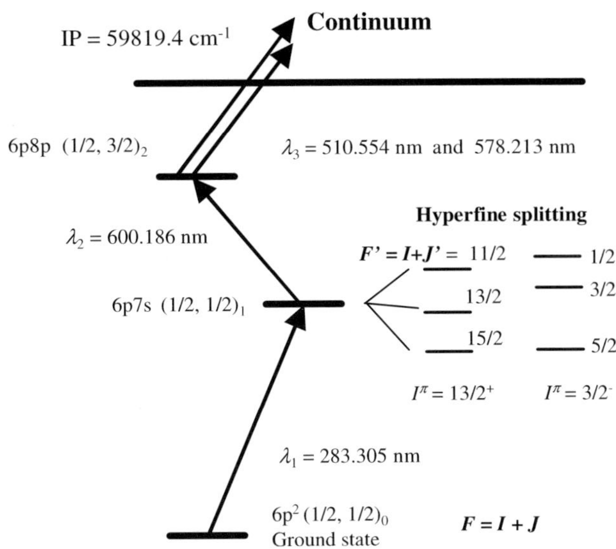

Fig. 1 The three-step laser ionisation scheme for Pb isotopes, applied in this work. Shown are electronic configurations for the levels involved and laser wavelengths for each step. On the right a schematic illustration of the hyperfine structure components in case of the 13/2$^+$ and 3/2$^-$ isomeric states in odd Pb isotopes is presented. More details are given in the text

transitions (λ_1=283.305 nm and λ_2=600.186 nm) was provided by two tunable pulsed dye lasers. The ultraviolet radiation (λ_1) was obtained by doubling of the fundamental dye laser radiation frequency using a non-linear BBO crystal. For the pumping of the dye lasers and the excitation of the final ionising transition (the third step) into the continuum, copper vapour lasers, operating at a pulse repetition rate of 11 kHz, were used. An ionisation efficiency of about 3% was measured in off-line tests with stable lead isotopes [9].

During the online experiment with radioactive nuclei the laser power of the first excitation step was reduced to avoid additional broadening of the optical line, so we estimate the on-line efficiency for Pb isotopes to be $\approx$ 1%. The laser ion source is not only a very effective in its normal use as an element-selective tool for producing intense ion beams, but it can be used as a very powerful atomic-spectroscopy tool due to the resonance character of the photoionization.

In contrast to other laser spectroscopic techniques, the scanning procedure is applied directly within the ion source. For atomic spectroscopy measurements, a narrow bandwidth laser (line width 1.2 GHz) was used for the first excitation step. By scanning its frequency over the resonance, together with simultaneous counting of the mass-separated photo-ions, the isotope shifts and hyperfine structure of the atomic spectral lines could be measured. From these data the change in nuclear mean square charge radii and the static nuclear moments (magnetic moments and spectroscopic quadrupole moments) have been determined.

Due to the low production yield for the very neutron-deficient lead isotopes, the direct ion counting was prohibited by a severe isobaric contamination of the ion beam. An isotope and isomer selective registration of the photo-ions was required and could be achieved by the synchronous measurement the α- and γ-ray spectra characteristic for the decay of the Pb isotopes and isomers in question. For this work, the radioactive Pb isotopes were either detected with a α-spectroscopy setup composed of a wind mill system and a passivated implanted planar silicon detector and a γ- or α-γ spectroscopy setup comprising a tape transport system, a planar Ge(HP) X-ray detector, two 70% efficiency coaxial Ge(HP) detectors and an silicon α-detector.

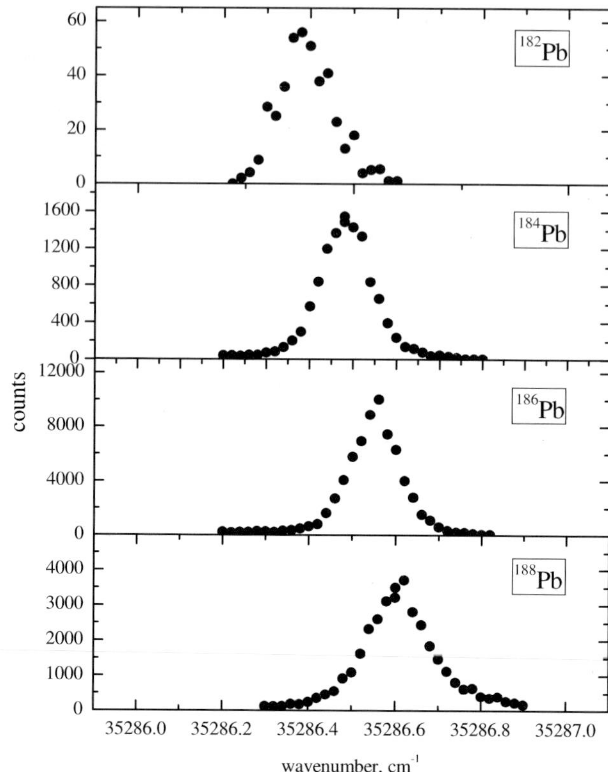

Fig. 2 Typical examples of the collected spectra (α-counts versus laser wavenumber) of the even-mass isotopes of Pb

During each laser scan an α- or γ-ray spectrum was recorded. In the data analysis the characteristic lines were integrated for every laser frequency setting. The optical photo-ionization spectra (number of photo-ions versus laser frequency) were obtained. Typical examples of the collected optical spectra are shown in Figs. 2 and 3. Isotopes $^{182-188}$Pb were studied with the α- detection, ^{189}Pb with the γ-ray and α-detection and ^{190}Pb with the γ-ray detection.

A multiparameter servo-control for the laser frequency scanning was implemented for this experiment. The new control system incorporates controls for the narrowband laser etalon and diffraction grating positions, a frequency read-out with feed back control and also communication to the data acquisition system using the RS232 or TCP/IP protocols. The long-term stability of the laser control system enabled laser scans in excess of two hours to be performed. This is essential for the measurement of the radioactive isotopes at low production yield.

Throughout the experimental period, measurements of radioactive Pb isotopes at the scanning frequencies were, at regular time intervals, combined with measurements at a fixed reference frequency in order to monitor the stability of the experimental system. For an absolute wavelength scale calibration, spectra of stable Pb isotopes were recorded regularly. For these isotopes, the photo-ion current was directly measured with a Faraday cup.

For heavy elements such as Pb, with large isotope shift and hyperfine structures, the Doppler limited RILIS spectroscopy is particularly advantageous. Using an optical setup with a high long term stability and isobaric-background free ion detection, an extremely high sensitivity can be obtained and is essential in such a study.

 Springer

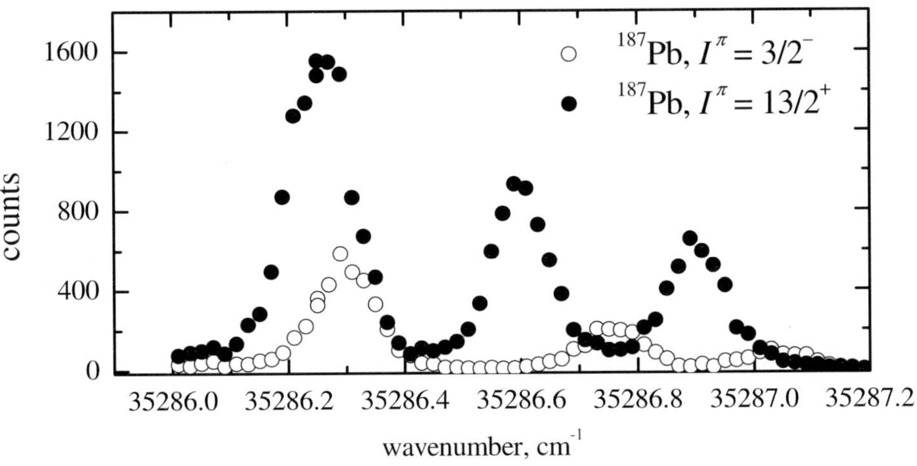

Fig. 3 Spectra for the high spin isomer ($I^{\pi} = 13/2^{+}$, $E_{\alpha} = 5.993$ MeV) and the low spin isomer ($I^{\pi} = 3/2^{-}$, $E_{\alpha} = 6.194$ MeV) of ^{187}Pb collected in a single scan

3 Results and discussion

The isotopic change of the charge radius $\delta\langle r^{2}\rangle_{A,A'}$ was evaluated through the standard formula:

$$\delta v_{A,A'} = F \cdot \lambda_{A,A'} + M\frac{A' - A}{A \cdot A'}, \tag{1}$$

where $\delta v_{A,A'}$ is the isotope shift between isotopes with mass numbers A and A', F is the electronic factor, M is the mass shift constant, $\lambda_{A,A'}$ is the nuclear parameter [10]:

$$\lambda = \delta\langle r^{2}\rangle + C_{2}\delta\langle r^{4}\rangle + C_{3}\delta\langle r^{6}\rangle + \ldots = C\delta\langle r^{2}\rangle. \tag{2}$$

The following values of constants were used [11]:
F=20.26 (0.18) GHz fm^{-2},
M=0.19(0.25) $\times$ N (here N is the normal mass shift constant),
C=0.93.
The magnetic dipole moments were evaluated from the scaling relations, based on the known magnetic moment μ_{0}, magnetic coupling constant A_{0} and nuclear spin I_{0} of a stable ^{207}Pb isotope (μ_{0}=0.592583(9), I_{0}=1/2, A_{0}=8,807.2(3.0)) [11, 12]:

$$\mu = \frac{A \cdot I \cdot \mu_{0}}{A_{0} \cdot I_{0}}. \tag{3}$$

This expression disregards any hyperfine anomaly, which was found to be less than $\approx 10^{-3}$ for $^{190-197}$Pb isotopes studied in [13].

In Fig. 4 our data are shown along with the data for the neighboring Hg (Z=80) and Pt (Z=78) chains [14–17]. The observed slope of the charge radii as a function of the mass follows the smooth trend of the heavier isotopes down to and even below the neutron mid-shell, where the excitation energy of the oblate 0^{+} state in the even isotopes reaches its minimum. This finding indicates a spherical shape for the Pb ground states. The charge radii follow rather closely the droplet model prediction. This is in contrast with the Hg isotopes

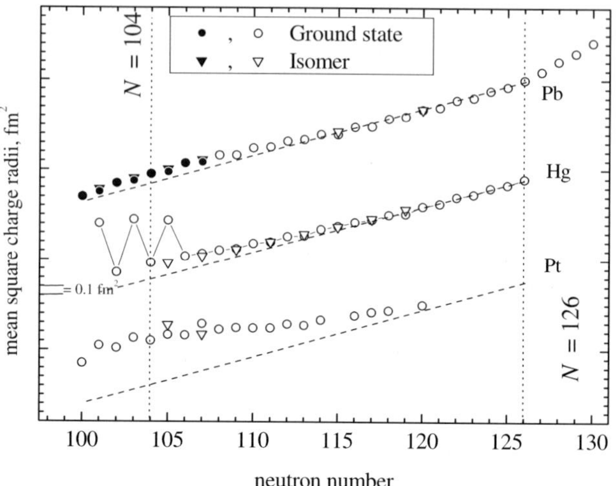

Fig. 4 Change of the mean square charge radii for Pb, Hg and Pt isotopes, compared to the predictions of the spherical droplet model. The *experimental error bars* are smaller than the symbol size. Our data are shown as *filled symbols*, the literature data are shown as *open symbols*. The distance between the different chains is chosen arbitrarily for ease of display. Each minor division on the vertical scale corresponds to 0.1 fm²

where a large odd–even staggering and shape isomerism set in at $N=105$. In the Pt series an overall deviation from the linear trend for heavier isotopes indicates the influence of deformation near the mid-shell. Such an effect is absent or at least considerably smaller in the Pb isotope chain, confirming their spherical nature.

The magnetic moments of the high spin isomers ($I^\pi = 13/2^+$) also follow a smooth trend from $A=197$ to $A=183$ with a slight down-sloping towards the Schmidt line.

4 Conclusions and outlook

Substantial improvements in the frequency stabilization for the RILIS laser system have been achieved. With its improved precision and reliability, the in-source laser spectroscopy technique at ISOLDE has been combined with synchronous nuclear decay spectroscopy. The high sensitivity and isomer selectivity enabled measurement of very short-lived neutron deficient Pb isotopes down to ^{183}Pb ($T_{1/2}=300$ ms) and ^{182}Pb ($T_{1/2}=55$ ms) which are produced at very low yield.

The laser spectroscopy investigation of the Pb isotope chain has been extended beyond the neutron mid-shell at $N=104$. The behavior of charge radii indicates that in this region the ground states of the Pb isotopes remain essentially spherical. Beyond mean-field calculations are underway to compare directly with the experimental data reported in this paper [18, 19].

The in-source laser resonance ionization spectroscopy technique has become a very efficient tool for atomic spectroscopy of rare isotopes approaching the driplines.

Acknowledgement This work was supported by German Ministry for Education and Research (BMBF) under contract No. 06 MZ 171.

References

1. Andreyev, A.N., et al.: Nature **405**, 430 (2000)
2. Wood, J.L., et al.: Phys. Rep. **215**, 101 (1992)

 Springer

3. Köster, U., et al.: Nucl. Instrum. Methods, B **204**, 347 (2003)
4. Alkhazov, G.D., et al.: Nucl. Instrum. Methods, B **69**, 517 (1992)
5. Barzakh, A.E., et al.: Eur. Phys. J., A **1**, 3 (1998)
6. Köster, U., et al.: Hyp. Int. **127**, 417 (2000)
7. Fedosseev, V.N., et al.: Nucl. Instrum. Methods, B **204**, 353 (2003)
8. Andreyev, A.N., et al.: Eur. Phys. J., A **14**, 63 (2002)
9. Köster, U.: Nucl. Phys., A **701**, 441 (2002)
10. Seltzer, E.C.: Phys. Rev. **188**, 1916 (1969)
11. Anselment, M., et al.: Nucl. Phys., A **451**, 471 (1986)
12. Lutz, O., Stricker, G.: Phys. Lett., A **35**, 397 (1971)
13. Dutta, S.B., et al.: Z. Phys., A **341**, 39 (1991)
14. Bonn, J., et al.: Z. Phys., A **276**, 203 (1976)
15. Ulm, G., et al.: Z. Phys., A **325**, 247 (1986)
16. Lee, J.K.P., et al.: Phys. Rev., C **38**, 2985 (1988)
17. Le Blanc, F., et al.: Phys. Rev., C **60**, 054310 (1999)
18. Bender, M., et al.: Phys. Rev., C **73**, 034322 (2006)
19. Bender, M. et al.: Phys. Rev., C **69**, 064303 (2004)

Hyperfine Interact (2006) 171:233–241
DOI 10.1007/s10751-006-9483-2

Experimental investigation of the stability diagram for Paul traps in the case of praseodymium ions

W. Koczorowski · G. Szawioła · A. Walaszyk ·
A. Buczek · D. Stefańska · E. Stachowska

Published online: 1 February 2007

Abstract The present paper describes an investigation of non-linear resonances of praseodymium ion clouds stored in a Paul trap as a function of the storage parameters. These have been observed in traps with different ring electrode diameters. In these different traps the resonances occur for different values of the operating parameters. Discrepancies with the approximation model for one ion have been found. The intensity of the fluorescence signal and the Doppler half width have been recorded as a function of one of the storage parameters: q. We use our results to optimize the fluorescence signal of the stored ions, which is especially useful in the case of the double-resonance method.

Key words Paul trap · non-linear resonances · laser spectroscopic investigations

1 Introduction

Ion storage in electromagnetic traps requires a special configuration of the dynamic electrical field, consisting of DC and AC components. The motion of individual ions in the radial plane and along the symmetry axis of the trap obeys the well-known Mathieu equation [1, 2]. According to this equation, the storage conditions are stable in all directions within certain ranges of the values of the DC and AC voltages, depending on the mass of the ions, the trap dimensions and the frequency of the AC voltage. This trapping area is usually visualized in a stability diagram. The trapping area is crossed by non-linear resonances which we calculated for one Pr^+ ion in the non-limited field approximation. We are especially interested in

W. Koczorowski (✉) · G. Szawioła · A. Walaszyk · A. Buczek · D. Stefańska · E. Stachowska
Chair of Quantum Engineering and Metrology, Poznan University of Technology,
Ul. Nieszawska 13 B, 60-965 Poznan, Poland
e-mail: Wojciech.Koczorowski@put.poznan.pl

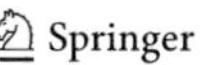

precision spectroscopy of rare-earth ions. To obtain spectroscopic data for these complex ions, with many hyperfine levels, we need however a large number of ions to obtain a large enough signal. We also need a buffer gas to depopulate the several metastable states, which could trap ions in a state no longer contributing to a signal. Theoretical calculations for the many ion case have been made [3–6], taking into account the space charge effect, but do not provide results which describe an actual experimental situation accurately enough. Also the presence of the buffer gas introduces further complications for which an accurate theoretical description not exists to our knowledge. Therefore studying experimentally the real instabilities is needed to optimize the spectroscopic signal. In this way we can obtain strong and well resolved signals, which are particularly important in investigations using the laser-microwave double resonance technique. We choose Pr^+ as a case study because it has only one stable isotope, relatively strong lines from the ground level and a low lying metastable level with good separations of the hyperfine structure (hfs) components. Knowledge of the non-linear resonance in a specific case could also be helpful to modify the number of ions and, in case of other rare earth ions, the isotope composition of the ion cloud [7].

2 Theoretical background

The basic trapping theory has been described in a few monographs see e.g. [8] and most recently in a paper by Madsen et al. [9]. In a three-dimensional Paul trap the potential is described by:

$$\phi_0(t) = U_0 - V_0 \cos(\Omega t) \tag{1}$$

where U_0 and V_0 are the DC and AC amplitudes respectively, and Ω is the AC frequency.

In order to describe the stability diagram and the operating point, two dimensionless parameters are defined, for both the radial (r) and axial (z) directions :

$$a_z = -2a_r = -\frac{8eU_0}{r_0^2 m\Omega^2} \tag{2}$$

$$q_z = -2q_r = -\frac{4eV_0}{r_0^2 m\Omega^2} \tag{3}$$

In Fig. 1 a region of the stability diagram of the Paul trap, calculated for Pr^+ with the one ion model, without interaction with the buffer gas, using a program provided by Drakoudis (Drakoudis, personal communication).

Within the stability diagram some curves are present, where the ions can resonantly absorb the energy of the electric field, called non-linear resonances. They result from the imperfections of the quadrupole field, i.e. higher-order contributions, caused mainly by the finite dimensions of the electrodes and the deviations from the hyperboloidal geometry, required for the experiment as explained in Section 3. The positions of the non-linear resonances on the stability diagram have been experimentally determined for up to 5,000 ions of one element, hydrogen, by Alheit et al. [10].

 Springer

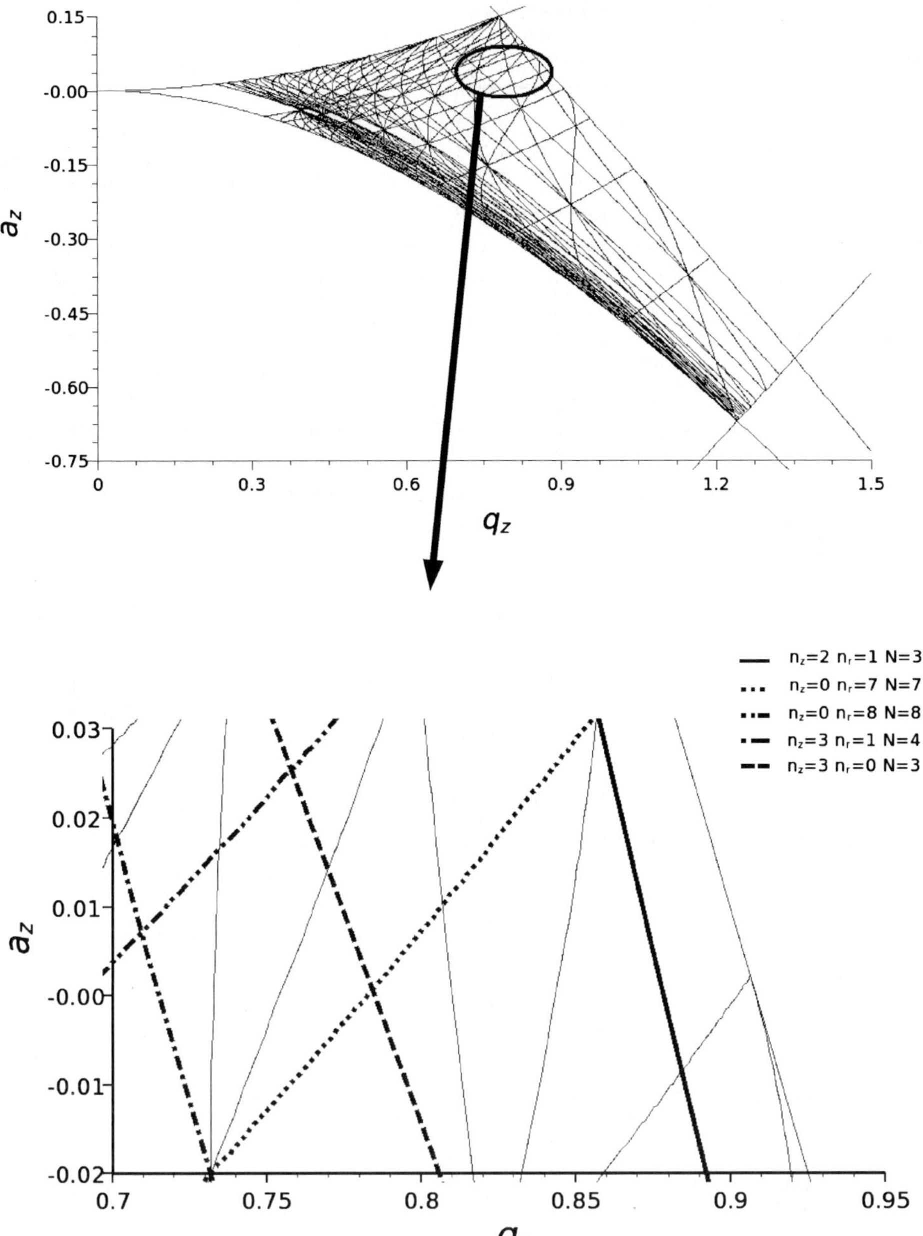

Fig. 1 Stability diagram for the Mathieu equation, including non-linear resonances; the *lower part* shows the area we experimentally investigated

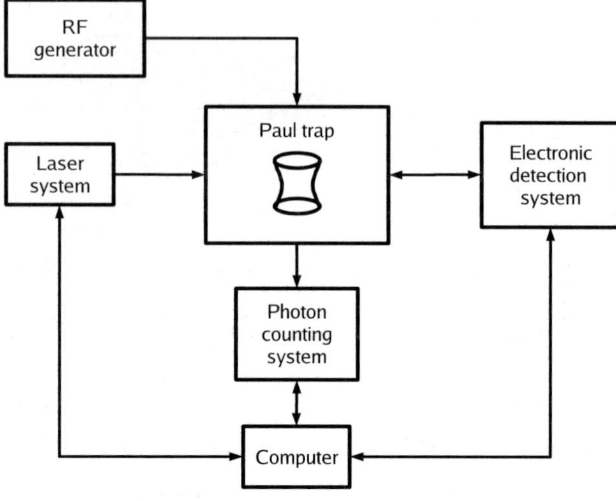

Fig. 2 Experimental setup for spectroscopic investigations in the Paul trap

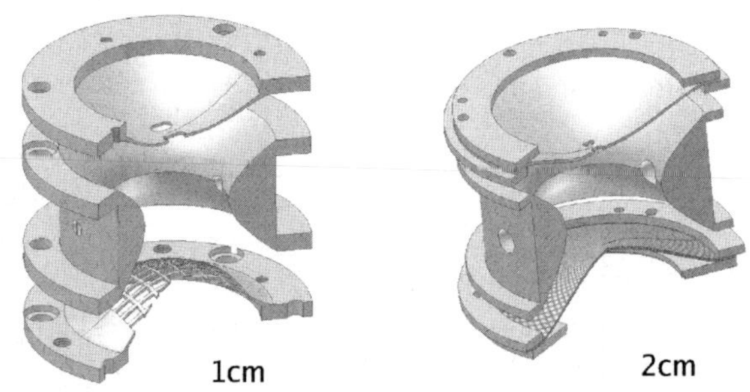

Fig. 3 The two different traps

Non-linear resonances are treated in detail in the monograph by Major et al. [8]. They are characterized by the integer mode numbers n_r, n_z; see the insert in Fig. 1, which shows the region we studied in more detail. The sum

$$|n_r| + |n_z| = N \tag{4}$$

determines the order 2N of a perturbing multipole.

3 Experiment

The detailed description of our experimental setup has been presented in our earlier papers [11, 12] and is shown in Fig. 2. The experimental work was done using two traps with different radius ($r_0 = 1$ cm and $r_0 = 2$ cm, see Fig. 3), which were mounted in two independent ultrahigh vacuum vessels. The axial dimensions have the same proportion in comparison to the radius of the trap. The smaller trap was build using stainless steel, while the larger one was made of OHFC copper. In the ring electrodes holes of resp. 2.7 and 8 mm were drilled to allow the laser beam to enter the trapping area. Furthermore two rectangular holes were made in each trap through which a tantalum wire with a Pr^+ sample could be introduced (each 3 mm by 5 mm). In

 Springer

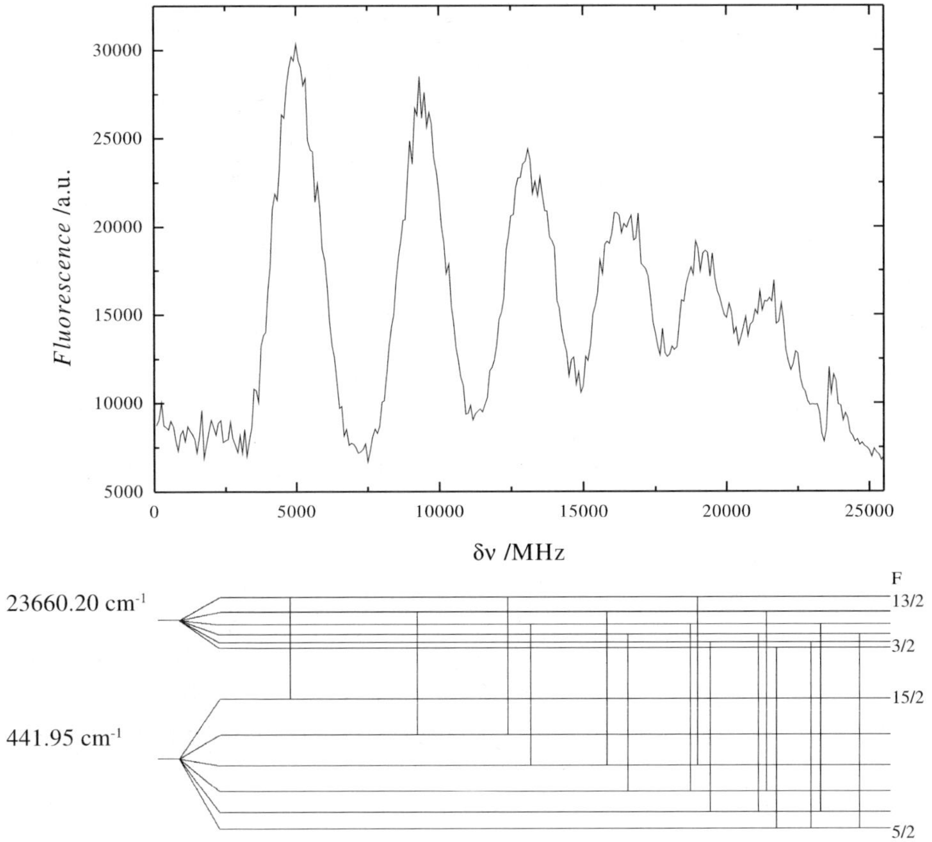

Fig. 4 Recorded LIF signal for praseodymium ions for the transition at 430.6 nm : $4f^36s^5I_5$ $(441.95\ \mathrm{cm^{-1}}) \rightarrow 4f^36p^5I_4$ $(23660.20\ \mathrm{cm^{-1}})$

the smaller trap one end cap electrode was formed by a mesh made from 0.5 mm wire with rectangular openings of 1.5 mm by 2.8 mm for collecting the laser induced fluorescence. In the larger one the openings were 2 mm by 2 mm. Deviations of the trap from an ideal quadrupole shape originate from the finite axial dimensions and these disturbances of the ideal shape of the electrodes and the mounting rings of the end caps.

We detected non-linear resonances by observing the variation in the laser induced fluorescence (LIF) of $^{141}Pr^+$ ions. Electronic detection is not well suited for a direct investigation of the non-linear resonances due to the relative large time delay, but it provides an independent estimate of the relative number of ions.

The fluorescence light has been recorded with a photomultiplier, an example of the Pr^+ signal is presented in Fig. 4. To study the non-linear resonances the laser frequency was tuned to the maximum of the strongest and well-isolated hyperfine component of the optical transition, and the parameter q was varied by changing the AC voltage. This method has been used to detect non-linear resonances in earlier

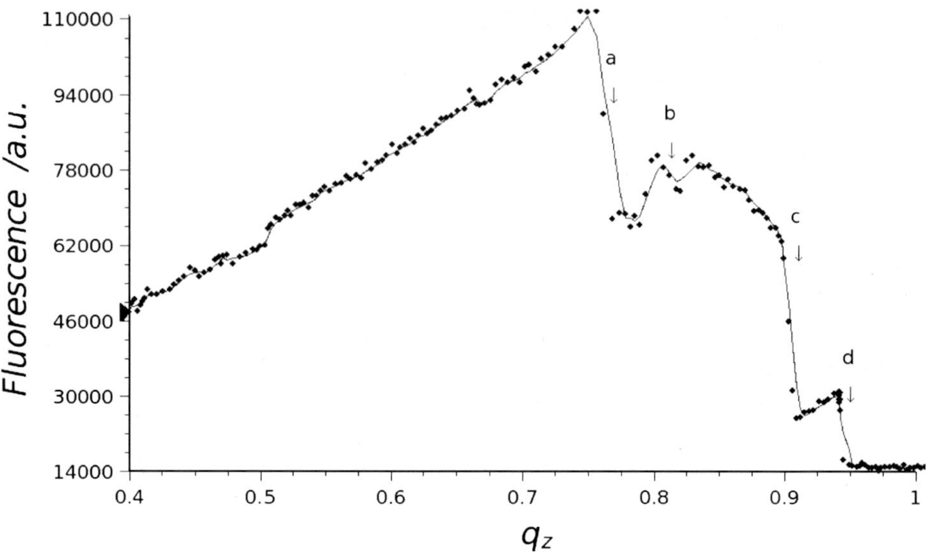

Fig. 5 LIF vs q_z at $a_z = -0.01525$ for the $r_0 = 2$ cm trap (laser frequency tuned to the *top* of the first hyperfine component of the *spectral line* in Fig. 4)

work see e.g. [13]. The AC amplitude has been varied in the range 250–1000 V, with an approximate accuracy of 1 V, while the DC voltage was kept constant.

4 Results

Crossing a non-linear resonance is indicated by an abrupt decrease of the fluorescence light intensity while varying the parameter q. An example of the recording is presented in Fig. 5. Different non-linear resonances also yield different degrees of fluorescence intensity increase after crossing the resonance q value. Compare resonance b with a and c in Fig. 5. We noticed that a sufficiently fast variation of the parameter q through a non-linear resonance results in a small or almost negligible loss of ions. This suggests that although at the point of the non-linear resonance the kinetic energy of the ions considerably increases, which results in a decreased fluorescence intensity due to the expansion of the ion cloud (and thus a decrease of the number of ions in the observation volume), the process is in most cases slow enough to be inverted, when stable storage conditions are restored.

The points in the stability diagram, where non-linear resonances have been experimentally observed for traps of ring electrode radius $r_0 = 1$ cm or $r_0 = 2$ cm, are shown in Fig. 6. In the larger trap we could observe more lines of non-linear resonances then in the smaller one. The lines of non-linear resonances we can construct from our measured points are clearly different from the ones calculated in the one ion approximation. A tentative identification is given in these figures too, which led us to conclude that higher order non-linearities are observed in the larger trap.

 Springer

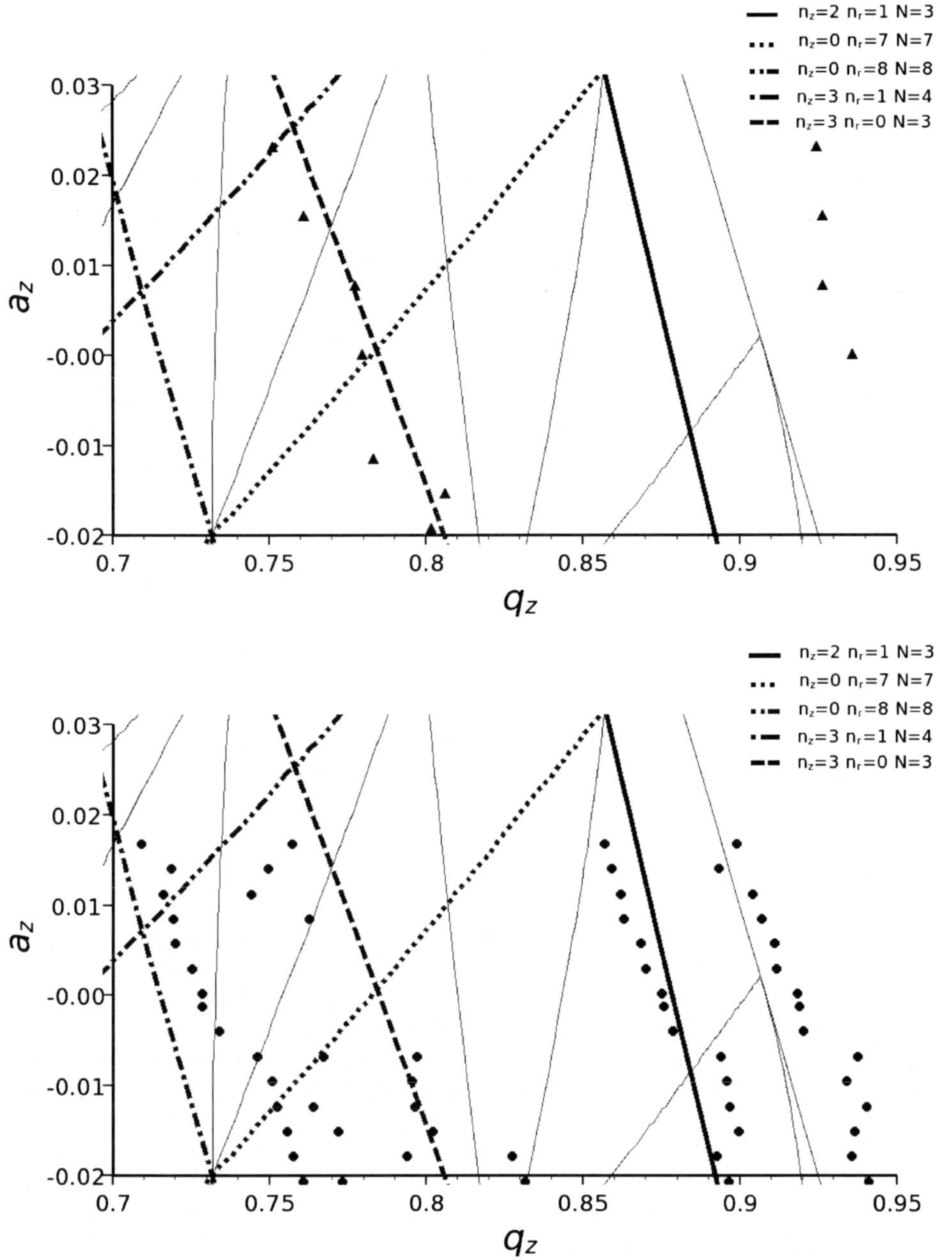

Fig. 6 Part of the stability diagram with theoretical lines of non-linear resonances and experimental points determined for an ensemble of praseodymium ions; upper figure for the trap with $r_0 = 1$ cm, lower figure for the trap with $r_0 = 2$ cm

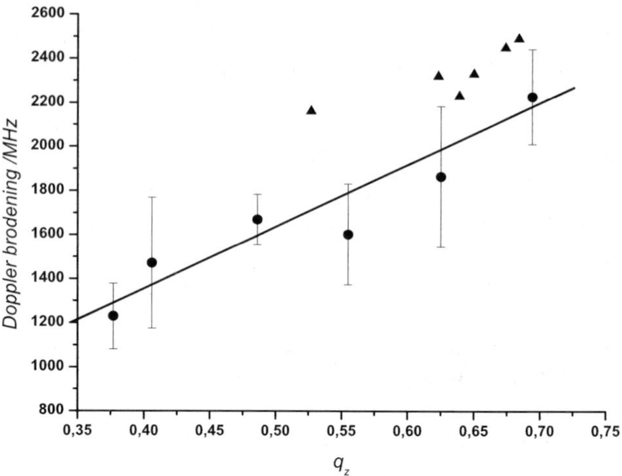

Fig. 7 Doppler broadening vs the storage parameter q_z. The • denote the results obtained with the larger trap; the points (▲), taken from [14], enable a comparison with results previously obtained in the smaller trap ($r_0 = 1$ cm)

The accuracy of determination of the positions of the non-linear resonances is however strongly limited by space charge effects, since the number of the ions in the cloud can only be controlled roughly. It is rather difficult to state unequivocally, but the experimental evidence suggests that the outermost points observed with a fluorescence intensity decrease resembling a non-linear resonance in fact fall on the border lines of the stability diagram (d in Fig. 5). After passing through these points a complete ion loss is observed and the fluorescence signal cannot be restored regardless of the variation rate of the parameter q. If this conclusion could be verified, the present investigation method could also be used to determine the borders of the stability diagram and to study its position and shape dependency on the number of stored ions, provided the latter could be controlled more precisely.

The gradual increase of the LIF signal intensity and half width with storage parameters q, before reaching a non-linear resonance can clearly be seen in Fig. 5; the same has been observed in the smaller trap. This intensity increase can be used to maximize the LIF-signal, in the case where the hfs is large compared with the Doppler broadening, which is the case in many rare earth spectra.

In Fig. 7 the dependence of the half widths of the observed LIF signal, dominated by Doppler broadening, is presented (only the regions far away from non-linear resonances have been considered, since the severe ion loss at the resonances considerably influences the line half width). The large error bars are due to difficulties in reproducing the experimental conditions in subsequent measurement runs (different numbers of stored ions in the observation volume). The errors are statistical, each point has been measured at least four times. In the case of close lying hfs components this effect could limit the possibility for maximizing the LIF signal.

5 Concluding remarks

Spectroscopic investigations allow to study the stability diagram with respect to inhomogeneities and instabilities of the storage process. The positions of the points, where such inhomogeneities occur, have been determined with an accuracy better

 Springer

than 0.5% for q and 0.1% for a. In both traps the values of the parameters q_z and a_z for which non-linear resonances occur differ significantly from those calculated for one ion in the non-limited field approximation. In the larger trap a considerably larger number of non-linear resonances has been observed, in particular such as characterized by a less abrupt decrease of the fluorescence intensity. Also the parameter values for which they appear are different.

When approaching a non-linear resonance the LIF, as well as the Doppler-width increase.

Our results allow simultaneous optimization of the LIF signal with respect to intensity, half width as well as reproducability. This is particularly important for spectroscopic investigations of the metastable levels of Pr^+ lying above 5,000 cm^{-1}, which have a low thermal population, using the laser-microwave double resonance method.

A still more accurate determination of the position of the non-linear resonances might allow a selective modification of the relative isotopic abundance in the ion cloud of the element under study, or a controlled decrease of the number of the stored ions [7].

In order to obtain a versatile optimization procedure, more systematic investigations of the stability diagram with respect to the storage efficiency are still required.

Acknowledgements We wish to thank dr. F.G. Meijer for reading and discussing the revision of the manuscript. This work was performed as part of the project *DS-63-029/06* of the Poznan University of Technology.

References

1. Mathieu, E.: J. des Mathémathiques Pure et Appliquées, 137–203 (1868)
2. McLachlan, N.W.: Theory and Applications of Mathieu Functions. Clarendon, Oxford (1947)
3. Fisher, E.: Z. Phys. **156**, 1–26 (1959)
4. Schwebel, C., Moller, P.A., Manh, P.-T.: Rev. Phys. Appl. **10**, 227–239 (1975)
5. Vedel, F., André J., Vedel, M., Brincourt, G.: Phys. Rev., A **27**, 2321–2330 (1983)
6. Meis, C., Desaintfuscien, M., Jardino, M.: Appl. Phys., B **45**, 59–64 (1988)
7. Alheit, R., Enders, K., Werth, G.: Appl. Phys., B **62**, 511–513 (1996)
8. Major, F.G., Gheorghe, V.N., Werth, G.: Charged particle physics. In: Springer Series on Atomic, Optical, and Plasma Physics, vol. 37. Springer, Berlin Heidelberg New York (2005)
9. Madsen, M.J., Hensinger, W.K., Stick, D., Rabchuk, J.A., Monroe, C.: Appl. Phys., B **78**, 639–651 (2004)
10. Alheit, R., Gudjons, Th., Kleineidam, S., Werth, G.: Rapid Commun. Mass Spectrom. **10**, 583–590 (1996)
11. Dembczyński, J., Stefańska, D., Szawioła, G., Furmann, B., Stachowska, E., Jarosz, A., Arcimowicz, B., Buczek, A., Koczorowski, W., Krzykowski, A., Kajoch, A., Elantkowska, M., Ruczkowski, J., Kowalkiewicz, W.: Act. Phys. Pol., A **92**, 517–526 (1997)
12. Stachowska, E., Szawioła, G., Buczek, A., Koczorowski, W., Furmann, B., Stefańska, D., Walaszyk, A., Dembczyński, J.: M & MS **13**, 207–217 (2006)
13. Gudjons, T., Seibert, P., Werth, G.: Appl. Phys., B **65**, 57–62 (1997)
14. Zdrojewski, S.: M.Sc. Thesis, Poznań (2004)

Hyperfine Interact (2006) 171:243–253
DOI 10.1007/s10751-006-9484-1

Correlation effects on the charge radii of exotic nuclei

M. Tomaselli · T. Kühl · D. Ursescu · S. Fritzsche

Published online: 1 February 2007
© Springer Science + Business Media B.V. 2007

Abstract The structures and distributions of light nuclei are investigated within a microscopic correlation model. Two particle correlations are responsible for the scattering of model particles either to low momentum- or to high momentum-states. The low momentum states form the model space while the high momentum states are used to calculate the G-matrix. The three and higher order particle correlations do not play a role in the latter calculation especially if the correlations induced by the scattering operator are of sufficient short range. They modify however, via the long tail of the nuclear potential, the Slater determinant of the (A) particles by generating excited Slater's determinants.

Key words nuclear models · distribution · charge radii

PACS 21.10.Ma · 21.60.-n · 21.10.Ft · 21.60.Fw

1 Introduction

Correlation effects in nuclei have first been introduced in nuclei by Villars [1], who proposed the unitary-model operator (UMO) to construct effective operators. The method was implemented by Shakin [2, 3] for the calculation of the G-matrix from hard-core interactions.

M. Tomaselli (✉)
Physics Department, TUD Darmstadt, 64289 Darmstadt, Germany
e-mail: m.tomaselli@gsi.de

M. Tomaselli · T. Kühl · D. Ursescu
Gesellschaft für Schwerionenforschung (GSI), 64291 Darmstadt, Germany

S. Fritzsche
Institut für Physik, Kassel University, 34132 Kassel, Germany

The UMO is based on the separation of the two body potential in a short and a long components. Within this separation the effective n-body Hamiltonian contains only the long component. The short-range component is considered up to the two body correlation and produces no energy shift in the pair state.

Non perturbative approximations of the UMO have been recently applied to even nuclei in [4, 5] which is treated here in more detail. The basics formulae of the Boson Dynamic Correlation Model (BDCM) presented in the above quoted paper have been obtained by solving the n-body problem in terms of the long range component of the two-body force. This component has the effect of generating a new correlated model space (effective space) which departs from the originally adopted one (shell model). The amplitudes of the model wave functions are calculated in terms of non-linear equation of motions (EoM).

By linearizing the systems of commutator equations, which characterize the EoM, we derive the eigenvalue equations for our model space. Within this correlated formalism we generate a model that includes not only the ladder diagrams of [6] but also the folded diagrams of Kuo [7].

The n-body matrix elements which define the eigenvalue equations are calculated exactly via the Cluster Factorization Theory (CFT) [8].

In this paper the BDCM model is applied to calculate the influence of the correlations on the charge distributions of the lithium isotopes and on the charge distribution of ^{6}He.

The value obtained for the charge radius of the correlated ^{6}He is slightly bigger than the radius calculated in other theories [9–13] and that derived within the isotopic-shift IS theory [14]. A charge radius which agrees with the radii calculated in the [9–13] and those calculated in the cluster models of [15–17] is on the other hand obtained by considering only two protons in the $s_{\frac{1}{2}}$. This non-correlated radius agrees also with the radius derived at Argonne within the IS theory [14]. Correlations have therefore the property to increase the charge radius of ^{6}He as observed for the isotopes of Lithium.

The calculations performed in [18] for the charge radii of the lithium isotopes, although in good agreement with those measured at GSI-TRIUMF [19–21] and analyzed with the help of [22], are always slightly larger than those measured. For the stable isotope ^{6}Li the calculated radius agrees with the value obtained with the electron scattering experiments of [23]. However, the charge radii calculated in the IS theory could also depend on the nuclear correlations. The calculations performed in [24] for the field shift (FS) of ^{7}Li show that the departure from a point nuclear approximation is a rather large effect. Additionally the higher order cross term contributions of [25] need to be considered.

A direct comparison between the calculated and the measured charge radii should therefore be performed after an accurate analysis of these two correcting factors.

2 Theory of two correlated particles

In order to describe the structures and the distributions of nuclei we start from the following Hamiltonian:

$$H = \sum_{\alpha\beta} \langle \alpha|t|\beta \rangle \, a_\alpha^\dagger a_\beta \; + \; \sum_{\alpha\beta\gamma\delta} \langle \Phi_{\alpha\beta}|v_{12}|\Phi_{\gamma\delta} \rangle \, a_\alpha^\dagger a_\beta^\dagger a_\delta a_\gamma \tag{1}$$

 Springer

were v_{12} is the singular nucleon-nucleon two body potential. Since the matrix elements $|\alpha\beta\rangle$ are uncorrelated the matrix elements of v_{12} are infinite. This problem can be avoided by taking matrix elements of the Hamiltonian between correlated states. In this paper the effect of correlation is introduced via the e^{iS} method. In dealing with very short range correlations only the S_2 part of the correlation operator need to be considered.

Following [2, 3] we therefore calculate an "effective Hamiltonian" by using only the S_2 correlation operator obtaining:

$$H_{eff} = e^{-iS_2} H e^{iS_2} = \sum_{\alpha\beta} \langle\alpha|t|\beta\rangle a_\alpha^\dagger a_\beta + \sum_{\alpha\beta\gamma\delta} \langle\Psi_{\alpha\beta}|v_{12}^l|\Psi_{\gamma\delta}\rangle a_\alpha^\dagger a_\beta^\dagger a_\delta a_\gamma$$

$$= \sum_{\alpha\beta} \langle\alpha|t|\beta\rangle a_\alpha^\dagger a_\beta + \sum_{\alpha\beta\gamma\delta} \langle\Psi_{\alpha\beta}|v|\Psi_{\gamma\delta}\rangle a_\alpha^\dagger a_\beta^\dagger a_\delta a_\gamma \qquad (2)$$

where v_{12}^l is the long component of the two body interaction (note that the v_{12}^l is in the following equations simply denoted as v). The $\Psi_{\alpha\beta}$ is the two particle correlated wave function:

$$\Psi_{\alpha\beta} = e^{iS_2} \Phi_{\alpha\beta} \qquad (3)$$

In dealing with complex nuclei however the S_i with $i = 3 \cdots n$ correlations should also be considered.

The evaluation of these diagrams is, due to the exponentially increasing number of terms, difficult in a perturbation theory.

We note however that one way to overcome this problem is to work with $e^{i(S_1+S_2+S_3+\cdots+S_n)}$ operator on the Slater's determinant by keeping the n-body Hamiltonian unvaried.

After having performed the diagonalization of the n-body Hamilton's operator we can calculate the form of the effective Hamiltonian which, by now, includes correlation operators of complex order.

Using (1), we can compute the commutator of the Hamiltonian with the operator $(a_{j_1}^\dagger a_{j_2}^\dagger)^J$ that creates a valence particle pair. By performing this calculation we shall retain the linear and the non-linear terms which are formed by coupling the shell model states with the particle-hole excitations of the core.

At this point in order to obtain a complete system of equations, we also have to calculate the commutator of the Hamiltonian (1) with the non-linear terms. With this calculation we introduce in the model space states which are formed by coupling the valence states with the two particle-hole excitations of the core.

The successive model equations are then formed by calculating the commutator with the operator obtained in the previous step.

The set of commutator equations above is suitable to be solved by means of a perturbation expansion. The perturbative solution of the system of commutator equations is however not easily obtainable due to the high number of diagrams one needs to calculate.

A non-perturbative solution of the system of commutator equation can be obtained within the linearization method, which consists by applying the Wick's theorem to the ((n+2) p-2h) terms and by neglecting the normal order terms.

This approximation is motivated by the consideration that the low lying spectra of nuclei the ((n+2) p-2h) terms are lying at much higher energy than that of the ((n+1) p-1h) states.

The linearized system of the commutator equations is then solved exactly in terms of the CFT which calculates the n-body matrix elements in an expedite and exact way.

In the following we give the basic formula of the method for a (nuclear) system with two valence particles. In second quantization, the two particle states are defined by:

$$\Phi_{2p} \longrightarrow \Phi^{J}_{j_1 j_2} = A^{\dagger}_1(\alpha_1 J)|0\rangle = \left[a^{\dagger}_{j_1} a^{\dagger}_{j_2}\right]^{J}_{M} |0\rangle, \tag{4}$$

where, for the sake of simplicity, we have omitted the isospin quantum numbers and where

$$\alpha_1 \leftrightarrow j_1 j_2 \tag{5}$$

has been introduced to ensure a compact index notation of the angular momenta of the two particles. In this notation, the operator product $a^{\dagger}_{j_1} a^{\dagger}_{j_2}$ just creates two *coupled* particles of single particle j_1 and j_2 *coupled* to the final J quantum number.

To derive the effect of the correlation on the two valence particles, we have, at this stage, to evaluate the next commutator

$$[H, A^{\dagger}_1(\alpha_1 J)]|0\rangle$$

$$= \left[\left(\sum_{\alpha} \epsilon_{\alpha} a^{\dagger}_{\alpha} a_{\alpha} + \frac{1}{2} \sum_{\alpha\beta\gamma\delta} \langle\alpha\beta|v(r)|\gamma\delta\rangle a^{\dagger}_{\alpha} a^{\dagger}_{\beta} a_{\delta} a_{\gamma}\right), \left(a^{\dagger}_{j_1} a^{\dagger}_{j_2}\right)^{J}\right]|0\rangle \tag{6}$$

which, after some operator algebra becomes

$$[H, A^{\dagger}_1(\alpha_1 J)]|0\rangle = \sum_{\beta_1} \Omega(2p|2p') A^{\dagger}_1(\beta_1 J)]|0\rangle$$

$$+ \sum_{\beta_2 J'_1 J'_2} \Omega(2p|3p1h) A^{\dagger}_2(\beta_2 J'_1 J'_2 J)]|0\rangle. \tag{7}$$

In (7) the $A^{\dagger}_1(\beta_1 J)$ operators are those of (4) and the $A^{\dagger}_2(\beta_2 J'_1 J'_2 J)$ are defined below:

$$\Phi_{3p1h} \longrightarrow \Phi^{J'_1 J'_2 J}_{j_1 j_2 j_3 j_4} = A^{\dagger}_2(\beta_2 J'_1 J'_2 J)|0\rangle = \left(\left(a^{\dagger}_{j_1} a^{\dagger}_{j_2}\right)^{J'_1} \left(a^{\dagger}_{j_3} a_{j_4}\right)^{J'_2}\right)^{J} |0\rangle. \tag{8}$$

In (8) we have used the additional convention:

$$\beta_2 \longrightarrow j'_1 j'_2 j'_3 j'_4 \tag{9}$$

and we have associated:

$$\begin{array}{l} J'_1 \text{ to the coupling of } j'_1 j'_2 \\ J'_2 \text{ to the coupling of } j'_3 j'_4 \end{array} \tag{10}$$

Having extended the commutator as in (7), we have also to calculate the commutator equation for the $A_2^\dagger(\alpha_2 J_1 J_2 J)$ operators as given below:

$$[H, A_2^\dagger(\alpha_2 J_1 J_2 J)]|0\rangle = \sum_{\beta_2 J_1' J_2'} \Omega(3p1h|3p'1h') A_2^\dagger(\beta_2 J_1' J_2' J)|0\rangle$$

$$+ \sum_{\beta_3 J_1' J_2' J_3'} \Omega(3p1h|4p2h) A_3^\dagger(\beta_3 J_1' J_2' J_3' J)|0\rangle, \qquad (11)$$

where we have introduced the (4p-2h) wave functions defined below:

$$\Phi_{4p2h} \longrightarrow \Phi_{j_1 j_2 j_3 j_4 j_5 j_6}^{J_1' J_2' J_{12}' J_3' J} = A_3^\dagger(\beta_3 J_1' J_2' J_3' J)|0\rangle$$

$$= \left(\left(\left(a_{j_1}^\dagger a_{j_2}^\dagger \right)^{J_1'} \left(a_{j_3}^\dagger a_{j_4}^\dagger \right)^{J_2'} \right)^{J_{12}'} \left(a_{j_5}^\dagger a_{j_6}^\dagger \right)^{J_3'} \right)^J |0\rangle, \qquad (12)$$

and where we have consistently extended the definition given in (5, 10):

$$\beta_3 \longrightarrow j_1 j_2 j_3 j_4 j_5 j_6 \qquad (13)$$

with:

$$J_1' \text{ associated to the coupling of } j_1 j_2$$
$$J_2' \text{ associated to the coupling of } j_3 j_4$$
$$J_3' \text{ associated to the coupling of } j_5 j_6 \qquad (14)$$

In the definition of $A_3^\dagger(\beta_3 J_1' J_2' J_3' J)$ the coupling of J_1' to J_2' to J_{12}' has been discarded from the notation. In (7, 11) the Ω's are the matrix elements of the Hamilton's operator in the model wave functions. The next step would then be the computation of the commutator of the Hamiltonian with the $A_3^\dagger(\beta_3 J_1' J_2' J_3' J)$ operators. Here we linearize these contributions by considering that in the study of the low energy spectrum and in the calculation of ground-state correlated distributions the $A_3^\dagger(\beta_3 J_1' J_2' J_3' J)$ terms have a small contribution. The linearization is performed by applying to the (4p2h) terms:

$$\sum_{\alpha\beta\gamma\delta} \langle \alpha\beta|v(r)|\gamma\delta\rangle a_\alpha^\dagger a_\beta^\dagger a_\delta a_\gamma \, A_3^\dagger(\beta_3 J_1' J_2' J_3' J) \qquad (15)$$

the Wick's theorem and to discard the normal order terms. Within this linearization approximation we generate non-perturbative solutions of the EoM from the commutator equations of (7, 11), i.e.: the eigenvalue equations for the mixed mode system:

$$\left[H, A_1^\dagger(\alpha_1 J) \right]|0\rangle = \sum_{\beta_1} \Omega(2p|2p') A_1^\dagger(\beta_1 J)|0\rangle$$

$$+ \sum_{\beta_2 J_1' J_2'} \Omega(2p|3p'1h') A_2^\dagger(\beta_2 J_1' J_2' J)|0\rangle, \qquad (16)$$

and

$$\left[H, A_2^\dagger(\alpha_2 J_1 J_2 J) \right]|0\rangle = \sum_{\beta_1} \Omega(3p1h|2p') A_1^\dagger(\beta_1 J)|0\rangle$$

$$+ \sum_{\beta_2 J_1' J_2'} \Omega(3p1h|3p'1h') A_2^\dagger(\beta_2 J_1' J_2') |0\rangle. \qquad (17)$$

Within the application of the GLA approximation we convert $(7, 11)$ in an eigenvalue equation for the configuration mixing wave functions (CMWFs) of the model. In fact, the linearization provides the additional matrix elements necessary to write the following identity:

$$\Omega(3p1h|3p'1h') = \langle j_1 j_2 j_3 j_4 | v(r) | j_1' j_2' j_3' j_4' \rangle, \tag{18}$$

and to introduce the off-diagonal matrix elements which couple the (2p) to the (3p1h) subspaces. Now, by writing $(16, 17)$ in the following matrix form:

$$\left(\begin{array}{c} \left[H, A_1^\dagger(\alpha_1 J) \right] |0\rangle \\ \left[H, A_2^\dagger(\alpha_2 J_1 J_2 J) \right] |0\rangle \end{array} \right)$$

$$= \left(\begin{array}{cc} E_{2p} + \Omega(2p|2p') & \Omega(2p|3p'1h') \\ \Omega(3p1h|2p') & E_{3p1h} + \Omega(3p1h|3p'1h') \end{array} \right) \left(\begin{array}{c} A_1^\dagger(\beta_1 J)|0\rangle \\ A_2^\dagger(\beta_2 J_1 J_2 J)|0\rangle \end{array} \right), \tag{19}$$

and by multiplying to the left with:

$$\left(\begin{array}{c} \langle 0| A_1(\alpha_1 J) \\ \langle 0| A_2(\alpha_2 J_1 J_2 J) \end{array} \right) \tag{20}$$

we generate the eigenvalue equation for the dressed particles:

$$\sum_{\beta_1 \beta_2 J_1' J_2'} \left(\begin{array}{cc} E_{2p} + \langle A_1(\alpha_1 J)|v(r)|A_1^\dagger(\beta_1 J)\rangle & \langle A_1(\alpha_1 J)|v(r)|A_2^\dagger(\beta_2 J_1' J_2' J)\rangle \\ \langle A_2(\alpha_2 J_1 J_2 J)|v(r)|A_1^\dagger(\beta_1 J)\rangle & E_{3p1h} + \langle A_2(\alpha_2 J_1 J_2 J)|v(r)|A_2^\dagger(\beta_2 J_1' J_2' J)\rangle \end{array} \right)$$

$$\cdot \left(\begin{array}{c} \chi_1(\beta_1 J) \\ \chi_2(\beta_2 J_1' J_2' J) \end{array} \right) = E \left(\begin{array}{c} \chi_1(\alpha_1 J) \\ \chi_2(\alpha_2 J_1 J_2 J) \end{array} \right) |0\rangle. \tag{21}$$

In (21) $E_{2p} = \epsilon_{j_1}^{HF} + \epsilon_{j_2}^{HF}$ and $E_{3p1h} = \epsilon_{j_1}^{HF} + \epsilon_{j_2}^{HF} + \epsilon_{j_3}^{HF} - \epsilon_{j_4}^{HF}$ are the Hartre-Fock energies while the χ's are the projections of the model states:

$$|\Phi_{2p}^J\rangle = \chi_1(\alpha_1 J) A_1(\alpha_1 J)|0\rangle + \chi_2(\alpha_2 J_1 J_2 J) A_2(\alpha_2 J_1 J_2 J)|0\rangle \tag{22}$$

to the basic vectors 2p, 3p1h. To conclude, although the (4p-2h) CMWFs are not active part of the model space, they are important for structure calculations. One may therefore associate the GLA approximation to a parameter which describes the degree of complexity of the model CMWFs. Within the first order linearization we obtain the EoM for the shell model while within the second and third order linearization approximations we derive the EoM of valence particles coexisting with the complex particle-hole structure of the excited states.

In this paper we solve (21) self-consistently. The solutions for the first iteration step are obtained by diagonalizing the eigenvalue (21). The first step of the iterative method generates the dynamic amplitudes for the two dressed particles, i.e. two particles coexisting with the 3p1h structures. With the calculated eigenvectors we recompute then the matrix elements $\langle j_1 j_2 | v(r) | j_1 j_2 j_3 j_4 \rangle$ and $\langle j_1 j_2 j_3 j_4 | v(r) | j_1' j_2' j_3' j_4' \rangle$ and we diagonalize again the eigenvalue equation. The iterations are repeated until the stabilization of the energies has been reached. Before performing the diagonalization of relative Hamilton's operator in the CMWFs base we have to eliminate the spurious center of mass components. In the BDCM this is performed, following the calculations of [26, 27], by calculating the percent weights of spurious

states in the model wave functions. These can be obtained by evaluating the energy of the center of mass according to the following equation:

$$E_R = \int dR \Psi^{\dagger dressed}(j_i j_j J)(R^2) \Psi^{dressed}(j_i j_j J)$$

$$+ 2 \sum_{ij} \int dr_i dr_j \Psi^{\dagger dressed}(j_i j_j J)(r_i \cdot r_j) \Psi^{dressed}(j_i' j_j' J). \tag{23}$$

In (23) the calculation of the integrals can be performed by using the CFT expansion for the (3p1h) states and by considering that for two particle states we have:

$$\langle j_i j_j J | (r_i \cdot r_j) | j_i j_j J \rangle$$

$$= \frac{4\pi}{3} \left[\hat{j}_i \hat{j}_j \right] \begin{pmatrix} j_i & 1 & j_j \\ -\frac{1}{2} & 0 & \frac{1}{2} \end{pmatrix} 2 \left\{ \begin{matrix} j_i & j_j & J \\ j_i & j_j & 1 \end{matrix} \right\} \langle l_i | r | l_j \rangle 2, \tag{24}$$

where:

$$\hat{j} = (2j + 1). \tag{25}$$

By diagonalizing the above operator in the model space we obtain the energy of the center of mass. The overlap with the model space give the degree of "spuriosity" of the different components. The model space which characterize the BDCM is formed by adding even particle to a closed-shell nucleus. The closed shell configuration can be described by a single Slater determinant and one can use the Hartree-Fock's theory to obtain the binding energy and the single-particle energies. Alternatively one can remark that for a closed shell nucleus (Z, N) the single particle energies for the states above the Fermi surface are related to the binding energies differences:

$$\epsilon_p^> = BE(Z, N) - BE^*(Z + 1, N), \tag{26}$$

and

$$\epsilon_n^> = BE(Z, N) - BE^*(Z, N + 1). \tag{27}$$

The single particle energies for the states below the Fermi surface are given by:

$$\epsilon_p^< = BE^*(Z - 1, N) - BE(Z, N), \tag{28}$$

and

$$\epsilon_n^< = BE^*(Z, N - 1) - BE(Z, N). \tag{29}$$

The BE are ground states binding energies which are taken as positive values, and ϵ will be negative for bound states. $(BE^* = BE - E_x)$ is the ground state binding energy minus the excitation energy of the excited states associated with the single particle states. Within this method, which recently has been reintroduced by B.A. Brown [28], we derive the single particle energies from the known spectra of neighbor nuclei (see Table 1).

The generalization of the previously defined formalism needed to calculate the charge radii of the $A = 7$ (one particle DCM), $A = 8$ (four particles BDCM) and $A = 11$ (three particles DCM) is not given explicitly in this paper, but will be presented shortly.

Table 1 Single-particle scheme and single particle energies (MeV) used to form the model CMWFs for the $A = 6$ isotopes

Hole	$1s_{1/2}$							
Energy	-20.58							
Hole/particle	$1p_{3/2}$							
Energy	1.43							
Particle	$1p_{1/2}$	$1d_{5/2}$	$2s_{1/2}$	$1d_{3/2}$	$1f_{7/2}$	$2p_{3/2}$	$1f_{5/2}$	$2p_{1/2}$
Energy	1.73	17.21	22.23	23.69	25.23	27.18	28.33	29.67

Table 2 Calculated charge radii for ^{6}He in fm compared with the results obtained in other theoretical models and with the radius derived within the IS theory

Charge radius of ^{6}He	Model
1.944 [9, 10]	No-core shell model
2.09 [12, 13]	Quantum Monte Carlo technique
2.25 This work	BDCM
2.39 This work	BDCM without the folded diagrams
2.06 This work	Two correlated $1s_{\frac{1}{2}}$-protons
1.99 [15]	Cluster
1.99 [16]	Cluster
1.99 [17]	Cluster
$2.054 \pm .014$[14]	Isotopic shift (Exp.)

2.1 Results

In order to perform structure calculations, we have to define a single particle base with the relative single-particle energies and to choose the nuclear two-body interactions. The single-particle energies of these levels are taken from the known experimental level spectra of the neighboring nuclei and given in Table 1. For the experimentally unknown single particle energies of the fp shells we use the corresponding energies for the mass $A = 9$ nuclei scaled accordingly the different binding energies. In this paper we perform as in [12, 13] calculations by assuming all levels as bound for the particle-particle interaction, we use the G-matrix obtained from Yale potential [29]. These matrix elements are evaluated by applying the e^S correlation operator, truncated at the second order term of the expansion, to the harmonic oscillator base with size parameter $b = 1.76$ fm. As also explained in [4, 5] the potential used by the BDCM is separated in low and high momentum components. Therefore, the effective model matrix elements calculated within the present separation method and those calculated by Kuo [30–33] are pretty similar. The separation method generates matrix elements, which are almost independent from the radial shape of the different potentials generally used in structure calculations.

The particle-hole matrix elements could be calculated from the particle-particle matrix elements via a re-coupling transformation. We prefer to use the phenomenological potential of [34]. The same size parameter as for the particle-particle matrix elements has been used. In Table 2 the calculated charge radii of ^{6}He are compared with the radii calculated by the other theoretical models and with the radius obtained by the IS theory. In the theoretical models quoted in this table the calculations of the charge distributions and of the charge radii are performed in terms of non- correlated

Table 3 Calculated charge radii for the Lithium isotopes in fm compared with the results obtained in other theoretical models and with the radius derived within the IS theory (Exp.)

Lithium	Exp.(GSI) [19]	Exp.+Theo. [35]	Theo. [9]	Theo. [12]	Theo. [36]	DCM+BDCM
	rms	rms	rms	rms	rms	
^{6}Li	2.51	2.47	2.22	2.54	-	2.55
^{7}Li	2.39	2.43	2.13	2.41	2.43	2.41
^{8}Li	2.30	2.42	2.13	2.26	2.34	2.40
^{9}Li	2.22	2.34	2.16	2.21	2.27	2.42
^{11}Li	2.47	3.01	-	-	2.57	2.67

operators. The correlations are included only in the derivation of the S_2 effective Hamiltonian.

For the stable ^{6}Li the calculated charge radius is equal to 2.55 fm, a value that reproduces well the charge radius of 2.55 fm obtained in [23] from the electron scattering experiments. The charge radii given in Table 3 for other lithium isotopes are however larger then those calculated in the other quoted theoretical models and then those obtained within the IS theory. Here also the main difference between the results obtained in the DCM and BDCM models and those of the other theoretical calculations has to be found in the treatment of the correlation operator. In [9, 10] the charge radii are calculated in the "no core shell model" which is based on exact solutions of the two particles Schrödinger's equation by considering large computational spaces. The calculations do not include however the S_3 correlations. [12, 13] presents charge radii evaluated within an accurate Quantum Monte Carlo Method. The Hamiltonian used includes two- and three-bodies forces in a two-body correlated mechanics. The calculation method of [36–38] is based on a microscopic cluster method in which few particles are interacting with the rest nucleus considered in its ground state. In our model the excitations of the core are associated to the S_3 correlation operator which increases the charge radius of the Lithium isotopes.

3 Conclusions and outlook

In this contribution we have investigated the effect of the microscopic correlation operators on the charge distributions of ^{6}He and of the Lithium isotopes. The microscopic correlation has been separated in short- and long-range correlations according the definition of Shakin [2, 3]. The short-range correlation has been used to define the effective Hamiltonian of the model while the long-range is used to calculate the structures and the distributions of exotic nuclei. As given in the work of Shakin, only the two-body short-range correlation need to be considered in order to derive the effective Hamiltonian especially if the correlation is of very short range. For the long range correlation operator the three body component is important and should not be neglected. Within the three body correlation operator one introduces in the theory a three body interaction which compensates for the use of the genuine three body interaction of the no-core shell model.

By using generalized linearization approximations and cluster factorization coefficients we can perform expedite and exact calculations.

Within the calculated correlated distributions we obtain charge radii slightly larger than those calculated for non-correlated distributions and derived by the IS experiments.

The application of the DCM [24] to the two and three electron energies and distributions of the Helium and Lithium atoms, respectively, could serve as future motivation for a reevaluation of the IS theory. From one side, the calculation of CM of the different isotopes, could help to obtain a non-perturbative formulation of the isotopic change of the electron transition energies. From the other side the field shift theory could include the correct isotopic variation. Since the derivation of the charge radii from the two photon experiments is influenced by the precision of the theoretical calculations, this new proposed method could contribute to evaluate with better precision the charge radii of exotic nuclei.

References

1. Villars, F.: Proceedings of the Enrico Fermi Int. School of Physics XXII. Academic, New York (1961)
2. Shakin, C.M., Waghmare, Y.R.: Phys. Rev. Lett. **16**, 403 (1966)
3. Shakin, C.M., Waghmare, Y.R., Hull, M.H.: Phys. Rev. **161**, 1006 (1967)
4. Tomaselli, M., Liu, L.C., Fritzsche, S., Kühl, T., Ursescu, D.: Nucl. Phys. **A738**, 216 (2004)
5. Tomaselli, M., Liu, L.C., Fritzsche, S., Kühl, T.: J. Phys. G: Nucl. Part. Phys. **30**, 999 (2004)
6. Brückner, K.A.: The Many Body Problem. Wiley, New York (1959)
7. Kuo, T.T.S., Osnes, E.: Folded-Diagrams Theory of the Effective Interaction in Atomic Nuclei, Springer Lecture Notes in Physics Vol. 366. Berlin (1991)
8. Tomaselli, M., Kühl, T., Ursescu, D., Fritzsche, S.: Prog. Theor. Phys. **116**, 699 (2006)
9. Navrátil, P., Barrett, B.R.: Phys. Rev. **C57**, 3119 (1998)
10. Navrátil, P., Caurier, E.: Phys. Rev. **C69**, 014331 (2004)
11. Stancu, I., Barrett, B.R., Navrátil, P., Vary, J.P.: Phys. Rev. **C71**, 044325 (2005)
12. Pieper, S.C., Wiringa, R.B.: Ann. Rev. Part. Sci. **51**, 53 (2001)
13. Pieper, S.C., Wiringa, R.B.: Phys. Rev. **C66**, 044310 (2002)
14. Wang, L.-B., Müller, P., Bailey, V., et al.: Phys. Rev. Lett. **93**, 142501 (2004)
15. Würzer, J., Hofmann, H.M.: Phys. Rev. **C55**, 688 (1997)
16. Funada, S., Kaneyama, H., Sakuragi, Y.: Nucl. Phys. **A575**, 93 (1994)
17. Varga, K., Suzuki, Y., Ohbayasi, Y.: Phys. Rev. **C50**, 189 (1994)
18. Tomaselli, M., Kühl, T., Nörtershäuser, W., Ewald, G., Sanchez, R., Fritzsche, S., Karshenboim, G.S.: Can. J. Phys. **80**, 1347 (2002)
19. Ewald, G., Nörtershäuser, W., Dax, A., et al.: Phys. Rev. Lett. **93**, 113002 (2004)
20. Bushaw, B.A., Nörtershäuser, W., Ewald, G., et al.: Phys. Rev. Lett. **91**, 043004 (2003)
21. Sánchez, R., Nörtershäuser, W., Ewald, G., Albers, D., Behr, J., Bricault, P., Bushaw, B.A., Dax, A., Dilling, J., Dombsky, M., Drake, G.W.F., Götte, S., Kirchner, R., Kluge, H.-J., Kühl, T., Lassen, J., Levy, C.D.P., Pearson, M.R., Prime, E.J., Ryjkov, V., Wojtaszek, A., Yan, Z.-C., Zimmerman, C.: Phys. Rev. Lett. **96**, 033002 (2005)
22. Yan, Z.-C., Drake, G.W.F.: Phys. Rev. **A66**, 042504 (2002)
23. Li, G.C., Sick, I., Whitney, R.R., Yearian, M.R.: Nucl. Phys. **A162**, 583 (1971)
24. Tomaselli, M.: Can. J. Phys. **83**, 467 (2005)
25. Aufmuth, P.: J. Phys. B: At. Mol. Phys. **15**, 3127 (1982)
26. Baranger, E., Lee, C.W.: Nucl. Phys. **22**, 157 (1961)
27. Unna, I., Talmi, I.: Phys. Rev. **112**, 452 (1958)
28. Brown, B.A.: Prog. Part. Nucl. Phys. **47**, 524 (2001)
29. Shakin, C.M., Waghmare, Y.R., Tomaselli, M., Hull, M.H.: Phys. Rev. **161**, 1015 (1967)
30. Jiang, M.F., Machleidt, R., Stout, D.B., Kuo, T.T.S.: Phys. Rev. **C46**, 910 (1992)
31. Lacombe, M., Loiseau, B., Richard, J.M., Vinh Mau, R., Côté, J., Pirès, P., de Tourreil, R.: Phys. Rev. **C21**, 861 (1980)
32. Machleidt, R.: Adv. Nucl. Phys. **19**, 189 (1989)
33. Bogner, S., Kuo, T.T.S., Coraggio, L., Covello, A., Itaco, N.: Phys. Rev. **C65**, 051301(R) (2002)

34. Millener, D.J., Kurath, D.: Nucl. Phys. **255**, 315 (1975)
35. Tanihata, I.: Phys. Lett. **B206**, 592 (1988)
36. Suzuki, Y.: Prog. Theor. Phys., Supp. **146**, 413 (2002)
37. Varga, K., Suzuki, Y., Taniata, I.: Phys. Rev. **C52**, 3013 (1995)
38. Varga, K., Suzuki, Y., Lovas, R.G.: Phys. Rev. **C66**, 041302 (2002)

Hyperfine Interact (2006) 171:255–259
DOI 10.1007/s10751-006-9506-z

Testing QED with resonance conversion

F. F. Karpeshin · M. B. Trzhaskovskaya

Published online: 9 February 2007

Abstract Various phenomena: hyperfine shift of $^{209}\mathrm{Bi}^{82+}$, nuclear decay of $^{229}\mathrm{Th}$ and $^{169}\mathrm{Yb}^{69+}$ are considered as cross-invariant channels of the internal conversion process from the viewpoint of testing QED. The way of accelerating nuclear decay in laser field via merging the nuclear and laser photons is discussed. Discovering this fundamental process would give a convincing test of both QED along with our experimental abilities.

Key words quantum electrodynamics test · resonance conversion (internal and bound) · nuclear decay · $^{209}\mathrm{Bi}^{82+}$ · $^{229}\mathrm{Th}$ · $^{169}\mathrm{Yb}^{69+}$

1 Introduction

Probing space and matter with strong electromagnetic field is a topic of fundamental interest. As long ago as in 1920s, Klein considered relativistic equation for an electron put in strong electrical field. He predicted the process of spontaneous generation of the e^+e^- pairs in the field (Klein's paradox) with critical electrical strength $F_0 = \frac{m^2c^3}{e\hbar} = 2 \cdot 10^{16}$ V/cm. At this strength, the non-linear QED effects reach their optimal value. Experimental investigation of this process was undertaken at GSI Darmstadt half centenary later. Another process of restoration of the spontaneously broken symmetry in strong field was predicted by Salam and Strathdee. The latter process was searched for by David et al. at the University of Bonn in muon capture by the nuclei. Interpretation of the results is not clear in the both cases. The problem remains to be a challenging task for experimentalists.

At the same time, elementary estimate shows that the electrical strength at a nuclear surface is $\varepsilon \sim eZ/R^2 \sim 10^{19}$ V/cm, that is overcritical value. This gives rise to strong

F. F. Karpeshin (✉)
Fock Institute of Physics, St. Petersburg State University, 198504 St. Petersburg, Russia
e-mail: Feodor.Karpeshin@pobox.spbu.ru

M. B. Trzhaskovskaya
Petersburg Nuclear Physics Institute, Leningrad district, 188300 Gatchina, Russia

polarisation of vacuum around the nucleus, which considerably affects the processes discussed below. This example gives an idea of complexity of the processes occurring in strong field, and illustrates the dynamics of the "Moscow zero," another paradoxical result. The example gives an idea how any charge ultimately disappears due to the polarisation of vacuum, and coolly reminds that QED is only asymptotic theory valid in the limit of comparatively weak field.

It is worthy to note that traditional tests of QED deal with static properties of atoms, such as energy shifts. But these are sensitive to the square of variation of the wave function. Advantage is the practically infinite time, which allows one to reach precision of many digits. Dynamical tests based on measuring the amplitude of a process depend linearly on the wave function and its variation. Therefore, they may be *much* more sensitive. This gives advantage of using lasers, storage rings and other facilities.

On the other hand, strength of the field of most powerful lasers is within $\sim 10^9$ V/cm. At first glance, this rules out any idea of affecting nuclear processes. There is one way. That is offered by the resonance phenomena. We shall focus on the resonance interaction mediated by the electron shell. This way is realised via bound (BIC), or resonance (RC) internal conversion (IC) [1–3]. The resonance conversion was observed in muonic atoms [4] and 45-fold ions of ^{125}Te [5]. Much experimental effort was spent without result because of negation of the theory of RC in studies of the 3.5 eV state in ^{229m}Th (see e.g. [6]). This story is discussed in refs. [7, 8] and other papers cited therein.

2 Atoms in electromagnetic field

We recall that typical nuclear sizes are within $R \sim 10$ fm and typical scale for the transition energies is of the order of ~ 1 MeV. Therefore, meaningful effect of the electromagnetic field on nuclei can be only expected for fantastic overcritical electrical strength of $\varepsilon \sim 10^{18}$ V/cm. Such a simple estimate helps to realise that all the photo-nuclear reactions are due to resonance effect. Only resonance quanta can be absorbed by the nuclei, with the energy which exactly equals separation energy of the nuclear levels.

In contrast, atomic size is by four orders of magnitude larger than the nuclear one. This means that it is much easier to affect the nucleus by electromagnetic field through mediation of the electron shell, which plays a role of resonator [1]. A gain of several orders of magnitude can be benefited from making use of the resonance properties of BIC. Further experimental study in the field of producing nuclear isomers and triggering their energy, constructing gamma laser must be directed in this way.

3 Tuning BIC

The idea is to arrange a two-photon resonance. Consider atom in an external field of a plane electromagnetic wave. Some atomic electron can go to an excited state by absorption of one or several photons from the field. The probability of multiphoton absorption increases drastically if the total energy of the absorbed photons approaches the difference of the electron levels. This effect is used by RIS – Resonance Ionisation Spectroscopy. Let us consider the two-photon resonance, and replace one photon of the field by the nuclear photon [1]. This is laser assisted nuclear BIC transition, as the electron makes a two-photon transition to an excited state, one photon being from the nucleus, and the other from the

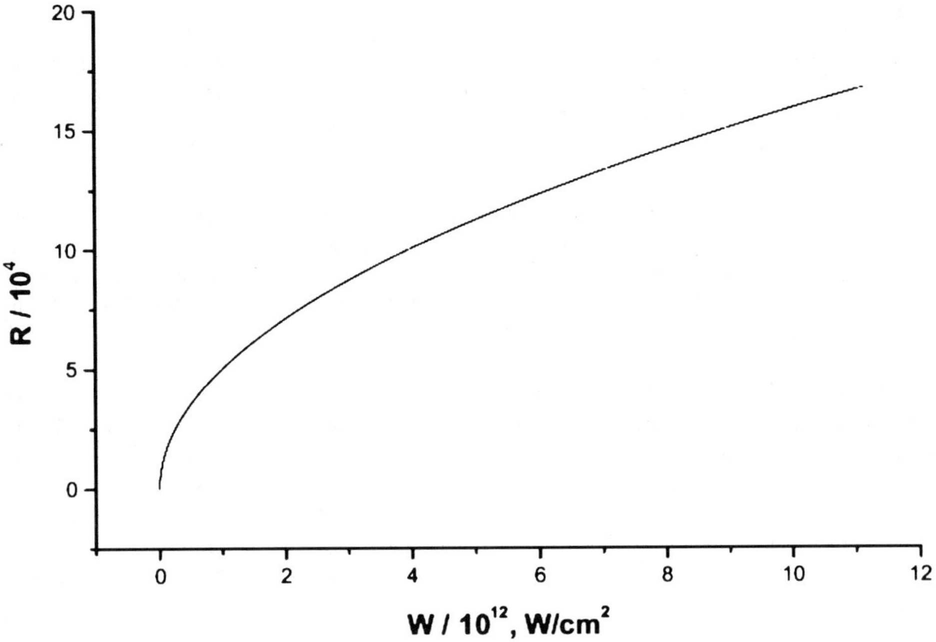

Fig. 1 Resonance conversion factor R calculated for the 70.882 keV $M1$ transition in hydrogen-like ions of ^{169}Yb against power of externally applied laser field

field. The both should arrive in the same time within the lifetime of the electron in the intermediate state. Necessary condition is then

$$\omega_n \pm \omega_l = \omega_a, \tag{1}$$

where ω_n, ω_l and ω_a stand for the nuclear, laser and electron frequencies, respectively.[1] The two signs in (1) correspond to either absorption, or induced emission of a photon of the field. The both probabilities are of the same value.

A very promising experiment can be undertaken in hydrogen-like ions of ^{169}Yb. The energy of the $M1$ 70.882 keV transition differs by about 10 eV from the energy of the atomic $1s \rightarrow 10s$ transition. Therefore, it is expected that the nuclear lifetime will be considerably shortened in the field of electromagnetic wave of such frequency [9, 10]. This effect is demonstrated in Fig. 1. The R value is already 50 at moderate electrical strength of 30 kV/cm. At higher intensities, as one can see in the Figure, the effect can achieve as much as four orders of magnitude. Such an experiment can be fulfilled on the GSI facilities.

4 Hyperfine splitting in ^{209}Bi^{82+}

The hyperfine shift (HFS) is given by the diagonal matrix element of the electron–nucleus interaction. The corresponding Feynman diagram is presented in Fig. 2. Generally, this diagram describes the internal conversion amplitude. If the QED amplitude is an analytical

---

[1] We use relativistic units $\hbar = c = 1$, hence nuclear and electron frequencies coincide with the corresponding transition energies

Fig. 2 Feynman graph for the HFS

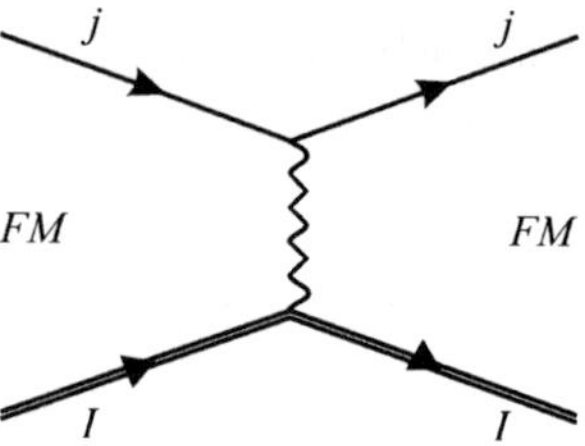

Table 1 Hyperfine splitting (hfs) calculated for the hydrogen-like ions of ^{209}Bi in various nuclear models: surface (*SC*) and volume (*VC*) nuclear currents, uniform and Fermi (*F*) charge distribution, without penetration (*NP*), with polarisation of vacuum ruled out (no p.v.) in comparison with experiment and other calculations

Nuclear model	SC	SC,F	SC, no p.v.	NP	VC	VC, F	Experim.	Ref. [12]
hfs	5.105	5.107	5.081	5.192	5.138	5.141	5.0841(8)	5.091

function of its kinematics variables, then this amplitude describes many processes along with the IC: electron scattering on nuclei, non-photonic positron annihilation on bound electrons with nuclear excitation, pair and resonance IC, and others [2, 3]. Specifically, diagram of Fig. 2 is a particular case of RC for the $M1$ interaction with the energy of the virtual photon $\omega=0$. It was shown by Sliv [11] in the general case of IC that the resulting matrix element depends on the nuclear model for the transition current. In the given case that means that HFS depends on the space distribution of the magnetisation inside the nucleus. This is in contrast with the magnetic moment itself which can be measured with high precision in a model independent way. And the $M1$ transitions are just most sensitive to the nuclear model.

The results of calculation of the hyperfine splitting for the hydrogen-like ions of ^{209}Bi in various nuclear models are presented in Table 1. The calculations are conducted by making use of the computer codes package RAINE [13]. Finite nuclear size, higher QED corrections for polarisation of vacuum and electron self energy are included. We do not consider corrections for the nuclear recoil energy. That is of the order of m/M, with m and M being the electron and nuclear masses, respectively, and hence contributes to the sixth decimal. As can be seen from Table 1, the latter value is hardly distinguishable in experiment in view of the uncertainty brought about by the nuclear model [2], which is at the level of 1%. This result was expected from the IC theory. Calculated values are in reasonable agreement with experiment and other calculations. What we can really know from comparison with experiment is a characteristic radius of distribution of the magnetic current in the nuclear volume. The surface distribution looks most plausible in view that magnetic moment is produced mainly by the last odd nucleon. In the case of s valence nucleons, or in the excited state of giant dipole resonance, a volume distribution can be expected, which will be immediately reflected in the enlarged HFS.

5 Conclusion

Acceleration of the nuclear processes with externally applied radiation is a challenging task at the contemporary stage of investigation. This is a problem of great fundamental interest

with many practical applications. This gives a nontrivial test of QED, and simultaneously of our experimental abilities.

Acknowledgment This work was supported by Russian Foundation for Basic Research under grant no. 05-02-17430.

References

1. Karpeshin, F.F.: Hyperfine Interact. **143**, 79 (2002)
2. Karpeshin, F.F.: Part. Nucl. **37**, 523 (2006)
3. Karpeshin, F.F.: Prompt Fission. Nauka, St. Petersburg (2006)
4. Zaretsky, D.F., Karpeshin, F.F.: Sov. J. Nucl. Phys. **29**, 151 (1979)
5. Karpeshin, F.F., Harston, M.R., Attallah, F., Chemin, J.F., Scheurer, J.N., Band, I.M., Trzhaskovskaya, M.B.: Phys. Rev. C **53**, 1640 (1996)
6. Irwin, G.M., Kim, K.H.: Phys. Rev. Lett. **79**, 990 (1997)
7. Karpeshin, F.F., Band, I.M., Trzhaskovskaya, M.B., Pastor, A.: Phys. Rev. Lett. **83**, 1072 (1999)
8. Karpeshin, F.F., Trzhaskovskaya, M.B.: Yad. Fiz. **69**, 596 (2006) (Engl. transl. Phys. At. Nucl. **69**, 571 (2006))
9. Karpeshin, F.F., Trzhaskovskaya, M.B., Gangrsky, Yu.P.: Zh. Exper. Teor. Fiz. **126**, 323 (2004) (Engl. transl. JETP **99** 286 (2004))
10. Gangrsky, Yu.P., Karpeshin, F.F., Trzhaskovskaya, M.B.: Izv. RAN. Ser. fiz. **68**, 149 (2004)
11. Sliv, L.A.: Zh. Exp. Teor. Fiz. **21**, 770 (1951)
12. Tomaselli, M.S., Schneider, M., Kankeleit, E., Kühl, T.: Phys. Rev. C **51**, 2989 (1995)
13. Band, I.M., Trzhaskovskaya, M.B., Nestor, C.W., Jr., Tikkanen, P.O., Raman, S.: At. Data Nucl. Data Tables **81**, 1 (2002)

Author Index to Volume 171 (2006)